MECHANICAL BEHAVIOUR OF MATERIALS

SOLID MECHANICS AND ITS APPLICATIONS
Volume 58

Series Editor: G.M.L. GLADWELL
Solid Mechanics Division, Faculty of Engineering
University of Waterloo
Waterloo, Ontario, Canada N2L 3G1

Aims and Scope of the Series

The fundamental questions arising in mechanics are: *Why?*, *How?*, and *How much?* The aim of this series is to provide lucid accounts written by authoritative researchers giving vision and insight in answering these questions on the subject of mechanics as it relates to solids.

The scope of the series covers the entire spectrum of solid mechanics. Thus it includes the foundation of mechanics; variational formulations; computational mechanics; statics, kinematics and dynamics of rigid and elastic bodies; vibrations of solids and structures; dynamical systems and chaos; the theories of elasticity, plasticity and viscoelasticity; composite materials; rods, beams, shells and membranes; structural control and stability; soils, rocks and geomechanics; fracture; tribology; experimental mechanics; biomechanics and machine design.

The median level of presentation is the first year graduate student. Some texts are monographs defining the current state of the field; others are accessible to final year undergraduates; but essentially the emphasis is on readability and clarity.

For a list of related mechanics titles, see final pages.

Mechanical Behaviour of Materials

Volume II: Viscoplasticity, Damage, Fracture and Contact Mechanics

by

DOMINIQUE FRANÇOIS
École Centrale de Paris,
Chatenay-Malabry, France

ANDRÉ PINEAU
École Nationale Supérieure des Mines de Paris,
Paris, France

and

ANDRÉ ZAOUI
École Polytechnique,
Palaiseau, France

KLUWER ACADEMIC PUBLISHERS
DORDRECHT / BOSTON / LONDON

A C.I.P. Catalogue record for this book is available from the Library of Congress.

ISBN 978-90-481-4974-2

Published by Kluwer Academic Publishers,
P.O. Box 17, 3300 AA Dordrecht, The Netherlands.

Sold and distributed in the U.S.A. and Canada
by Kluwer Academic Publishers,
101 Philip Drive, Norwell, MA 02061, U.S.A.

In all other countries, sold and distributed
by Kluwer Academic Publishers,
P.O. Box 322, 3300 AH Dordrecht, The Netherlands.

Published with the help of the French Ministry of Culture

This is a translation of the original French work
Comportement Mécanique des Matériaux,
Hermès, Paris, ©1993
Translated from French by Jack Howlett

Printed on acid-free paper

MECHANICAL BEHAVIOUR OF MATERIALS

FOREWORD

Man discovered a long time ago that quenching steel would increase its hardness; the reason for this was found much more recently, and the understanding, together with the finding of ways of exploiting the interactions between martensitic transformations and dislocations, precipitates and texture, has led to the development of new steels, new heat treatments and new alloys with unusual properties. The art of metallurgy had provided many recipes rooted in empiricism: the introduction of scientific thinking has made it possible to improve these, and the science of metals thus founded has opened the way to the wider subject of materials science. In the same way the practical problems of construction have led to the growth of solid mechanics as a branch of applied mathematics. For a long time the constitutive equations needed in materials science remained crude idealisations of actual behaviour. The pioneers in this field could correspond equally well with their peers about metallurgy (or alchemy) as about mechanics (or astrology); later, scientists have become more and more specialised, and there is now little overlap between materials science and solid mechanics.

As technical equipment of ever greater sophistication has become available, the risk of catastrophes, of a scale that can affect the environment and kill many people, has increased; and safety has become a major concern. Economic considerations press for longer lifetimes and smaller safety factors; these generate strong incentives to use more realistic constitutive equations and better failure criteria in the calculations, and the computer now makes this possible. Materials design has become much more of a practical possibility, and materials can be produced with better and more reliable properties.

All this shows that establishing relations, as quantitative as possible, between the microstructure of materials and their macroscopic properties is nowadays essential. Thanks to fruitful cooperation between materials scientists and solid-mechanics specialists, recent research has led to promising achievements in this direction; but the number of training programs which cover both fields has remained low. It was the awareness of the need for advanced courses here that led us, some ten years ago, to create in France what is called a Diplome d'Études Approfondies (DEA) - Advanced Studies Diploma - with the title Mécanique et Matériaux - Mechanics and Materials; and the present books stem from the notes provided for the courses. The need was probably greater in France than in English-speaking countries, where the famous book of McClintock and Argon, Mechanical Properties of Materials, was already much in use. This, however, was published in 1966 and so did not deal with recent developments; and this gave us the incentive to embark on these books, even though we felt that we could not match McClintock and Argon.

The organisation of the two volumes follows the main classes of mechanical behaviour: the first deals with elastic and plastic deformations and the second with viscoplastic, followed by treatments of damage mechanisms, fracture mechanics and contact mechanics. Throughout we attempt to describe the physical processes that are responsible for the kinds of behaviour studied, the way in which the constitutive equations can represent the behaviour and how they relate to the microstructures. We follow each chapter with a set of exercises, to which we give either the solutions or hints on how these are to be obtained. Understanding the subject matter requires a good knowledge of solid mechanics and materials science; we give the main elements of these fields in a set of Annexes at the end of the first volumes.

Whilst the books are addressed primarily to graduate students, they could possibly be used in undergraduate courses; and we hope that practising engineers and scientists will find the information they convey useful. We hope also that English-speaking readers will be interested in the aspects of French culture which our treatment will undoubtedly display.

The authors are very grateful to all their colleagues, in particular those who teach in the DEA Mécanique et Matériaux, for their contributions and encouragement; and wish to thank all those people who have provided photographs to illustrate the book. We also thank Professor Gladwell of the University of Waterloo, Canada, for his final proofreading. The English translation was done by Dr. Jack Howlett, whose frequent questions and suggestions have helped to improve many paragraphs significantly. We have found cooperation with him very stimulating and we thank him for his excellent work.

MECHANICAL BEHAVIOUR OF MATERIALS

VOLUME II: VISCOPLASTICITY, DAMAGE, FRACTURE and CONTACT MECHANICS

CONTENTS

MECHANICAL BEHAVIOUR OF MATERIALS

CHAPTER 1 ELASTOVISCOPLASTICITY

1.1 INTRODUCTION

The type of behaviour to be discussed in this chapter has in common with elastoplasticity the fact that a permanent deformation remains after the load has been removed. On the other hand, viscosity now prevents the occurrence of instantaneous plastic deformations: time controls the inelastic deformation and we have what is called *rate-dependent plasticity*. Apart from this, viscoplastic behaviour is very similar to ordinary plasticity, already treated in Chapter 3 of Volume 1; but it is distinguished from this by the separation of the deformation into an elastic part and an inelastic viscoplastic part, say $\varepsilon = \varepsilon^e + \varepsilon^V$, with, in general, absence of reversibility, even delayed, for the inelastic part.

Viscoplastic behaviour occurs especially in metals and alloys when at high temperatures, not less than one-third of the melting point. It is then accompanied by diffusion phenomena with different characteristic distances: those related to dislocations are seen in the *creep* which results from the competition between work-hardening and recovery, or at greater distances those related to grain boundaries in *diffusional creep*. Low-temperature viscoplasticity also is possible, related to thermal activation of the plastic deformation. We define these various terms at the appropriate points later in the chapter. As well as to metals, viscoplasticity theory applies to resins and polymers when these are loaded beyond the point at which their behaviour can continue to be viscoelastic, and also, in studies of the evolution of mechanical behaviour over very long periods, to soils, rocks and ice.

As in the study of plasticity, we make the simplifying assumptions of linear elasticity, small deformations and absence of damage. This chapter gives:

- a description of the tests that enable the basic features of viscoplastic behaviour and the associated phenomena to be seen,

- an account of the physical models of viscoplasticity (thermal activation, creep). Particular attention is paid to the field of validity of each equation derived, so that the limits of the models can be predicted - for example, when results are to be extrapolated to longer times or higher temperatures,

- an account of phenomenological mechanical models of viscoplasticity, bringing out the physical significance of each term introduced; here we discuss the theories of plasticity and viscoplasticity in combination, and also, for reasons of computational convenience, the use of viscoplasticity rather than time-independent plasticity,

- a discussion of methods for reinforcing against creep, showing how, once the deformation mechanisms have been understood, means for guarding against excessive deformation and rupture can be found.

Only viscoplasticity is considered here, damage being left to the next chapter.

1.2 TYPICAL EXPERIMENTAL RESULTS

The basic features of viscoplasticity appear under steady loading, with either the force or the deformation held constant after the load is applied - corresponding to creep and relaxation experiments respectively. The importance of such tests is their simplicity, but in general they do not show the complete behaviour since either the stress or the plastic deformation remains practically constant each time. We must therefore consider more elaborate tests, with changes made incrementally and showing changes in the strain rate, or with the load increasing in steps. One of the most important requirements for a good characterisation of the mechanism is the separation of the effects of deformation from those of time on the state of the material.

1.2.1 One-dimensional response

1.2.1.1 Creep

Creep tests are carried out on cylindrical or flat test pieces under constant load; the time for a test can be anything from a few hours to several years, according to the application. The results are given as a time-deformation curve (Fig. 1.1). To the extent that the load is usually applied as a dead weight the stress is not constant throughout the test, but this is usually ignored and the test characterised by the initial value of the axial stress, $\sigma_0 = F/S_0$, where F is the applied force and S_0 the initial cross-section of the test piece. There are three main stages in the response to the load, more or less clearly distinguished according to the material and the temperature:

primary, during which the strain rate falls; this corresponds to an increase in the resistance of the material,

secondary, in which the strain rate remains constant,

tertiary, when the rate increases; significant mechanical damage appears in this stage, related to cavitation for example, or to a softening of the material, induced by localisation of the strain on the scale of the microstructure.

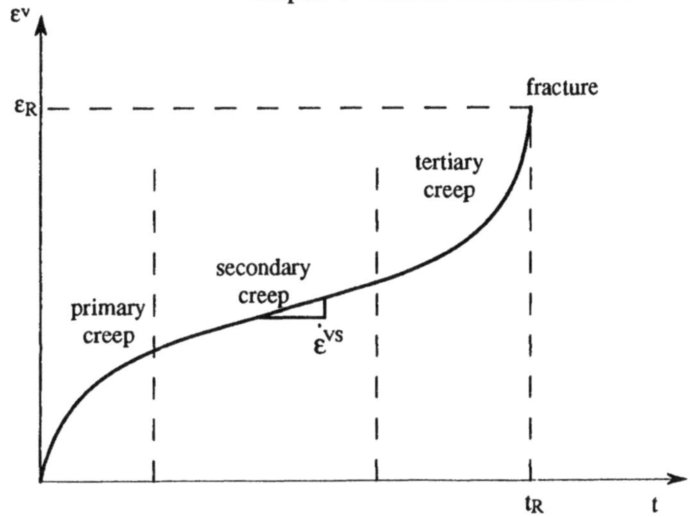

Fig. 1.1 The three phases in a creep test.

At low temperatures it is usually the primary and secondary creep that dominates; as the temperature rises the secondary phase becomes established sooner, and tertiary creep becomes more important.

The results of a set of tests can be presented as a body by plotting all the curves together on the same sheet; but it can be useful to plot lines of constant deformation in the time-stress plane, showing, for each initial stress, the time required to reach a certain deformation. Fig. 1.2, for example, shows these lines for deformations of *0.2%*, *1%* and *2%*. This has the advantage of making it easy to compare different materials, or to assess the effect of temperature: thus material A will have a "*50°C* creep difference" from material B if their curves are separated by this amount.

At temperatures below $0.3T_f$ (T_f being the melting point) only primary creep occurs, with the reduction in strain rate with time given by either a power or a logarithmic law, e.g.:

$$\varepsilon = At^{1/3} \qquad \text{(Andrade's law)} \qquad (1.1a)$$

$$\varepsilon = A\log(1 + t/t_0) \qquad \text{(logarithmic creep)} \qquad (1.1b)$$

where A and t_0 depend on the material.

These expressions should not be confused with the constitutive equations, since they cannot give a correct description of the way the deformation changes in response to a real variation of loading, such as partial or total unloading. We shall discuss later how work-hardening can be expressed, in terms of viscoplastic deformation for example.

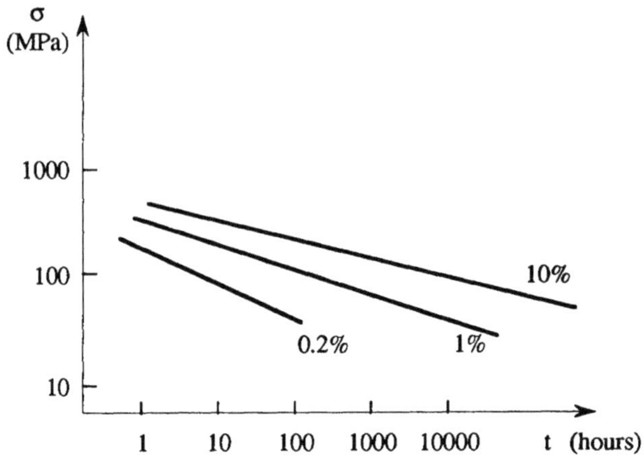

Fig. 1.2 Curves of constant deformation in creep.

Another way to present the results, restricted to secondary creep, is to show on a log-log plot of strain rate vs. time the minimum rate measured in each test. In any small interval of stress the points will lie on a straight line, enabling the tests to be interpreted with the aid of Norton's law:

$$\dot{\varepsilon}^{vs} = \left(\frac{\sigma}{K}\right)^{M} \tag{1.1c}$$

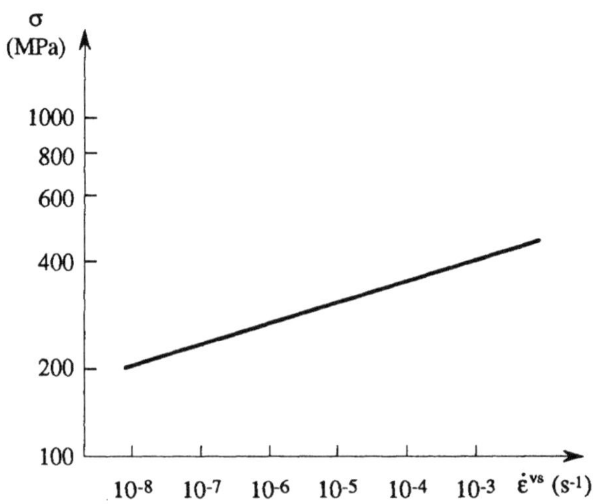

Fig. 1.3 Interpretation of creep test by Norton's law.

Whilst there is only one relation that is valid for describing the steady state, and which, in particular, takes no account of work-hardening, (1.1c) is often used in calculations as though it were an established constitutive equation. The coefficient M, the numerical value of which decreases with increasing temperature, is "Norton's exponent"; Table 1.1 gives the values for a number of pure metals. Physical models, as we shall show in our later discussion (§1.3 below), predict values of *1* for diffusional creep and *4-5*, according to the mechanism involved, for dislocation creep. However, these models are valid only over a limited range, and for low temperatures an internal stress has to be introduced; applying the rule as it is leads to exponents that can reach *40-50*, particularly in the case of complex alloys that are not simple solid solutions.

Metal	M	Q kcal/mole	Self-diffusion energy kcal/mole
Al	4.4	34	34
Cu	4.8	48.4	47.1
Au	5.5	48 ± 5	41.7
Ni	4.6	66.5	66.8
Pb	4.2	24.2 ± 2.5	24.2
Ta	4.2	114 ± 4	110
Cd	4.3	19 ± 2	19.1
Zn	6.1	21.6	24.3

Table 1.1 Values of Norton's exponent M, Q and self-diffusion energy for pure metals (Mukherjee et al (1969) p.155).

It can be useful to keep to the same type of law, modified to include both stress and temperature; such a law, involving an activation energy and the temperature (in °K) is

$$\dot{\varepsilon}^{vs} = \left(\frac{\sigma}{K}\right)^M \exp\left(-\frac{Q}{kT}\right) \tag{1.2}$$

As Table 1.1 shows, the activation energy is the same as the energy of *self diffusion* (see Vol. I, Annex 2) in the case of pure metals. For more complex materials the term cannot be given a precise physical meaning: Q can depend, for example, on the stress applied. Nevertheless, it continues to be used, in particular to establish time-temperature equivalences; these have a bearing on lifetime calculations and can be helpful in using the results of short-time tests at high temperatures to estimate times to failure at lower temperatures for which when the relevant times are very long, beyond the reach of experiment. If we accept the Monkman-Grant law (Ch. 2 §2.12) that states that the lifetime t_R is a power function of the rate of steady-state creep, that is, $\dot{\varepsilon}^{vs} t_R^{\alpha} = \text{constant}$, we find $t_R = A \exp(Q/kT)$. A parametric representation is then possible provided that A and Q do not both depend on the stress: this is the case, for example, for diffusional creep, where only A varies with stress and the parameter $P' = \log t_R$

- *const./T* can be used to represent the creep data. If only Q varies with stress the Larson-Miller parameter $P = T(\log t_R + const.)$ can be used, as shown in Fig. 1.4: see Larson and Miller (1952) and McClintock and Argon (1966) p.639.

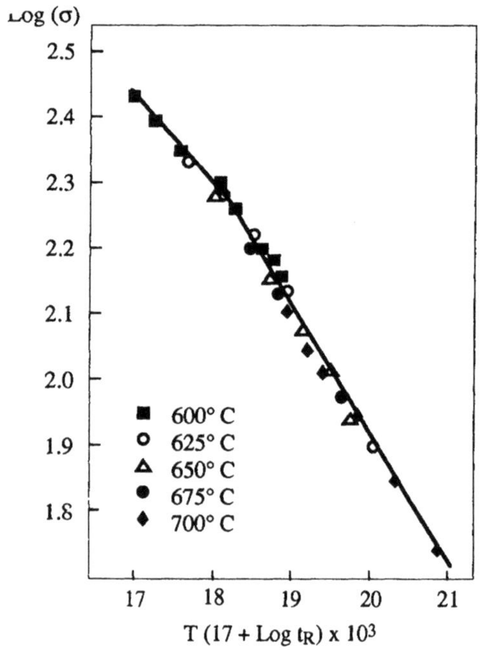

Fig. 1.4 Larson-Miller plot for the austenitic stainless steel Z03CND17-12, containing 2.5% molybdenum.

1.2.1.1 Tensile curves

In contrast to the case in which the plasticity does not depend on time, no single stress-strain curve represents tensile loading at different rates. The "rate effect" is usually to increase the stress for a given deformation, as shown in Fig. 1.5. There is a tendency to saturation at very high and at very low rates, suggesting limiting cases of instantaneous plasticity at very high rates and at zero rate. If we define a *critical stress* $\sigma_c(\varepsilon^v, \alpha)$, the stress corresponding to the zero-rate case, where α is the conventional representation of the work-hardening variables, and a *viscous stress* $\sigma_v(\dot{\varepsilon}^v, \alpha)$, which depends on the viscoplastic strain rate and is zero at zero rate, we can write the stress-strain relation for different strain rates v as

$$\sigma = \sigma_c(\varepsilon^v, \alpha) + \sigma_v(\dot{\varepsilon}^v, \alpha) \tag{1.3}$$

The existence of this "viscous stress" can be made evident by making a temporary change in the rate of loading in the course of a test. As Fig. 1.6 shows, there is a tendency to rejoin the undisturbed curve after the return to the original rate.

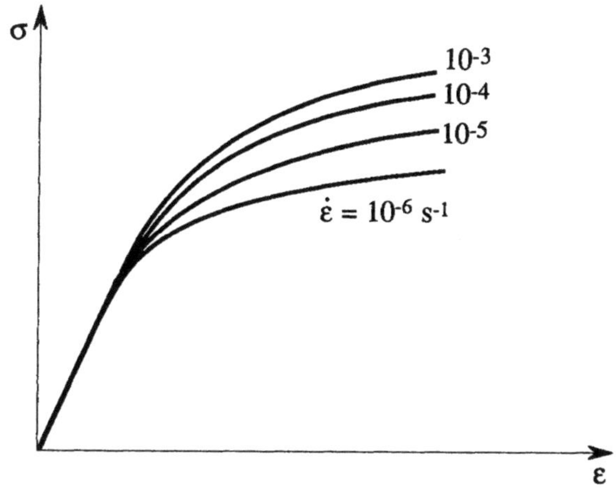

Fig. 1.5 Effect of rate of application of the load in a tensile test.

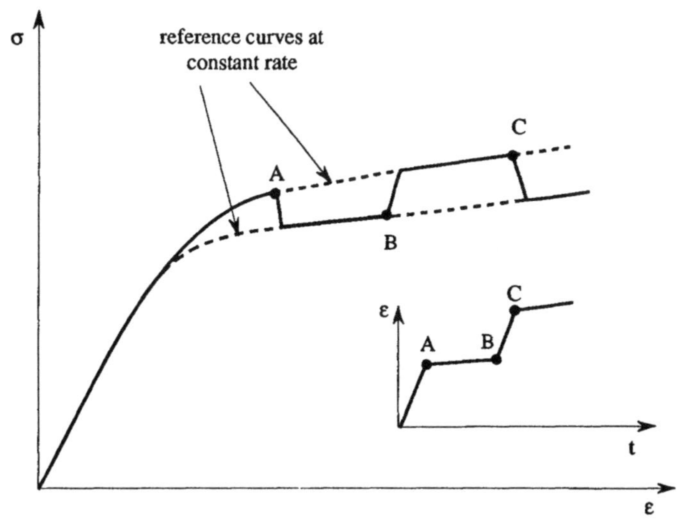

Fig. 1.6 Effect of change of loading rate in a tensile test.

The observations given above do not apply generally. Some materials show an "inverse strain rate effect", peculiar to themselves, in certain regions of temperature and loading rate, in which the resistance to deformation *increases* with decreasing strain rate. These effects are usually associated with instabilities in the modes of deformation. These are shown at the macroscopic level by the Portevin-Le Chatelier effect, characterised by drops in the work-hardening curve when the strain rate is imposed by a servomechanism powerful enough to hold this constant (Fig. 1.7).

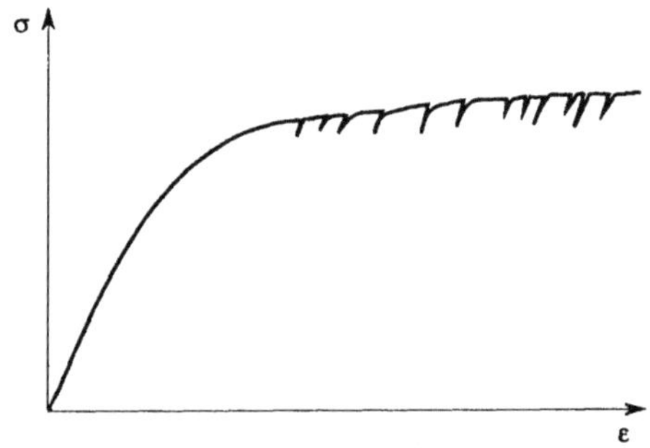

Fig. 1.7 Portevin-Le Chatelier effect in a tensile test under strain control.

If it is the loading rate that is imposed, the curve has broad plateaux so long as nothing keeps the strain rate constant when instabilities occur as is the case in Fig. 1.8: see also I-3.3.8.5 and I - Fig. 3.65.

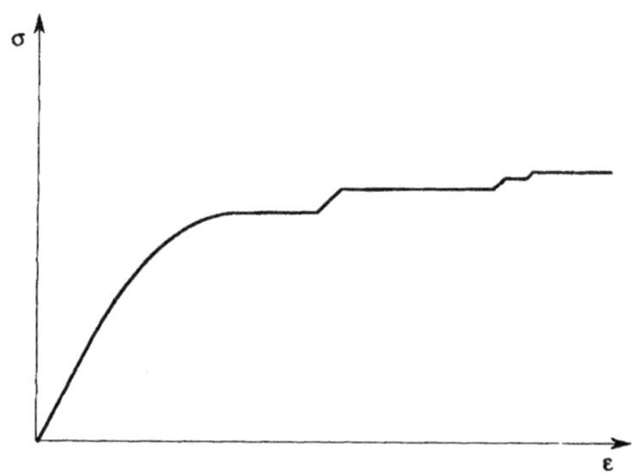

Fig. 1.8 Portevin-Le Chatelier effect in a tensile test under stress control.

1.2.1.3 Relaxation

In a relaxation test a load is applied, generally at a controlled strain rate, followed by holding the deformation constant. During the period of constant total deformation the stress falls while the viscoplastic deformation continues to increase (Fig. 1.9): this confirms that there is a fundamental difference from time-independent plasticity, since the points on the curve are *outside* the elastic region, if indeed there is one. During the relaxation the rates of change of stress and viscoplastic deformation are related by

$$\dot{\sigma} = - E\dot{\varepsilon}^{v} \tag{1.4}$$

were E is Young's modulus in the direction of the traction.

In the case where, in the absence of servo control of some local value, it is not the deformation but a displacement of the test piece that is held constant, the flexibility of the cross-head must be taken into account, and an equivalent, smaller modulus E^* used which comes from outside the working area of the test piece.

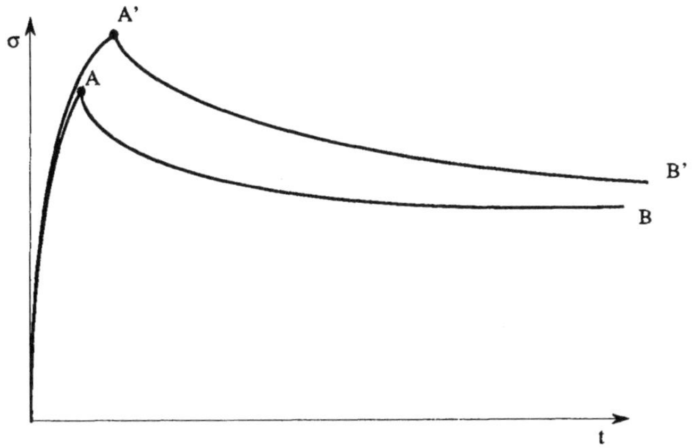

Fig. 1.9 Relaxation test.

It is often more difficult here than in the case of creep to carry out tests of very long duration, mainly because of the difficulty of controlling the temperature. Even so, relaxation tests have the advantage of enabling a wide range of viscoplastic strain rates to be covered, typically from 10^{-2} to 10^{-10}-10^{-12} sec^{-1}. The residual value reached when the fall in the stress has practically stopped (rates of less than 10^{-10} sec^{-1}) corresponds to the upper limit σ_c of the elastic region, introduced in §1.2.1.2. Classically, for a given material, this limit is smaller the higher the temperature, but the value is often high enough for it to be unreasonable to work without an elastic region. The different forms obtained for the relaxation curves according to the laws derived from creep tests are the subject of Exercise 1; in most cases, simple laws that do not involve internal stresses are not able to describe both creep and relaxation simultaneously.

1.2.1.4 Stress discontinuities in creep

The application of sudden changes to the applied stress during primary creep enables the work-hardening in this period to be investigated. The results of such tests can be studied from a plot on a time-viscoplastic strain diagram. The two classical assumptions are that the best indicator of the state of work-hardening is provided, (i) by the time and (ii) by the deformation. This leads to deriving curves for different values of stress by means of translations parallel either to the time axis (*time hardening*) or to the deformation axis (*strain hardening*): see Fig. 1.10, also Exercise 1. The second assumption is the one most often used, the first not having any firm foundation - a problem that arises is what to do if the test is interrupted? It is expressed in terms of Norton's law, modified to allow the constant K to vary with the viscoplastic stress:

$$\dot{\varepsilon}^{v} = \left(\frac{\sigma}{K(\varepsilon^{v})} \right)^{M} \tag{1.5}$$

It is standard to take $K(\varepsilon^v)$ as a power function.

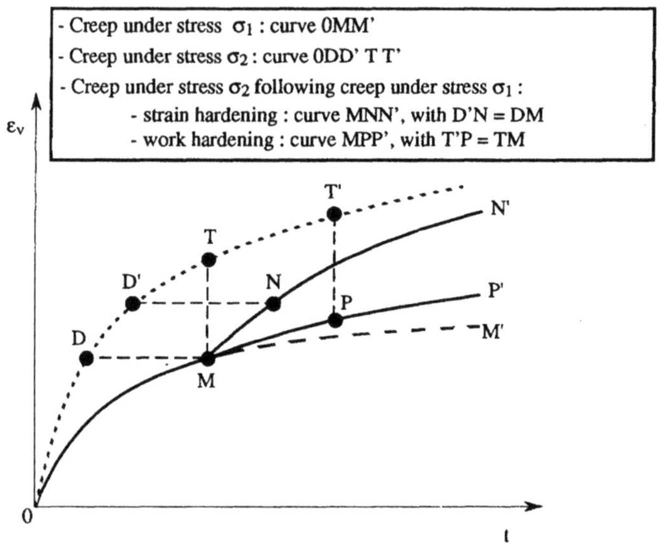

Fig. 1.10 Comparison of the time-hardening and strain-hardening assumptions.

In the *dip test*, used to locate the elastic region, the loading path is more elaborate. It starts with the material in the secondary creep phase, under the stress σ_0, and it is the variables corresponding to this load state that are to be investigated. As Fig. 1.11 shows, a sequence of unloadings of greater or lesser amplitude is made, returning after each to the original stress σ_0 so as not to depart too far from the steady state being studied.

The following effects are normally seen:

- after a small unloading (A in the diagram) viscoplastic flow occurs, possibly after a latency interval: this is *creep hesitation*,

- after a moderate unloading (B) there is no further flow, and the stress state is inside the elastic region,

- after a large unloading compressive viscoplastic flow may occur, the applied stress remaining positive (i.e. tensile), as in case in (C) in the diagram. This shows that the elastic region has been crossed and the effective stress that determines the strain rate has changed sign.

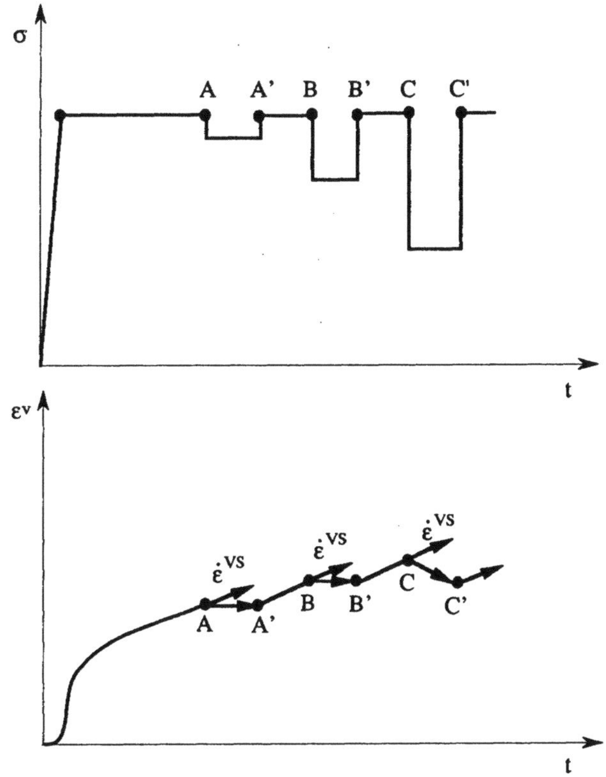

Fig. 1.11 The various stages in a dip test.

Tests of this type give an indication that only laws that involve a threshold term can give a correct representation of the behaviour of a material; Norton's law can be modified with this aim in view by replacing σ by σ - σ_s:

$$\dot{\varepsilon}^{vs} = \left(\frac{\sigma - \sigma_s}{K}\right)^M \tag{1.6}$$

Thus kinematic work-hardening occurs in viscoplasticity as in time-independent plasticity (see I - 3.2.2.2) The formulations that lead to (1.5) and (1.6) respectively are in fact complementary rather than alternative, since both may be required to model general viscoplastic behaviour.

1.2.1.5 Recovery

At high temperatures viscoplasticity is accompanied by other temperature-dependent phenomena that evolve with time; recovery is one of these.

Static recovery is the name given to the phenomena associated with thermally-activated rearrangements at the microstructure level, dislocations in particular, which occur during annealing (I-§3.3.8.3). Recovery usually results in the partial or complete loss of the work-hardening built up in the course of the deformation; to account for this a "forgetting" term must be included in the work-hardening laws, as a function of time. The effect is seen at very low strain rates, even zero.

Dynamic recovery, on the other hand, is directly related to the deformation process and appears in the course of this when the rate is high, say from *1* to *100* sec^{-1}. In particular, it occurs during torsion testing of solid cylindrical bars (I-1.3.4), undertaken to assess the ease with which a metal can be worked, in which the deformations can be *100%* or more. Results of such tests are given in Fig. 1.12 for different strain rates, increasing from bottom to top. The peak of the curve corresponds, in the case of materials with low stacking-fault energy such as stainless steels, to a recrystallisation, and in those for which this is high, such as aluminium alloys, to a polygonisation in the grains (I-§3.3.8.3); the oscillations after the peak are due to alternations between work-hardening and recrystallisation or polygonisation.

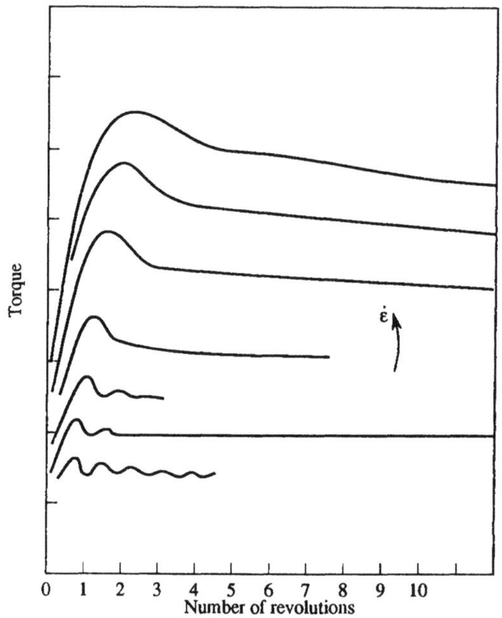

Fig. 1.12 Dynamic recovery during torsion tests

Detailed interpretation of such tests is difficult because new dislocation-free material is formed continuously, so that the state of the work-hardening becomes non-uniform through the test piece. Nevertheless, the test enables different materials to be classified relative to one another.

1.2.1.6 Ageing

There can be other phenomena at the microstructure level, for example increases in the volume fraction of precipitates already formed, or new precipitation, of carbides for example. In contrast to recovery, these can lead to hardening, which becomes evident in a tensile test, for example, if the test piece is held for a time at high temperature and zero load (point B in Fig. 1.13). If the loading is then re-applied the effect of the past history of the temperature changes can be to cause the stress/strain curve to rise above the form it would have had without the ageing. There can also be softening effects, the subject of Exercise 2, in which an attempt is made to form a link between internal stress and precipitation and which show that in certain cases tertiary creep is very simply related to a change in the stress-strain behaviour and not to damage.

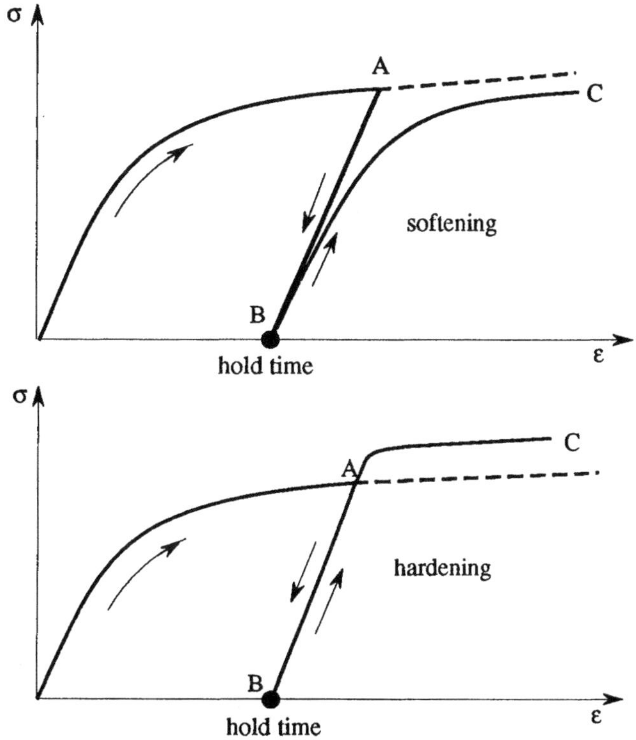

Fig. 1.13 Ageing effect in a tensile test.

1.2.1.7 Cyclic tests

Viscoplastic materials can be characterised with respect to their behaviour under cyclic loading under the same headings as can materials that obey the laws of instantaneous plasticity (Volume I §3.2.1). The cyclic work-hardening curve can be defined as before, but now depends on the loading rate. The *Bauschinger effect* is seen in the stress-strain loops, which can be more rounded than those for a plastic material to the extent that positive viscoplastic flow continues as the tensile load is reduced (Fig. 1.14). Among tests currently in use, in particular for calibrating the laws of rupture, are those of cyclic relaxation and cyclic creep.

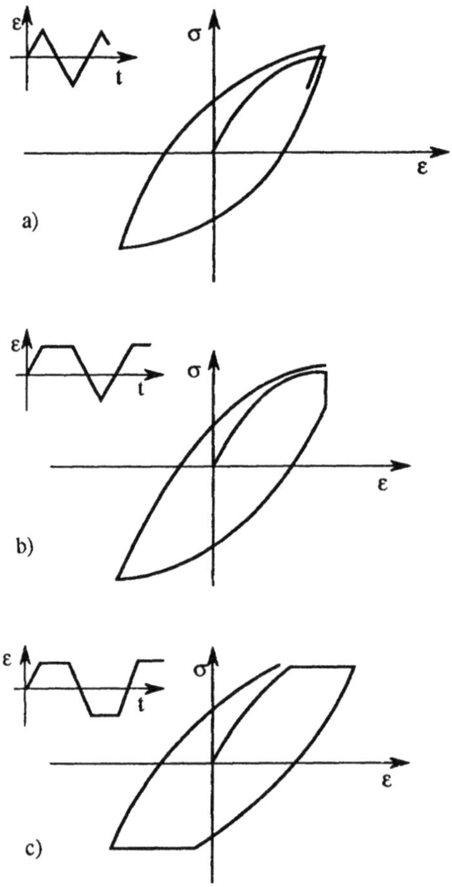

Fig. 1.14 Viscoplastic stress-strain cycles. (a) low-cycle fatigue tests (b) relaxation test with intervals at constant tension (c) creep test with intervals at constant tension and constant compression.

1.2.2 Multi-axial loading

To the extent that single-axis tests raise doubts concerning the concept of an elastic region it is reasonable to ask what might be retained from the formalism of time-independent plasticity in viscoplastic behaviour. Some answer can be derived from results of experiments with multi-axial loads. In plasticity experiments the delineation of the elastic region, either initially or after the first loading, can be made by recording the start of plastic flow in certain directions in a loading plane, for example σ_{11}-σ_{12} or σ_{11}-σ_{22}. In the corresponding tests on a viscoplastic material the strain rates are measured, and the test is in fact one of partial unloadings in multi-axial creep. Fig. 1.15 is for a biaxial creep test in the σ_{11}-σ_{12} plane in which, after arriving at a stable secondary creep regime, the loads are reduced by varying amounts over short intervals (Fig. 1.15a) and the magnitude and direction of the visco-plastic strain rate are measured. The results plotted in Fig. 1.15 show that the vector representing the direction of the flow is in general normal to the equipotentials in the stress plane located with respect to a region displaced from the origin. The magnitude of the flow vector is greater, the further the working point is from the centre of this region. These experimental data fit very well into a schema that uses at the same time the von Mises criteria, kinematic work-hardening and the normality law.

1.2.3 Summary

The experiments we have been describing show that for a viscoplastic material the stress has to be treated as the sum of a critical stress for plastic flow and a "viscous" stress. The first, which can vanish in certain conditions, in particular at high temperature, depends primarily on the deformation (more generally, on the work-hardening) and possibly also on the time over which the time-dependent recovery mechanisms are active. The second depends primarily on the strain rate and possibly on the deformation itself (in formulations of strain-hardening type) (§1.2.1.4). Thus there are two ways of describing the hardening of a viscoplastic material: as modifying the elastic region (*additive* hardening, since the changes raise the stress threshold); or as reducing the rate of flow outside the elastic region (*multiplicative* hardening, since the work-hardening terms multiply the factor that normalises the rate of visco-plastic flow). In models the work-hardening that affects the critical stress for plastic flow will follow a plastic-type law and can be isotropic or kinematic, linear or non-linear. That which affects the "viscous" stress will, by its very nature, be isotropic. As in plasticity, kinematic work-hardening has to be used in order to model cyclic loading correctly.

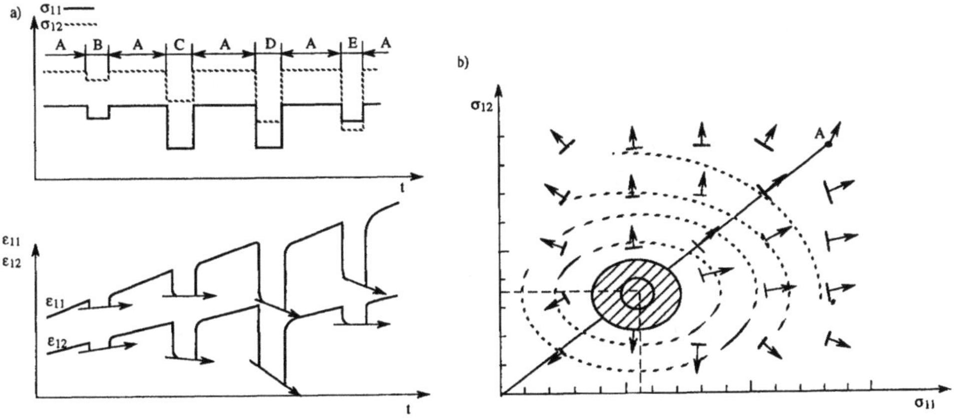

Fig. 1.15 Traction-torsion test with partial unloadings (a) Principle of the test (b) Equipotentials and flow directions in the σ_{11} - σ_{12} plane.

The problem of activating visco-plastic flow is simpler than the corresponding problem in plasticity, since now there are only two regimes to be considered: elastic, when the working point is inside the elastic region, and visco-plastic otherwise. Further, for any given point outside the elastic region the visco-plastic strain rate is completely independent of the rate of change of stress and depends only on the instantaneous values of the stress tensor components and of the work-hardening variables. There is no longer a consistency condition. Thus there is one equation lacking for the evaluation of the strain rate, which gives more freedom in defining this.

The choice between the descriptions we have given will be guided by a consideration of the physical mechanisms we describe next, in particular those concerning the threshold and the law of viscosity.

1.3 THE PHYSICAL MECHANISMS RESPONSIBLE FOR VISCO-PLASTICITY

In this section we look at the physical mechanisms that produce visco-plastic behaviour in crystalline materials. At low temperatures, $T/T_f < 0.3$, it will in general be only thermal activation of plasticity that is responsible for the rate effects seen in a number of metal alloys. The dislocations meet obstacles in the course of their movements, which they are enabled to surmount not only by the slip forces acting on them but also by the thermal agitation; in this temperature region diffusion is too slow for it to be possible for them to do this by climbing At higher temperatures, $T/T_f > 0.5$, it is diffusion that is responsible for viscoplasticity, whether over short or long distances (dislocation or diffusional creep respectively). We consider the two temperature regions in succession.

1.3.1 Low-temperature activation of plastic and visco-plastic deformation

1.3.1.1 Introduction

When carrying out tensile tests on certain metals to measure the yield strength *Rp* at very low temperatures ($T/T_f < 0.2$) we find that this is strongly dependent on the temperature and on the strain rate; this is shown in Fig. 1.16. More generally, the flow stress, measured for example in a tensile test, is a function of strain rate, as indicated in §1.2.3 above (see Fig. 1.5). The best example of this is provided by metals and alloys with BCC structures; in this case the cause of such behaviour is the dislocations' overcoming of the lattice or Peierls-Nabarro forces (Volume I §3.3.8.1) under the influence of the thermal activation. We shall show later that other obstacles may be surmountable by the dislocations under the combined effect of this and the applied stress, but first we shall give the general formalism for the thermal activation of plastic deformation.

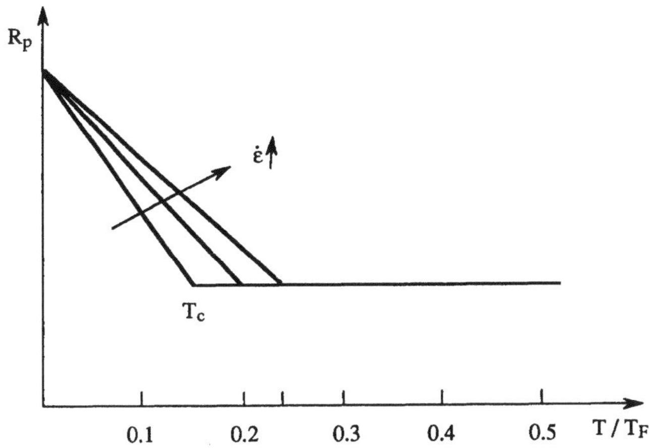

Fig. 1.16 Diagram showing that at low temperatures ($T/T_f < 0.2$) the yield strength varies with both temperature and strain rate. Above a critical temperature T_c is the "athermal" region.

1.3.1.2 Thermal activation of plastic deformation

(a) *Basic mechanisms.* Let ρ_m be the density of mobile dislocations with Burgers vector $\vec{b}$ and v their mean velocity From I-§3.3.3.3 the corresponding strain rate is

$$\dot{\varepsilon}^v = \rho_m b v \tag{1.7}$$

This important relation holds whether the displacement of the dislocations is controlled by slip or by climb. An equivalent form which is better suited to the activation formalism is

$$\dot{\varepsilon}^v = N_m \frac{\Delta A}{V} b \nu \tag{1.8}$$

where N_m is the number of activation sites, that is, those regions in the crystal where a segment of a dislocation is obstructed by an obstacle which the combination of the applied stress and the thermal agitation may enable to surmount; $b\Delta A/V$ is the elementary strain produced when the dislocation crosses a potential barrier that is opposing its free movement (ΔA is the area swept out and V is the volume of the crystal); and ν is the frequency of activation, that is, the number of times per second that a site is activated and the dislocation succeeds in surmounting the obstacle.

Two cases arise, according to the distance λ travelled by the dislocation in crossing the potential barrier (Fig. 1.17). If this is a small multiple of the length b then ΔA is of the order of λl, where l is the length of the liberated segment, and it is possible for the segment to jump back to its original location. But if it is large, corresponding to the distance between precipitates, 10^2 - $10^3 b$, return is not possible.

Over the years 1955-60 the results of the theory of chemical reaction have been applied to the basic physical phenomena underlying the deformation of crystalline solids; this has required a number of simplifying assumptions, which we now describe.

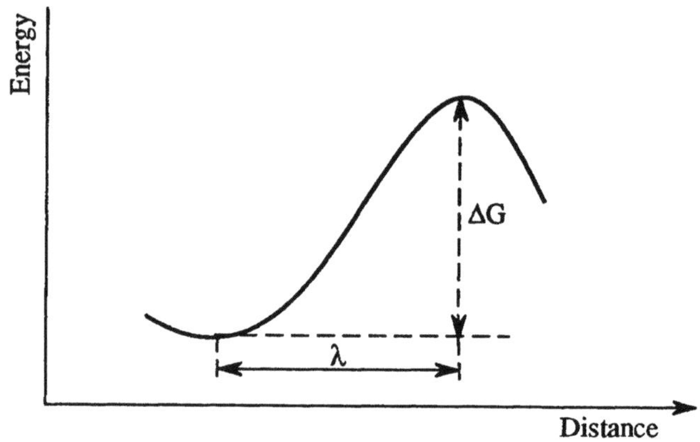

Fig. 1.17 Energy-distance diagram, activation energy.

(b) *General form of expressions relating dislocation rate to stress and temperature*
Taking first the case in which the return jump is possible, the effective frequency will be the algebraic sum of the forward (v^+) and backward (v^-) frequencies:

$$\begin{aligned}\bar{v} &= v^+ - v^- = v_0 \exp\left\{-\frac{Q_0 - \sigma bA}{kT}\right\} - v_0 \exp\left\{-\frac{Q_0 + \sigma bA}{kT}\right\} \\ &= 2v_0 \exp\left\{-\frac{Q_0}{kT}\right\} \sinh\frac{\sigma bA}{kT}\end{aligned} \tag{1.9}$$

Here Q_0 is, to a first approximation, the height of the potential barrier: we shall give the precise significance of this quantity later, and for A, the area swept by the dislocation. v_0 is the jump frequency. If the probability of return is very low we have

$$\bar{v} = v_0 \exp\frac{-Q_0}{kT} \exp\frac{\sigma bA}{kT} \tag{1.10}$$

Substitution of these results in (1.8) enables us, under certain conditions, to give the general form for the relation $\dot{\varepsilon}^v = f(T,\sigma)$.

(c) *Internal stress and effective stress*
As we indicated above when giving some typical experimental results , it is often found necessary to introduce the concept of a threshold; metal physicists usually express this as an internal stress, which we wrote as σ_c, so that the "effective" (σ_{eff}) or "viscous" (σ_v) stress is written

(c) *Internal stress and effective stress*
As we indicated above when giving some typical experimental results , it is often found necessary to introduce the concept of a threshold; metal physicists usually express this as an internal stress, which we wrote as σ_c, so that the "effective" (σ_{eff}) or "viscous" (σ_v) stress is written

$$\sigma_v = \sigma_{eff} = \sigma - \sigma_c \tag{1.11}$$

We shall return to this relation when we come to consider very general models for viscoplastic behaviour; for the moment we need only say that σ_c is an internal stress, at each point in the crystal the sum of the friction stresses and all the long-range (that is, varying like $1/r$) stresses associated with crystalline defects in the solid - that is, mainly the effects of the other dislocations. Thus the internal stress depends only on the temperature, through the intermediary of the elastic moduli, and on the sub-structure and thus on the level of the deformation. It oscillates in space, with a wavelength of the order of the dimensions of the sub-structure (cells or sub-grains) - much too long for the barriers formed by the maxima of the stress field to be surmountable by means of the thermal fluctuations. We denote by σ_s the (non-zero) mean value of this stress.

Table 1.2 classes various obstacles according to their range

Long range	Short range	Activation area (short range obstacles)
Frank network	Peierls valleys	$10b^2$
Attractive junctions	Repulsive junctions	$100b^2$
Precipitates	Solute atoms	$100b^2$

Table 1.2 Some long range and short range obstacles.

Alternatively, we could proceed as in Exercise 3, where, starting from a simplified model consisting of a pair of parallel bars , we try to construct a physical macroscopic image of this idea of an internal stress.

The force-distance diagram, as Fig. 1.18, can be helpful in representing the surmounting of an obstacle by thermal activation; from this:

- if $\sigma < \sigma_s$ the dislocation cannot move at all between the obstacles,

- if $\sigma > \sigma_s$ the dislocation can experience a certain displacement and take up a position of stable equilibrium in contact with an obstacle, which will exert on it a short-range returning force. The stress that contributes to the surmounting of the obstacle is $\sigma_{eff} = \sigma - \sigma_s$; this provides a part of the energy required (area I in Fig. 1.18), the remainder (area II) coming from the thermal activation,

- if the temperature is zero a stress $\sigma = \sigma_s + \sigma_B$ is required for crossing the barrier,

- if the temperature is above a critical value T_c, so that thermal activation alone can provide all the energy represented by the area under the curve, we are in the "athermal" region of Fig. 1.16, with $\sigma = \sigma_s$,

- there is an intermediate temperature T at which the combination of effective stress and thermal agitation is just sufficient for the obstacles to be surmounted at the frequency which allows the strain rate to be equal to the rate prescribed.

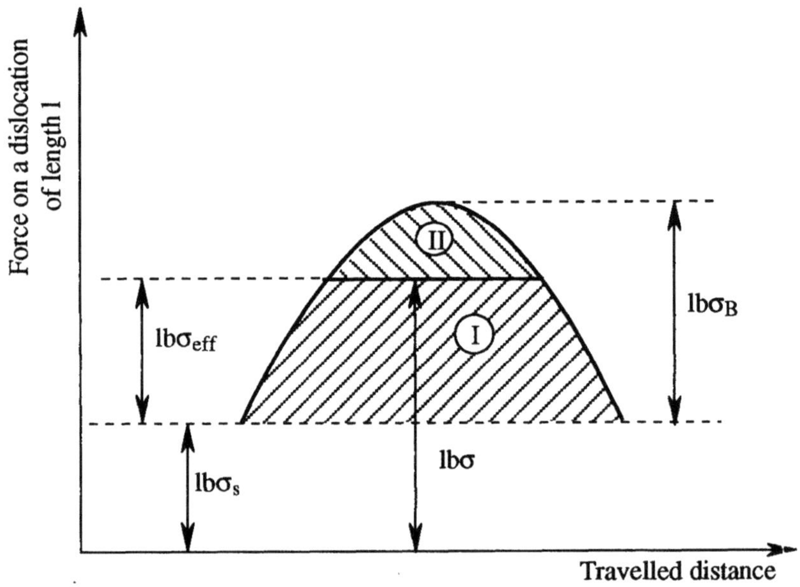

Fig. 1.18 Force-distance diagram, illustrating the concepts of internal stress and viscous or effective stress. Shaded area I represents the energy provided by the effective stress, area II that provided by the thermal activation.

(d) *Thermal activation magnitudes*

Two things are necessary to enable us to take advantage of the relations derived from thermodynamics and to compare the values measured in experiments in thermal activation with those predicted by formal theory. First, the theory must be based firmly on the laws of thermodynamics; and second, the treatment must go from the elementary activation event of a dislocation surmounting an obstacle to the behaviour on the macroscopic scale. We shall take up these two points in order, but in this section we shall give most attention to the first.

To a first approximation the change in Gibbs free energy between the initial position of the dislocation (State 1), corresponding to a stable equilibrium under the action of the stress $\sigma_{eff} = \sigma - \sigma_s$, and its final position (State 2), corresponding to an unstable equilibrium when it has crossed the barrier locally, can be written

$$\Delta G_A = \int_1^2 (\sigma_B - \sigma_{eff})bl\, dx \tag{1.12a}$$

Thus, at least in principle, we can compute ΔG_A if we know the form of $\sigma_B = f(x)$; but we must know also the way the length $l(x)$ changes during the activation.

Since $\sigma_{eff} = \sigma - \sigma_s$ and σ is constant, (1.12a) can be written

$$\Delta G_A = \int_1^2 [\sigma_B(x,T) + \sigma_s(x,T)]bl(x)\, dx \; - \sigma b\int_1^2 l(x)\, dx \tag{1.12b}$$

in which the second term gives the work done by the applied stress.

If we take σ and T as independent variables and σ_s as a constant we can write

$$d(\Delta G) = \left(\frac{\partial\, \Delta G}{\partial \sigma}\right)_T d\sigma + \left(\frac{\partial\, \Delta G}{\partial T}\right)_\sigma dT \tag{1.13}$$

from which we can define the activation entropy by

$$\Delta S_A = -\left(\frac{\partial\, \Delta G_A}{\partial T}\right)_\sigma \tag{1.14}$$

Now $\partial\Delta G/\partial\sigma$ has the dimensions of volume and we should prefer to define an activation *area*, to which it is often possible to give a precise physical meaning; so we define

$$A^* = -\frac{1}{b}\left(\frac{\partial \Delta G_A}{\partial \sigma}\right)_T \tag{1.15}$$

The activation enthalpy is $\Delta H_A = \Delta G_A + T.\Delta S_A$ (1.16)

which after substituting the value for ΔS_A gives

$$\Delta H_A = \frac{\partial\left(\frac{\Delta G}{T}\right)_\sigma}{\partial\left(\frac{1}{T}\right)} \tag{1.17}$$

We should note here that if in this elementary process the length l of the segment of the dislocation is independent of the distance x travelled, then the activation area $A^* = l.\Delta x$ has a very precise physical meaning: it is the area swept by the dislocation in moving from its initial position to that corresponding to the maximum of ΔG_A. In general, however, $\Delta x = f(\sigma)$; but in spite of this restriction it is convenient to write (1.12) in the simpler form

$$\Delta G_A = \int_1^2 [\sigma_B + \sigma_s] bl\, dx \; - \sigma bl \Delta x \tag{1.18}$$

In fact, using relations given by the theory of reaction rates, we get, for the case in which there is no possibility of a return jump,

$$\nu = \nu_0 \exp\left\{ -\frac{\int_1^2 (\sigma_B + \sigma_s) bl\, dx}{kT} \right\} \exp\frac{\sigma b A^*}{kT} \tag{1.19}$$

The integral in the numerator is the change in Gibbs free energy $\Delta G = \Delta H_0 - T\Delta S$; so

$$\nu = \nu_0 \exp\frac{\Delta S}{k} \exp\left(-\frac{\Delta H_0}{kT} \right) \exp\frac{\sigma b A^*}{kT} \tag{1.20}$$

and using (1.8) we have for the macroscopic strain rate

$$\dot{\varepsilon}^v = N_m \frac{\Delta A}{V} b \nu_0 \exp\frac{\Delta S}{k} \exp\left(-\frac{\Delta H_0}{kT} \right) \exp\frac{b\sigma A^*}{kT} \tag{1.21}$$

If there is possibility of a return jump this becomes

$$\dot{\varepsilon}^v = 2 N_m \frac{\Delta A}{V} b \nu_0 \exp\frac{\Delta S}{k} \exp\left(-\frac{\Delta H_0}{kT} \right) \sinh\frac{b\sigma A^*}{kT} \tag{1.22}$$

Three important remarks must be made at this stage.

1. The empirical relation (1.2), used to describe stationary creep, can often be written

$$\dot{\varepsilon}^v = \dot{\varepsilon}_0 \exp\left(-\frac{Q}{kT} \right) \tag{1.23}$$

which enables an apparent activation energy Q to be defined as

$$Q = -k \frac{\partial \log(\dot{\varepsilon}^v / \dot{\varepsilon}_0)}{\partial (1/T)}$$

If now in (1.21) and (1.22) we put

$$\dot{\varepsilon}_0 = N_m \frac{\Delta A}{V} b v_0 \exp\frac{\Delta S}{k} \tag{1.24}$$

and assume that the terms in $\dot{\varepsilon}^v$ are independent of T, we have $Q = \Delta H_0 - \sigma b A^*$. This shows that in the empirical relation (1.23) the apparent activation energy is indeed a function of the applied stress.

2. Norton's law, equation (1.1c) involves an exponent M; in rheological studies the parameter

$$m = 1/M = \left(\frac{\partial \log \sigma}{\partial \log \dot{\varepsilon}^v}\right)_T \tag{1.25}$$

is often used in investigations of rate sensitivity. If $\dot{\varepsilon}_0$ is independent of σ then m is directly related to the activation area and

$$m = b\sigma A^*/kT \quad \text{and} \quad A^* = \frac{kT}{b}\left[\frac{\partial \log(\dot{\varepsilon}^v / \dot{\varepsilon}_0)}{\partial \sigma}\right]_T \tag{1.26}$$

3. To give a physical meaning to the quantities m and Q just introduced we return to the simplifying assumptions that l is constant and $\dot{\varepsilon}^v$ is independent of T and σ, the first of which may in some respects seem inconsistent with the fact that in general A^* is a function of σ. However, in the process of deriving the law of macroscopic deformation from the elementary mechanisms the real problems have not been attacked; in particular, much more attention needs to be given to:

- finding a more faithful representation of the passage from the macroscopic stress state to the actual stresses acting on the dislocation,

- relating the deformation rate in a slip plane to the mean over the many events,

- expressing the macroscopic deformation as the mean over many slip planes.

For all these reasons we can only claim to measure mean values, given by

$$\overline{\Delta G_A} = -k\left[\frac{\partial \log(\dot{\varepsilon}^v / \dot{\varepsilon}_0)}{\partial (1/T)}\right]_\sigma, \quad \overline{A} = \frac{kTm}{b\sigma} = \frac{kT}{b}\left[\frac{\partial \log(\dot{\varepsilon}^v / \dot{\varepsilon}_0)}{\partial \sigma}\right]_T \tag{1.27}$$

1.3.1.3 *Nature and resistance of obstacles*

The obstacles met by a dislocation in the course of slipping can be put into two main categories, shown diagrammatically in Figs. 1.19 and 1.20:

- discrete, either already existing, such as precipitates or solid-solution atoms, or induced by the deformation, such as trees in the forest of dislocations,
- diffuse, such as lattice forces, which are overcome by the dislocation making *kinks* (I-3.3.8.1).

The strain rate is imposed either by the kinetics of surmounting discrete obstacles (Fig. 1.19) or by that of development and propagation of the kinks (Fig. 1.12). Table 1. gives the orders of magnitude of the energies and stresses ($\sigma_0 = \sigma_B + \sigma_s$) necessary to surmount the obstacle in the absence of thermal activation, that is, at absolute zero.

Obstacle force	Energy Q_0	σ_0 (0°K)	Examples
Strong	$\approx \mu b^3$	$\approx \mu b/l$	precipitates, by-passed by the dislocations
Medium	0.2 to μb^3	$\approx \mu b/l$	dislocation of the forest small sheared precipitates irradiation defects
Weak	$\leq 0.2\mu b^3$	$<< \mu b/l$	lattice forces solid solution

Table 1.3 Energies and stresses needed to surmount various obstacles

We now show how the formalism of thermal activation of plastic deformation can be used to treat three examples concerning obstacles of the two main classes.

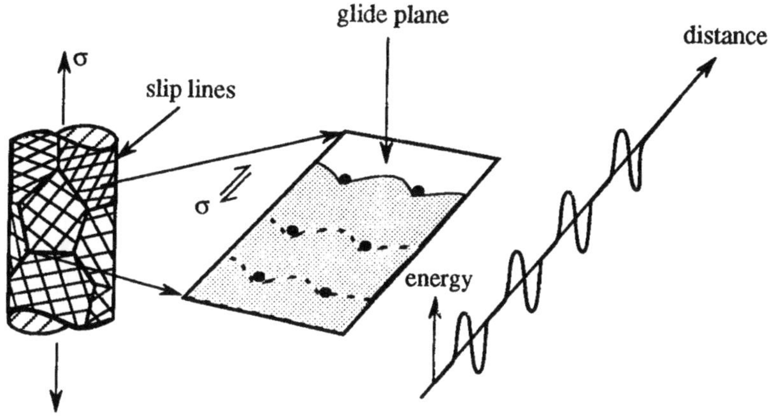

Fig. 1.19 Discrete obstacles in a slip plane, opposing the passage of a dislocation

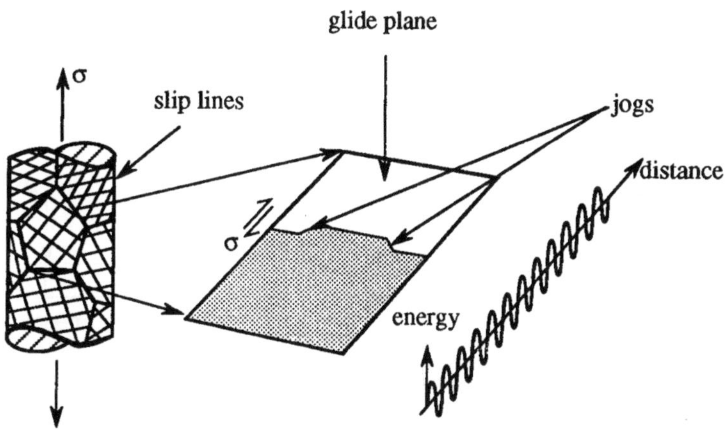

Fig. 1.20 Diffuse obstacles formed by kinks in a dislocation which crosses the Peierls valleys.

(a) *pre-existing obstacles:* precipitates
Hardening by precipitation provides an example of pre-existing obstacles (see Exercise 4). We recall that if these are perfectly coherent with the matrix then the dislocation can negotiate them by shearing if they are small or bypassing them if they are above some critical size, of the order of t/γ_A where t is the line tension of the dislocation and γ_A is the anti-phase energy corresponding to the shearing of the particles (I - 3.3.8.5).

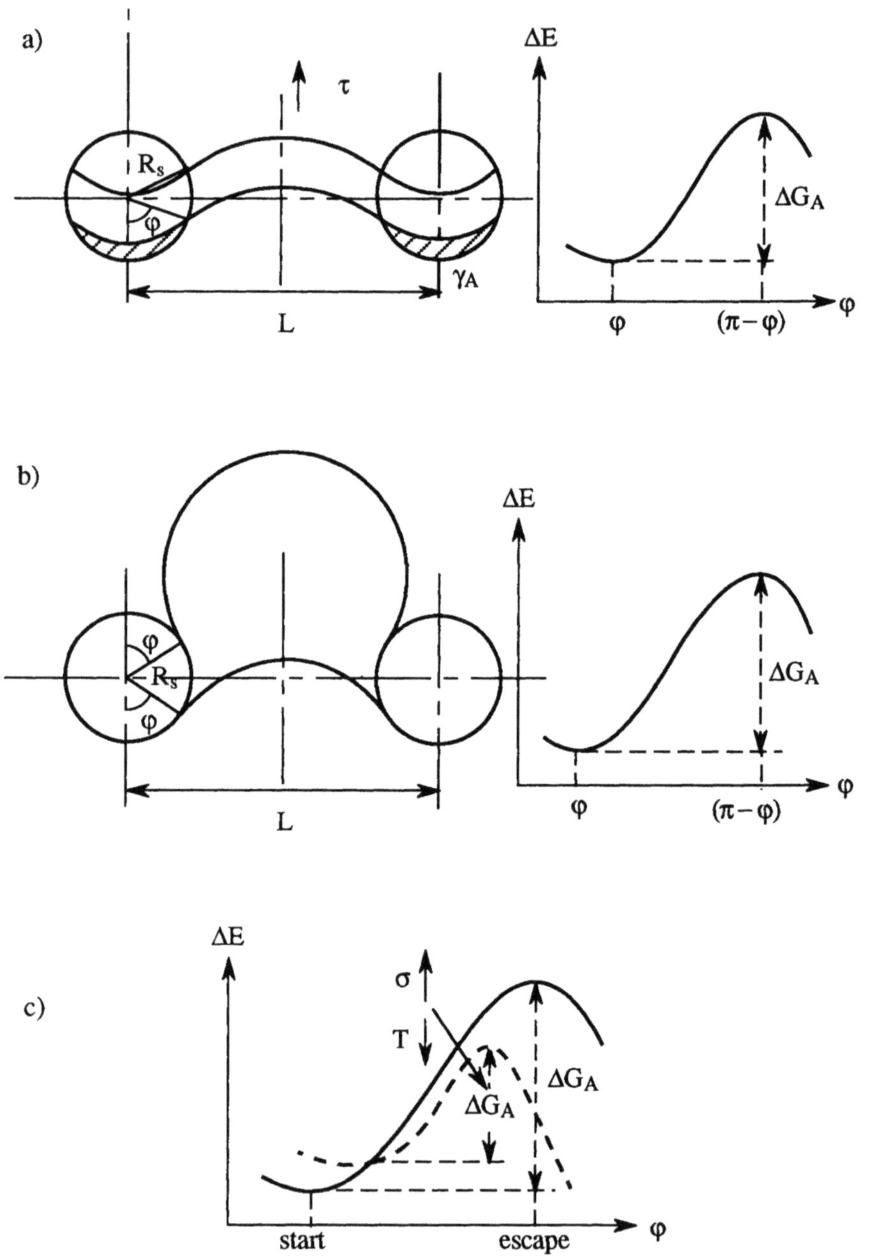

Fig. 1.21 Equilibrium positions for the crossing of two precipitates by a dislocation line by (a) shearing (b) bypassing. Fig. (c) shows the positions of the corresponding energy extrema and the effect of temperature (and thus of the applied stress for a given strain rate) (see also Exercise 4).

Whether the precipitates are sheared or bypassed, it is easy to show from the equilibrium equation that the dislocation has two equilibrium positions: one corresponding to the start (minimum on the energy curve, the other at the break-out (maximum of the energy curve) (Fig. 1.21 a and b). The way in which the dislocation crosses the energy jump ΔG_A, which it must when the stress increases (and therefore when the temperature decreases) is indicated in Fig. 1.21c. The exact expressions giving the total energy and the form of the curves of the diagram are (Exercise .4):

- for shearing $\Delta E = t(\Delta l_m + \Delta l_p) + \gamma_A \Delta S_p - \sigma b(\Delta S_m + \Delta S_p)$ (1.28)

- for bypassing $\Delta E = t\Delta l_m - \sigma b \Delta S_m$ (1.29)

where Δl_m, Δl_p are the increases in length of the part of the dislocation in the matrix and the precipitate respectively, relative to some arbitrarily-chosen reference position, and ΔS_m, ΔS_p are the increases in the areas swept out by this dislocation.

Since for obstacles of this type the probability of a return jump is very low we can use (1.10) to determine the strain rate: thus

$$\dot{\varepsilon}^v = N_m \frac{\Delta A b}{V} v_0 \exp\left(-\frac{Q_0 - \sigma b A^*}{kT}\right) \tag{1.30}$$

where N_m is a constant and $\Delta A \approx L\lambda$. The values of Q_0 and A are easily determined for the two modes from the expressions already given (§1.3.1.2). Thus for shearing under high stress, meaning in the region of absolute zero, it can be shown that Q_0 has the expected form, that is

$$Q_0 = \pi R_s^2 \gamma_A \tag{1.31}$$

and $A^* = 2LR_s \sqrt{1-(\sigma/\sigma_0)^2}$ where $\sigma_0 = 2\gamma_A/R_s$ (1.32)

The height of the activation barrier can be written

$$\Delta G_A = Q_0[1 - (\sigma/\sigma_0)]^{3/2} \tag{1.33}$$

enabling us to express the deformation rate in terms of temperature and stress, thus:

$$\dot{\varepsilon}^v = \dot{\varepsilon}_0^v \exp\left[-\frac{Q_0}{kT}(1-\sigma/\sigma_0)^{3/2}\right] \tag{1.34}$$

where $\dot{\varepsilon}_0^v$ is constant.

This relation, with the others of the same type developed in this section, is commonly used to represent the variation of the yield strength ($\sigma = R_p$) with temperature and stress.

(b) *discrete obstacles, deformation-induced* : trees in the forest of dislocations.
In Volume I- §3.3.8.2 we studied the hardening caused by attractive and repulsive boundaries between dislocations. We showed that if there is no thermal activation a critical stress σ_0 has to be applied to a dislocation in the middle of the forest to make it slip, where

$$\sigma_0 = \alpha\mu b\sqrt{\rho}\ \text{, with } \alpha \text{ approximately between } 1/3 \text{ and } 1/4 \tag{1.35}$$

The energy required to surmount the junctions is of the order of $\gamma\mu b^3$ with $1/5 < \gamma < 1$.

In the Orowan equation (1.7) the density of obstacles, in contrast to (a) above, is an increasing function of the deformation. With the mean velocity of the dislocations written as

$$\bar{v} = \beta b v_0 \exp\left(-\frac{Q}{kT}\right) \quad \text{where } \beta \text{ is a numerical factor} \tag{1.36}$$

and the relation between the density of mobile dislocations and applied stress as

$$\rho_m = \alpha'(\sigma/\mu b)^2 \quad \text{where } \alpha' \approx 1 \text{ is a constant,} \tag{1.37}$$

the strain rate can in general be written

$$\dot{\varepsilon}^v = \alpha'\left(\frac{\sigma}{\mu b}\right)^2 b^2 \beta v_0 \exp\left(-\frac{Q}{kT}\right) \tag{1.38}$$

The activation energy is a decreasing function of the applied stress, as in the case of pre-existing discrete obstacles; if we assume this to be linear, that is

$$Q = Q_0(1 - \sigma/\sigma_0) \tag{1.39}$$

we obtain

$$\dot{\varepsilon}^v = \dot{\varepsilon}_0 \exp\left[-\frac{Q_0}{kT}\left(1 - \frac{\sigma}{\sigma_0}\right)\right] \quad \text{where } \dot{\varepsilon}_0 = \alpha'\beta v_0(\sigma/\mu)^2 \tag{1.40}$$

(c) diffuse obstacles: low-temperature visco-plasticity controlled by lattice resistance. We first recall that at low temperatures ($T/T_f < 0.2$) the velocity of movement of dislocations in most metallic materials, particularly with BCC structures, is controlled by the surmounting obstacles of another type, those associated with the atomic structure (the Peierls-Nabarro forces). The energy of a dislocation varies with its position, as shown in Fig. 1.20; it advances by propagating simple or double kinks (I-Fig. 3.36) and in general the velocity with which it moves from one valley to another is controlled by the rate at which diversions are generated. As before, the activation energy can be written

$$Q = Q_0 \left[1 - \left(\frac{\sigma}{\sigma_0} \right)^p \right]^q \tag{1.41}$$

where $Q_0 \cong 0.2\mu b^3$, σ_0 is the elastic limit at absolute zero and p and q are close to 1; according to Frost and Ashby (1982) the choices $p = 3/4$, $q = 4/3$ give the best agreement.

The density of kinks is an increasing function of the strain; if we assume that it varies in the same way as that of trees in the forest we find that the deformation rate is

$$\dot{\varepsilon}^v = \dot{\varepsilon}_0 \exp\left[-\frac{Q}{kT} \left(1 - \frac{\sigma}{\sigma_0} \right)^{3/4} \right]^{4/3} \tag{1.42}$$

where $\dot{\varepsilon}_0 = \dot{\varepsilon}_0'(\sigma / \mu)^2$, $\dot{\varepsilon}_0'$ a constant

Comments

1. The relations we have just derived do not apply in the case of the very high strain rates (above $10^4\ sec^{-1}$) that are found in certain conditions of shock or in certain forming operations. In such cases the mobility of the dislocations is controlled by phonon or electron drag.

The velocity with which a dislocation moves can be related to the force F exerted on it through the intermediary of a mobility, so that

$$v = F/B = \sigma b/B \tag{1.43}$$

where B is the drag coefficient; substituting in the Orowan equation we get

$$\dot{\varepsilon}^v = \rho_m \frac{b^2 \mu}{B} \frac{\sigma}{\mu} \tag{1.44}$$

Fig. 1.22 Elastic limit in shear τ_p as a function of strain rate, at very high rates, for a mild steel.

At very high rates the ratio ρ_m/B is practically independent of stress and of temperature, so that this relation becomes an expression of Newtonian viscous flow (Fig. 1.22).

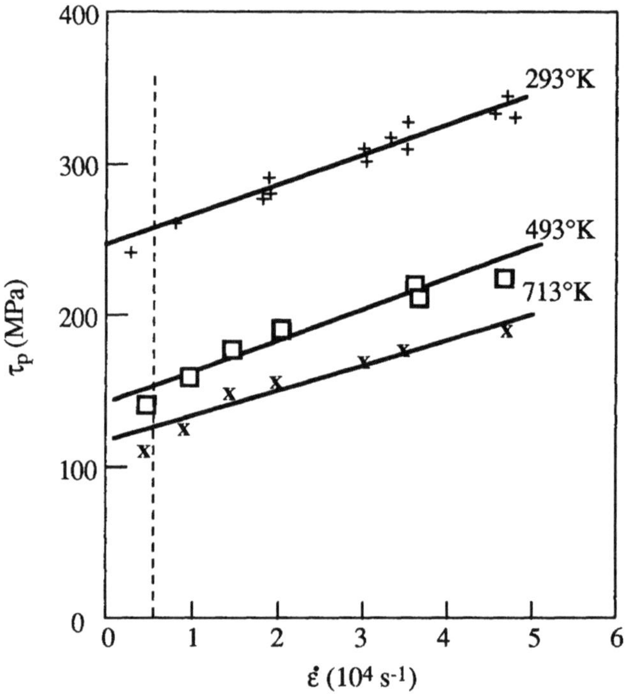

Fig. 1.22 Yield strength in shear τ_p as a function of velocity, for a mild steel at very high strain rates (> $10^4\ s^{-1}$) (Campell and Ferguson (1970)).

2. As we have just shown, this type of low-temperature visco-plasticity is observed especially in materials with covalent bonds or with BCC crystal structures. Fig. 1.23 gives another way of representing these effects, with the shear elastic limit plotted against strain rate for various temperatures. Three regions can be distinguished, as indicated:

(I) low temperature (< $293\ K$) and low-to-moderate rates (< $10^3\ sec^{-1}$); this is the region studied in the current paragraph.

(II) very high rates (> $10^4\ sec^{-1}$), in which the elastic limit is effectively a linear function of the rate; this is the region of phonon drag.

(III) a region in which the elastic limit varies very little with rate and temperature, corresponding to the athermal plateau of Fig. 1.16. The activation energy is now independent of temperature and applied stress and is sufficiently high for the thermal activation energy kT to be comparable to what is needed to surmount the obstacles.

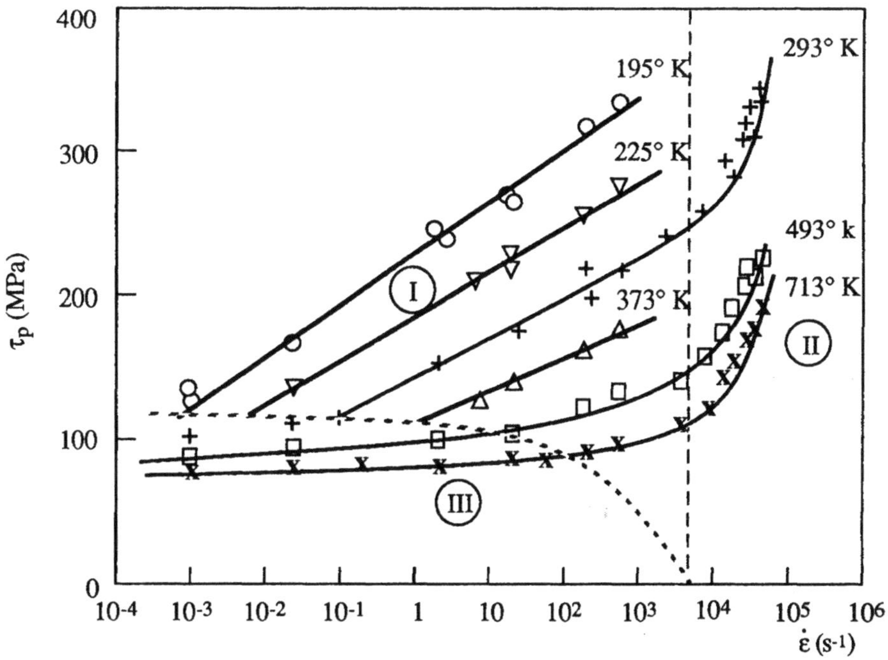

Fig. 1.23 Effect of temperature and strain rate on shear elastic limit for a mild steel (Campbell & Ferguson, *loc. cit.*)

1.3.2 Physical models of high-temperature viscoplasticity

As we stated in the introduction to §1.3, at high temperatures ($T > 0.5T_f$) viscoplasticity involves diffusion phenomena: short-range *dislocation creep* controlled by the movements of the dislocations and longer-range *diffusional creep* over distances of the order of the grain size, when the deformation is caused by movement of the material due to trans- or intergranular diffusion. We shall study these two modes of deformation in order, together with the physical models which have been suggested for their interpretation. These models are based mostly on simple creep experiments - which is why the terms dislocation creep and diffusional creep are used.

It is beyond the scope of this chapter to describe in detail all the many models that have been suggested for creep. We therefore limit our treatment to introducing the concepts underlying two main classes of models: *recovery creep* and *creep controlled by dislocation climb*; further, we shall consider only steady-state creep.

Diffusional creep, which acts at the higher temperatures and lower stresses, enables viscous deformation to occur without the influence of dislocations; inelastic deformation is brought about by transport of material along grain boundaries (*Coble creep* see Coble (1963)) or, at still higher temperatures ($T > 0.8T_f$), within the grains (*Herring-Nabarro creep* see Herring (1950), Nabarro (1947), 1952)). The behaviour is thus perfectly viscous and the threshold constant σ_s in Equation 1.6 is practically zero.

1.3.2.1 Dislocation creep

(a) *Recovery creep.* Many models have been based on the original idea of Bailey and Orowan, according to which a stationary state is reached when the work-hardening that normally results from the accumulation of deformations and the increase in density of dislocations is compensated for by the recovery, a phenomenon regulated primarily by time and diffusion (I-3.3.8.3).

When the diffusion is fairly rapid the obstacles to slip can vanish, which means that in Fig. 1.18 we can set $\sigma_B = 0$. The only barrier that remains is that due to the internal stress, and therefore $\sigma = \sigma_s$. The deformation is no longer activated thermally in the strict sense of that term, but the rate depends on that of the recovery, which is activated thermally through the intermediary of the diffusion.

If the creep strain rate is to remain constant so must the internal stress σ_s; thus

$$\frac{d\sigma_s}{dt} = \left(\frac{\partial \sigma_s}{\partial \varepsilon}\right)_t \dot{\varepsilon}^v + \left(\frac{\partial \sigma_s}{\partial t}\right)_\varepsilon = 0 \tag{1.45}$$

If we define $r = -(\partial\sigma_s/\partial t)_\varepsilon$, $h = (\partial\sigma/\partial\varepsilon)_t$ for the recovery and work-hardening rates respectively we have

$$\dot{\varepsilon}^v = r / h \tag{1.46}$$

The models proposed for this region differ from one another mainly in the way r and h are evaluated and in the assumptions made in obtaining their values.

An approach due largely to McLean (1962) shows that creep-recovery models lead naturally to the empirical equation 1.1c. His original model assumes that the internal stress arises solely from a three-dimensional network of dislocations of mean length l; this could correspond to the cells observed in creep at high stress. The internal stress is then inversely proportional to l, that is

$$\sigma_s = \mu b/l \tag{1.47}$$

Further, it is assumed that the deformation results from the slipping of dislocations in this network of cells, so that the Orowan relation can be written

$$\varepsilon = \rho_m bl \tag{1.49}$$

If we assume that the density of mobile dislocations (ρ_m) is proportional to the overall dislocation density we can write $\rho_m = \alpha/l^2$ where α is the constant of proportionality; and taking this together with (1.47) we get

$$h = (\partial\sigma_s/\partial\varepsilon)_t = \mu/\alpha = constant \tag{1.49}$$

in which it is assumed that α itself is constant. Various approaches have been made in order to determine the velocity of recovery, of which the following is the simplest. A network of cells will tends to increase its size so as to reduce the total length of dislocations and therefore the stored energy. Friedel (1964) assumes that this growth is brought about by the climbing of the arcs of dislocations which exchange vacancies. The flow of vacancies is stimulated by the difference in free energy of those in equilibrium in the neighbourhood of the different arcs.

The recovery rate shown by the increase of cell size with time (dl/dt) will be proportional to the diffusion coefficient D_v and the driving force $\sigma\Omega/kT$. where $\Omega \approx b^3$ is the volume occupied by a vacancy, and this speed will increase with decreasing cell size. Thus we can write

$$dl/dt = D_v(1/l)\sigma_s b^3/kT \tag{1.50}$$

so that

$$r = -\partial\sigma_s/\partial t = -(\partial\sigma_s/\partial l)(dl/dt) = (D_v/kT)(\sigma_s^4 b/\mu^2) \tag{1.51}$$

from which, with (1.46) we find

$$\dot{\varepsilon}^v = \alpha \frac{D_o b}{kT}\frac{\sigma^4}{\mu^3}\exp\left(-\frac{Q_D}{kT}\right) \tag{1.52}$$

where Q_D is the self-diffusion energy.

(1.52) can be written $$\frac{\dot{\varepsilon}^v kT}{D_v \mu b} = \alpha\left(\frac{\sigma}{\mu}\right)^4 \quad \text{(Dorn's form)} \tag{1.53}$$

Models of this type lead to laws of creep that are in reasonably good agreement with results found in experiments with pure metals, but as we have remarked previously, the exponent in the stress equation can differ from the value *4*, and other models give values between *4* and *5*. However, it is important to bear in mind that the agreement of the value of m deduced from a model with that found by experiment is never of itself a proof of the validity of the model.

(b) *Creep controlled by dislocation climb*. The mechanism of dislocation climb can enable recovery to take place. One model in particular, due initially to Weertman, is based on the questionable assumption that this mechanism acts through the intermediary of a configuration of dislocations such as in Fig. 1.24, in which there are two pile-ups of edge dislocations of opposite signs, emitted by sources S and S', a distance *h* apart and resting against a barrier which can be either a grain boundary or another pile-up of dislocations. The stresses at the head of the pile-up, in contrast to the case of low-temperature plasticity, can be relieved by the climb and annihilation of the dislocations there. When a dislocation at the head has climbed the source can become active again and emit a new dislocation. Thus as in the general model of recovery creep there is competition between the production and the loss of dislocations.

An estimate of the corresponding strain rate can be derived from the equation for the dislocation climb

$$\dot{\varepsilon}^{v} = bANc/h \tag{1.54}$$

where A is the area swept by the dislocation in moving between the source and the head of a pile-up of length L, approximately L^2, N is the number of sources per unit volume and c is the mean velocity of climb.

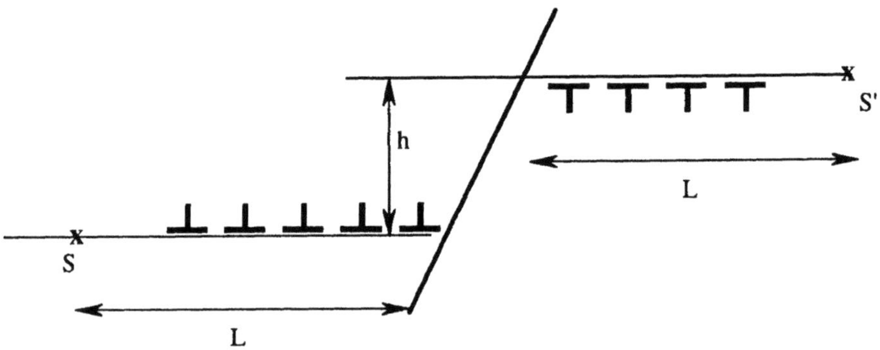

Fig. 1.24 Mechanism of recovery creep, by climb of dislocations located in pile-ups.

We know that if lattice friction can be neglected the local stress at the head of the pile-up is (I-3.85)

$$\sigma_l = \frac{\pi(1-\nu)L}{\mu b}\sigma^2 \approx \frac{2L\sigma^2}{\mu b} \tag{1.55}$$

The component of this stress normal to the slip plane favours the climb. Vacancies will be created or destroyed according as it is a tension or a compression; if C_0 is the equilibrium concentration of vacancies in the neighbourhood of the pile-up this is expressed by

$$C_S = C_0 \exp\left(\pm\frac{2L\sigma^2 b^2}{\mu kT}\right) \tag{1.56}$$

A flow of vacancies will be set up between the pile-ups of opposite signs, giving a velocity of climb

$$c = \frac{1}{AN}\frac{2D_v\sigma^2 b}{\mu kT} \tag{1.57}$$

Proof of these relations follows the same lines as for diffusional creep, which we shall give in §1.3.2.2 below.

The height h of the climb between the pile-ups is derived from the force needed to separate a pair of dislocations of opposite signs, a distance h apart: this is (I-3.46 and Fig. 3.32 in Volume I):

$$\sigma = \frac{\mu b}{4\pi(1-\nu)h} \tag{1.58}$$

from which we find

$$\dot{\varepsilon}^{v} = \frac{2D_v\sigma^2 b^2}{h\mu kT} = 8\pi(1-\nu)\frac{D_v\sigma^3 b}{\mu^2 kT} \tag{1.59, 60}$$

If we assume that both the number of sources and the sizes of the pile-ups remain constant, this gives a creep law with a smaller exponent than is normally found in pure metals. Weertman suggests that such an assumption is not valid, and that L^2 is proportional to σ, giving an exponent 4.4, which is closer to what is observed; but there is no theoretical basis for this. Further, a serious objection to this model is that the dislocation pile-ups on which it is based have never been observed in materials subjected to creep.

1.3.2.2 Diffusion creep

At high temperatures ($T > 0.7T_f$) and low stresses the relation between deformation rate and applied stress is often found to be linear, as indicated in the introduction to this chapter. The deformation then results not from movement of the dislocations but from transport of material by diffusion, either across the grains (Herring-Nabarro model) or between the grains (Coble model). The activation energy for the Coble model is less than that for volume diffusion, consequently Coble creep will occur at lower temperatures than will Herring-Nabarro. These purely diffusive modes of transport are very important for the study of the compatibility of deformation between the grains of a polycrystal and intergranular slip; and further, they are the origin of the deformation of superplastic alloys. We now consider the two in turn.

(a) *Herring-Nabarro creep.* Herring (1947, 1952)) suggested that since in a stress field that is not purely hydrostatic the concentration of vacancies will differ on surfaces having different orientations, there will be a concentration gradient that can generate a flow of vacancies and consequentially a flow of material in the opposite direction. We now give a simple form of the model based on this and later improved by Herring.

Consider first a cubic single crystal, of side d, subjected to shear (Fig. 1.25); we assume that it contains no dislocations and that the free surfaces are therefore the only sources and sinks of vacancies. Creating a vacancy on a surface in compression (BC) is equivalent to moving an atom from the interior to this surface; if the atom is represented by a small cube of side b, the atomic volume Ω can be taken as approximately b^3. To leave the surface the atom must do work against the stress σ, and the energy thus dissipated will be $\sigma b^2 \times b = \sigma b^3$, which is equivalent to saying that the energy of creation of a vacancy on BC is increased by σb^3.

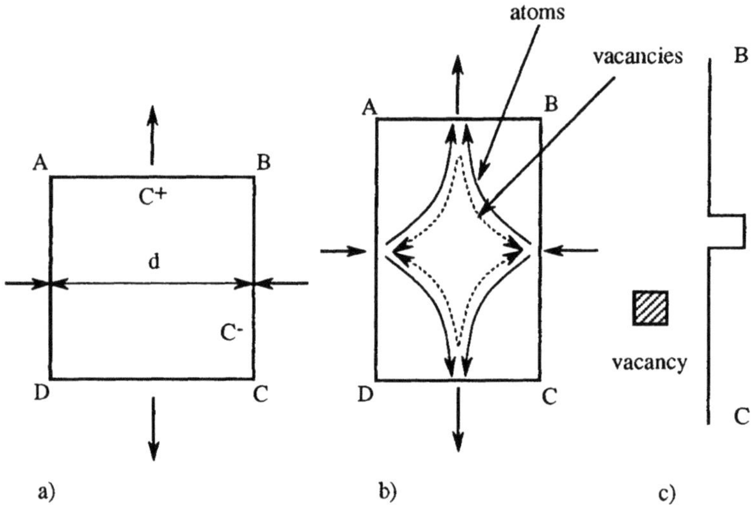

Fig. 1.25 Herring-Nabarro model: transport of vacancies from faces in tension (AB, CD) to faces in compression (AD, BC).

The corresponding atomic fractions will be
- on the faces in tension, AB, DC:

$$n^+ = \exp\left(-\frac{\Delta G_f - \sigma b^3}{kT}\right) = n_0 \exp\frac{\sigma b^3}{kT} \tag{1.61a}$$

-on the faces in compression, AD, BC

$$n^- = \exp\left(-\frac{\Delta G_f + \sigma b^3}{kT}\right) = n_0 \exp\left(-\frac{\sigma b^3}{kT}\right) \tag{1.61b}$$

where $n_0 = \exp(-\Delta G_f/kT)$ (1.62)

is the atomic fraction in thermal equilibrium in the absence of applied stress and ΔG_f is the Gibbs free energy of creation of vacancies, of the order of the sublimation energy of the solid.

This results in a gradient in the concentration of vacancies that causes a flow of vacancies across the body of the grains and of atoms in the opposite direction. If the faces are perfect sources and sinks for vacancies then, in the steady state, the number transported across a face in tension, whose area is d^2, will be

$$\Phi = -Jd^2 \tag{1.63}$$

where, from Fick's law

$$J = -D_L \,\mathrm{grad}C \tag{1.64}$$

with D_L the diffusion coefficient for the vacancies and grad C denoting the concentration gradient across the faces; we can write

$$\mathrm{grad}C = \alpha(C^+ - C^-)/d \tag{1.65}$$

where α is a numerical factor, close to *1*.

The concentration C is related to the atomic fraction n by $C = n/b^3$ (1.66)

and the diffusion coefficient D_v to D_L by $D_v = D_L n_0 = D_L C_0 b^3$ (1.67)

from which relations we find

$$\Phi = \frac{2\alpha D_v d}{b^3} \sinh\frac{\sigma b^3}{kT} \tag{1.68}$$

The escape of an atom from the face AB corresponds to an elementary extension which can be calculated by regarding the atom, of volume b^3, as spread over the surface, of area d^2:

$$\varepsilon = (b^3/d^2) \times (1/d) = b^3/d^3 \tag{1.69}$$

Thus the deformation rate corresponding to the flux Φ is

$$\dot{\varepsilon}^v = \Phi\frac{b^3}{d^3} = \frac{2\alpha D_v d}{d^2} \sinh\frac{\sigma b^3}{kT} \tag{1.70}$$

which for small values of σ becomes

$$\dot{\varepsilon}^v = \frac{2\alpha D_v}{d^2}\frac{\sigma b^3}{kT} \tag{1.71}$$

Extending this simplified calculation to the case of a polycrystal presents a number of problems: we have to take into account first the fact that the grain boundaries can act as perfect sources and sinks, and second the actual shape of the grains. However, all cases lead to expressions of the form

$$\dot{\varepsilon}^v = B\frac{D_v \sigma \Omega}{d^2 kT} \tag{1.72}$$

where the factor B is a function of the geometry and Ω ($\approx b^3$) is the atomic volume. Herring gives the value $B = 16$ for spherical grains.

The most important results of this model appear to be the following:

(i) diffusional creep can be expressed by a law of Newtonian viscous flow type

(ii) the rate of this creep increases with increasing temperature ($D_v = D_0\exp(-Q_0/kT)$ and with decreasing grain size (like $1/d^2$))

(b) *Coble creep.* Coble (1963) has proposed a model for creep in polycrystals in which the rate is controlled by diffusion, not through the grain bodies but along the grain boundaries; this can be faster, since the intergranular diffusion energy is only about half that of volume diffusion. The bases for the model are the same as for Herring-Nabarro (Fig. 1.25); we give a simplified treatment here, referring the reader to Exercises 5 and 6 for a fuller treatment.

Consider a grain in the form of a cube of side d. The vacancies and the atoms flow along the boundaries and if δ_j is the width of the grain boundary the flux of vacancies is

$$\Phi = \alpha D_L \frac{\Delta C}{d}(4\delta_j d) \tag{1.73}$$

Introducing the intergranular diffusion coefficient $D_j = D_L C_0 b^3$ and substituting the expression already found for ΔC:

$$\Delta C = 2C_0 \sinh\frac{\sigma b^3}{kT} \approx 2C_0\frac{\sigma b^3}{kT} \tag{1.74}$$

we find

$$\Phi = 8\alpha\delta_j D_j \frac{\sigma}{kT} \tag{1.75}$$

Proceeding as before we find for the strain rate

$$\dot{\varepsilon}^v = 8\alpha\delta_j D_j \frac{\sigma b^3}{kTd^3} \tag{1.76}$$

In general, this can be written

$$\dot{\varepsilon}^v = B'\frac{D_j\delta_j\sigma\Omega}{kTd^3} \tag{1.77}$$

where B' is a constant and δ_j, the effective width of the boundaries for diffusion, is of the order of $10b$ for metals. The difference between the Herring-Nabarro and Coble models lies in the effects of the grain size and of the activation energy. Experiments with copper have shown creep rates varying like $1/d^3$ at low temperature ($550°C$) and like $1/d^2$ at higher temperatures ($840°C$).

(c) *Compatibility of intergranular deformation:* sliding at grain boundaries

The Herring-Nabarro and Coble models are based on simplifying assumptions concerning the shape of the grains and the homogeneity of their deformation. However, these may not hold to better than a first approximation, in particular because a set of perfectly spherical grains is not compact. Further, while the deformation is occurring the grains must remain in contact so long as there is no intergranular damage. This compatibility condition for the deformation accounts for the development of intergranular-type stresses, as in plasticity, which at high temperatures can be largely relieved by viscoplasticity, with the grain boundaries gliding with respect to each other.

The importance of intergranular glide as a form of viscoplastic deformation is made clear by the displacement of reference marks, as shown in Fig. 1.26. If we define this intergranular deformation ε_j as $\Delta u_l/d$, where Δu_l is the displacement of a reference mark and d is the grain size (an approximation that in particular does not take into account the orientation of the boundary plane with respect to the plane of observation) we can use as a measure of the importance of this deformation mode the ratio

$$\lambda = \varepsilon_j/\varepsilon_T \qquad \text{where } \varepsilon_T \text{ is the total deformation.} \qquad (1.78)$$

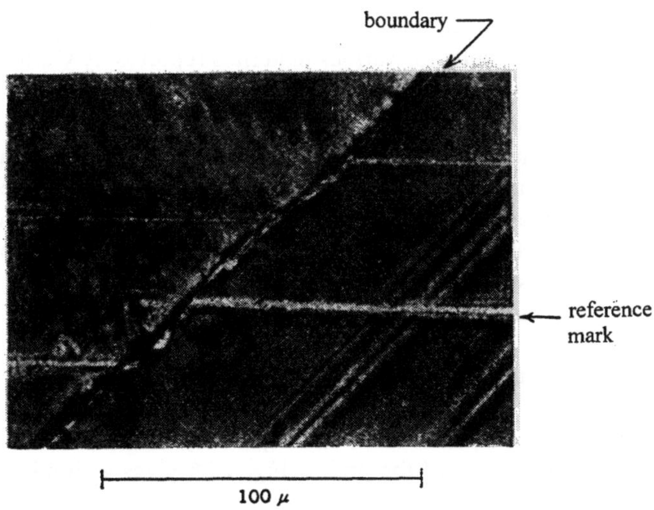

Fig. 1.26 Intergranular glide shown by displacement of reference marks in a bi-crystal of lead. (Garofalo ((1970) p.131).

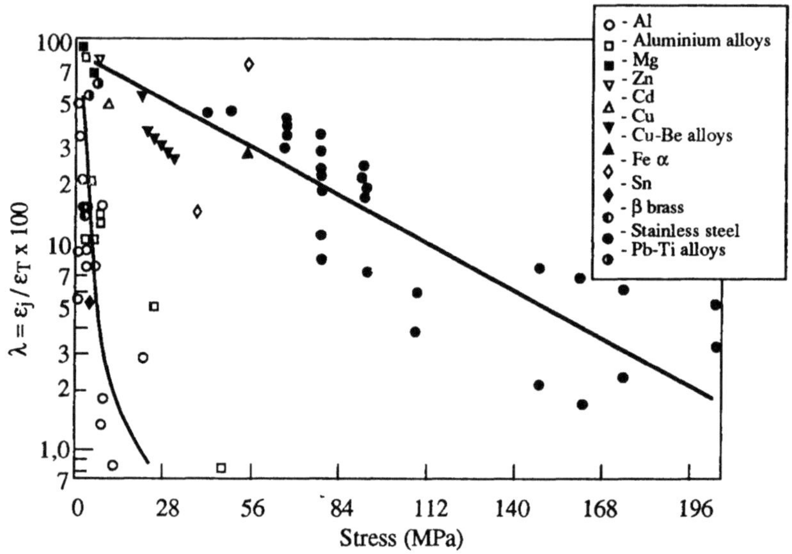

Fig. 1.27 Effect of applied stress in creep tests on the ratio of boundary-slide deformation (ε_j) to total deformation (ε_T), for various metals and alloys. (Garolfalo (1970) p.142).

The value of λ varies greatly with stress and less rapidly with temperature and grain size. Fig. 1.27 gives some results for various materials reported by Garofalo; these can be put into two groups, the first consisting of aluminium and its alloys, β-brass and tin, the second of copper, copper-beryllium, α-iron and stainless steel. In the first, λ decreases very quickly as the stress, and consequently the strain rate, increases; the contribution of the intergranular slip to the total deformation falls fairly quickly to below 5%. In the second the rate of decrease is much less. In both groups the deformations combine to form sub-grains, but again much more quickly in the first than in the second. These results suggest that there is a relation between intergranular slip and the changes to the microstructure associated with intergranular deformation.

Intergranular slip and diffusional creep are closely related (Raj and Ashby (1971) . The resultant deformation can be described as due either to diffusional creep of the grains accommodated by slip at the boundaries, or reciprocally to slip at the boundaries accommodated by diffusion creep of the grains. Fig. 1.28 gives a simple illustration of this.

In Fig. 1.28a the tensile stress favours transport of material over the surfaces AA', BB', which must be accommodated by slip along the boundary AB. In 1.28b the shear stress favours intergranular sliding which must be accommodated by transport of material over AB. The general situation, as in 1.28 c, will be intermediate between these extremes, necessitating both sliding and diffusion: both contribute to the deformation and each provides the accommodation required by the other.

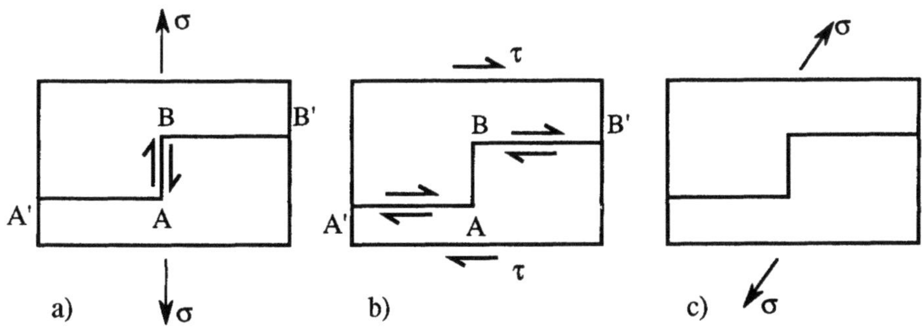

Fig. 1.28 Boundary slip in a bi-crystal, enabling compatibility of deformation to be achieved.

Raj and Ashby have made a detailed study of a non-planar boundary (Fig. 1.29), considering the case in which the incompatibilities (or internal stresses) created by the slip are accommodated either elastically or by diffusion. They showed that the way in which this accommodation is made controls the strain rate. In this work they considered the polycrystal as a two-dimensional compact set of hexagons.

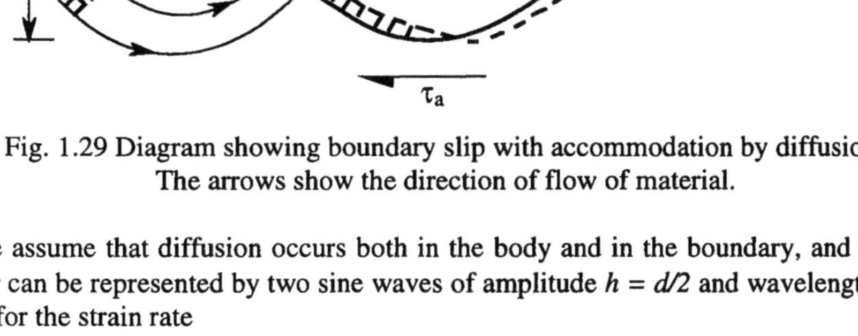

Fig. 1.29 Diagram showing boundary slip with accommodation by diffusion. The arrows show the direction of flow of material.

If we assume that diffusion occurs both in the body and in the boundary, and that the latter can be represented by two sine waves of amplitude $h = d/2$ and wavelength λ, we find for the strain rate

$$\dot{\varepsilon}^{v} = C\frac{\sigma\Omega}{kT}\frac{1}{d^{2}}D_{v}\left[1+\frac{\pi\delta_{j}}{\lambda}\frac{D_{j}}{D_{v}}\right] \tag{1.79}$$

where C is a numerical constant, of value about 40

The form of this result will be seen to combine volume diffusion creep (Herring-Nabarro) with boundary diffusion creep (Coble): this is because the hexagonal arrangement gives $\lambda = d$.

1.3.2.3 Creep diagrams (Figs. 1.30 - 1.35)

It is helpful to gather together all the experimental information and all the theoretical models concerning viscoplastic behaviour and present this in the form of diagrams or charts, as been done by Ashby for a number of polycrystalline materials (see Frost and Ashby (1982)). Fig. 1.30 shows one such presentation, for pure polycrystalline nickel with a grain size of $100\ \mu m$. In this diagram the dimensionless stress variable $\bar{\sigma} / \mu$, where $\bar{\sigma} = \left(\frac{1}{2} s_{ij} s_{ij}\right)^{1/2}$ is the equivalent stress, is plotted as ordinate against the temperature, as T/T_f, as abscissa, and the various regions that we have described are identified. One such region is that defined by high stresses and low temperatures, corresponding to low-temperature viscoplasticity, which in the case of pure nickel, an FCC material, is of small area. At the other extreme of low stress and high temperature are two regions corresponding to diffusional creep: Coble creep in the lower part of the temperature range ($T/T_f < 0.8$), Herring-Nabarro in the upper part ($T/T_f > 0.8$). The rest of the plane corresponds to dislocation creep.

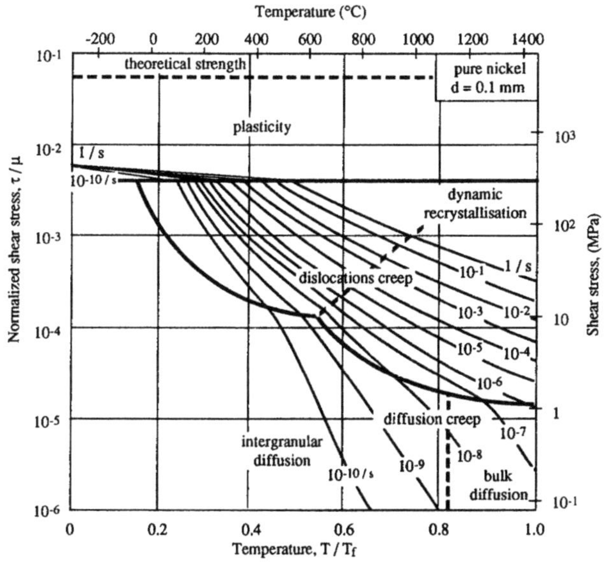

Fig. 1.30 Stress-temperature diagram for pure nickel of grain size *100 μm.*

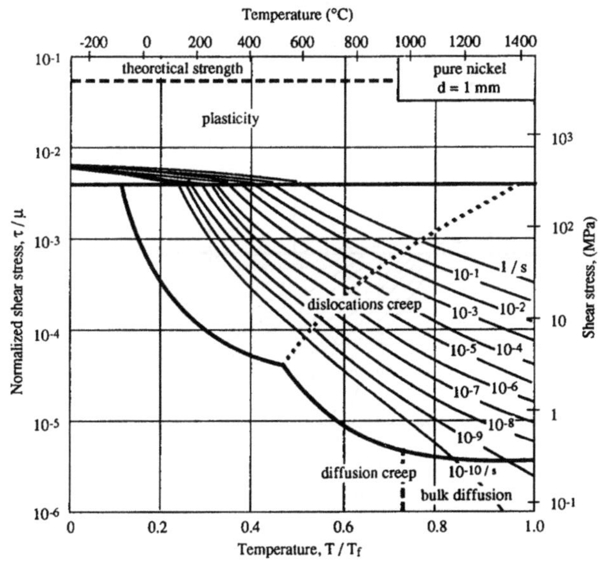

Fig. 1.31 Deformation diagram for pure nickel of grain size *1 mm.*

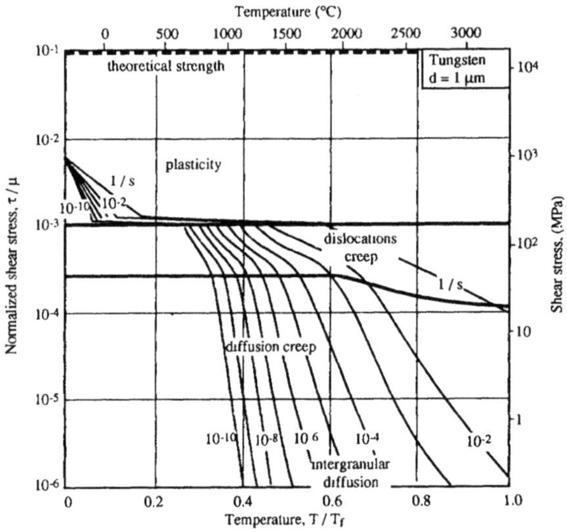

Fig. 1.32 Deformation diagram for pure tungsten of grain size *1 μm.*

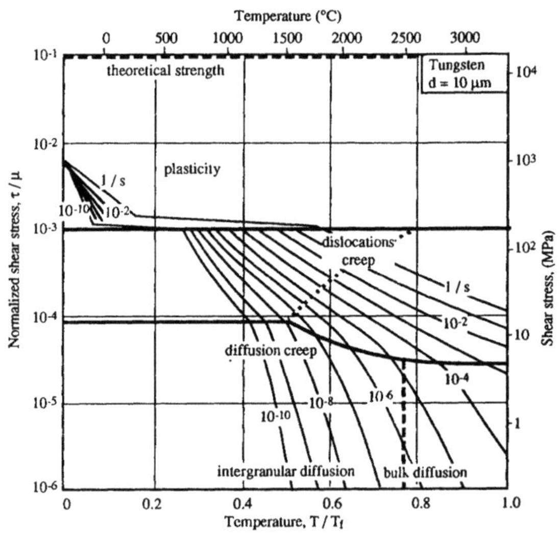

Fig. 1.33 Deformation diagram for pure tungsten of grain size *10 μm.*

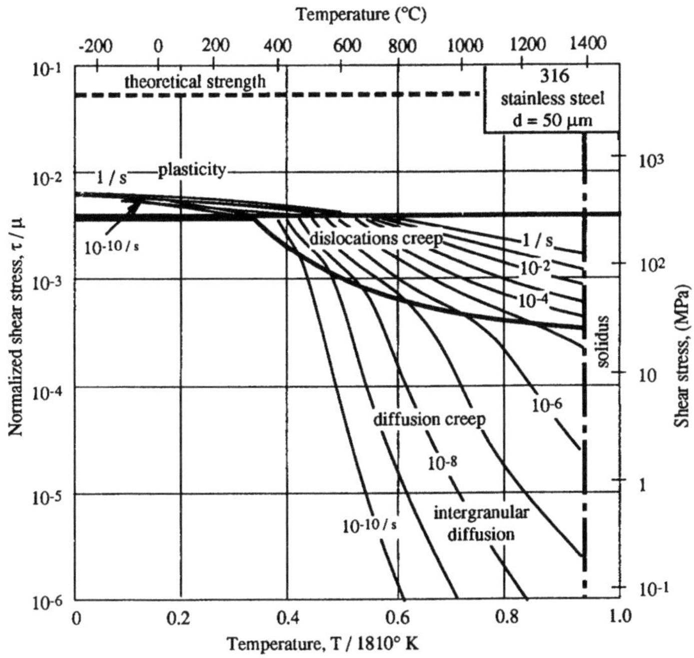

Fig. 1.34 Deformation chart for austenitic stainless steel 316 (American standard) Z 05 CND 17-13, grain size *50 μm*, molybdenum content *2.5%*.

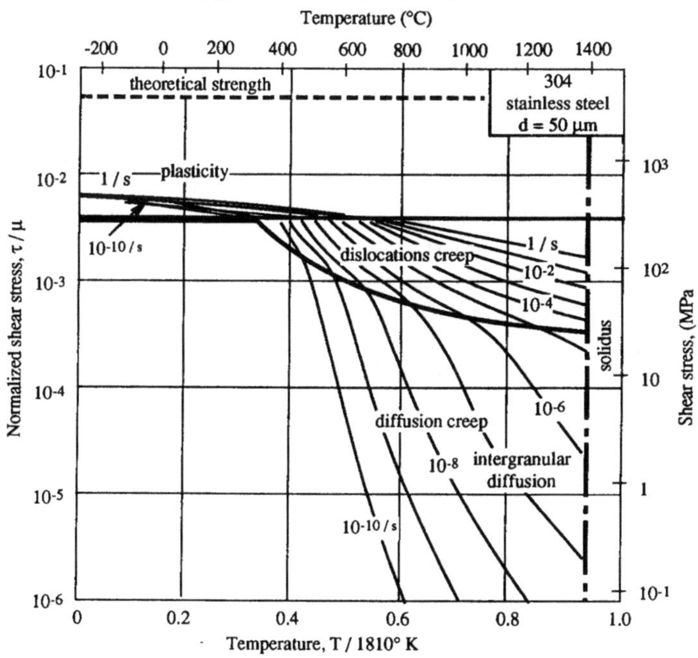

Fig. 1.35 Deformation diagram for austenitic stainless steel 304 (American standard) Z 05 CND 17-13, grain size *50μm*, no molybdenum content).

At high stresses and temperatures the diagram shows a region of dynamic recrystallisation, which we have not considered in this chapter; it was mentioned in the general introduction, and in I-§3.3.8.3. In the condition shown in Fig, 1.30 nickel recrystallises as it is being deformed, hence the term "dynamic".

In addition to showing the regions in which the different mechanisms are dominant the diagram has curves of constant deformation rate, where this is defined by $\dot{\gamma} = \left(2\dot{\varepsilon}_{ij}\dot{\varepsilon}_{ij}\right)^{1/2}$. Care must be taken when using these to avoid attributing too great a precision to them, since the experimental observations and the models derived from them are necessarily not perfect; but they are very useful for several purposes, for example:

- finding the order of magnitude of the strain rate for given stress and temperature,

- identifying the type of mechanism that will be operating in given conditions and, consequently, the type of law that will govern the creep. This can remove the need to make risky extrapolations from data obtained under conditions in which one mechanism is operating (dislocation creep, say) to conditions in which a different one (diffusional creep, say) dominates.

These deformation diagrams have been especially well determined for pure metals. Thus comparison of Figs. 1.30 and 1.31 will show the effect of grain size on the behaviour of pure nickel: this makes it very clear that whilst at low temperatures reducing the grain size will increase the resistance to deformation, the reverse is true at high temperatures.

Figs. 1.32 and 1.33 are for pure tungsten, a BCC metal very important for uses at high temperatures. Qualitatively, the diagrams for this metal and for nickel are quite similar; the main difference is at low temperatures ($T/T_f < 0.15$), where the yield strength increases much faster with decreasing temperature for a BCC metal than for FCC, because of the lattice (Peierls) forces. At high ($T/T_f > 0.5$) or even moderate temperatures the resistance of BCC metals such as tungsten is lower than that of FCC metals such as nickel, the difference arising from the faster diffusion in the less dense BCC structure.

Similar diagrams are available for a number of metallic alloys of great practical importance. Good examples are Figs 1.34 and 1.35 for austenitic stainless steels. The 304 and 316 steels are of essentially the same composition, *15%* Cr and *10-12%* Ni, differing in that 316 has also about *2.5%* Mo, the effect of which is to reduce the creep rate in the dislocation creep region by a factor of over *10*.

Frost et al (1982) give data for materials other than metal alloys, including oxides of various types, and ice.

1.3.2.4 Superplasticity

(a) *Introduction*

A material is said to be *superplastic* if it can sustain elongation to fracture without breaking - a rather imprecise definition, in that the "elongation to fracture" can vary considerably (ε_R = *100-1000%*). It is a very familiar property in the case of a number of non-crystalline materials - in glass-blowing, for example, or certain resins; and is seen in metals and metallic alloys (Al-Zn, Pb-Sn, Bi-Sn, titanium alloys, etc) which have a very fine microstructure and are in general polyphase when deformed under certain conditions of temperature and strain rate.

For a long time superplasticity remained a laboratory curiosity without any industrial applications. During the past twenty years, however, it has found practical uses, such as shaping certain alloys that are difficult to work (isothermal forming of some nickel-based superalloys manufactured by powder metallurgy), producing complex shapes (hemispheres in titanium alloys) or direct fabrication of assemblies (diffusion welding of aluminium- or titanium-alloy sheets, or forming hollow turbine blades with honey-comb reinforcement by superplastic inflation.)

(b) *Phenomenological description*

We recall that for materials deformed at low temperature the stability of the deformation of a tensile test piece is controlled by the rate of work-hardening, characterised by the coefficient $n = \partial \log \sigma / \partial \log \varepsilon$. But in the region of the temperature-strain rate diagram where the material is superplastic the value of this coefficient is very small, and stability results from the material's great sensitivity to the strain rate. This can be characterised by another coefficient, $m = \partial \log \sigma / \partial \log \dot{\varepsilon}$, whose variation with strain rate can be seen from Fig. 1.36; it is small (< *0.2*) at low and at high rates, reaching its greatest values, between *0.3* and *0.9*, at intermediate rates, where the behaviour is superplastic.

This sensitivity to deformation rate can be explained qualitatively as follows. If the deformation becomes localised in a neck the rate increases locally, resulting in an increase in the stress. This increase is greater, the greater the value of m; and its effect is to slow the development of the necking, resulting in the production of diffuse neckings and large elongations before rupture occurs. With the behaviour law written as

$$\sigma = K(\dot{\varepsilon}^v)^m \tag{1.80}$$

it is easily shown that as $m \to 1$ the rate of reduction of the cross-section area of a test piece under uniaxial tension P tends to P/K, and that this increases with decreasing m (see Chapter 2, on Damage, and Exercise 2 at the end of that chapter.)

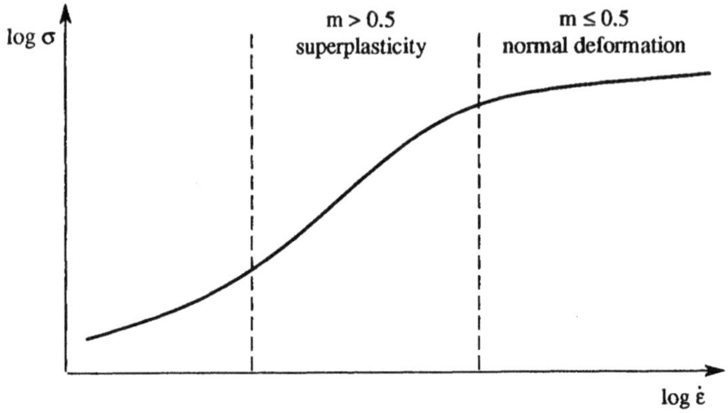

Fig. 1.36 Sensitivity of flow stress to strain rate, showing superplastic region.

(c) *Microstructure parameters and superplasticity in metallic alloys*
Superplastic behaviour occurs only at high temperatures (> $0.5T_f$), and usually at low strain rates (< 10^{-3} sec^{-1}); this makes it difficult to develop fabrication methods that make use of it. The activation energy that controls the deformation is small, in general less than the bulk diffusion energy and close to that corresponding to intergranular diffusion Q_j (approx. $\frac{1}{2}Q_v$) (see the tables of values at the end of this volume.)

The microstructure of a superplastic material must fulfil certain conditions. Above all, the grain size must be very small. As Fig. 1.37 shows for a number of aluminium alloys, an increase in grain size shifts the superplastic region very much towards small strain rates. Observations have shown that the variation of rate with grain size can be expressed as

$$\dot{\varepsilon}^v \approx d^{-\beta} \text{ with } \beta = 2 \text{ or } 3 \qquad (1.81)$$

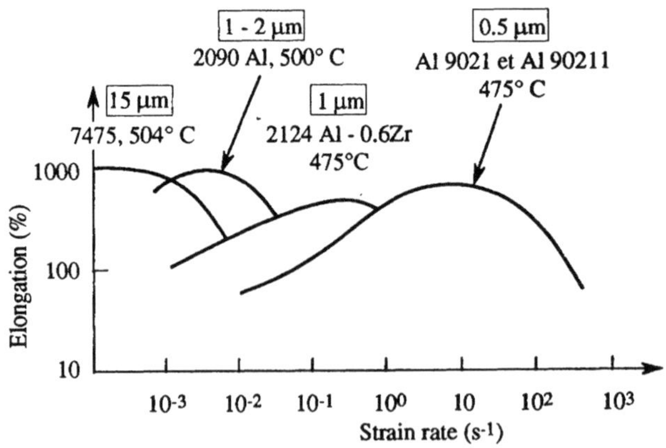

Fig. 1.37 Effect of grain size on strain rate for four alloys; showing that in recently developed alloys of the 9xxx series, with very small grain size (< $1 \mu m$), superplasticity can be achieved at relatively high rates ($\approx 10\ s^{-1}$).

Fig. 1.38, from V.N. Perevezentsev et al (1992), gathers together the effects of grain size, temperature, intergranular diffusion coefficient and deformation rate on stress for a number of materials. These results show that the rate-sensitivity coefficient *m* is close to *0.5* and that to a first approximation the rate is inversely proportional to the square of the grain size. Generally, to obtain such small grain sizes (< *1 micron*), which must remain stable during superplastic deformation, alloys consisting of approximately equal volumes of two phases (eutectic or eutectoid alloys) must be used; each phase will inhibit the growth of grain size in the other.

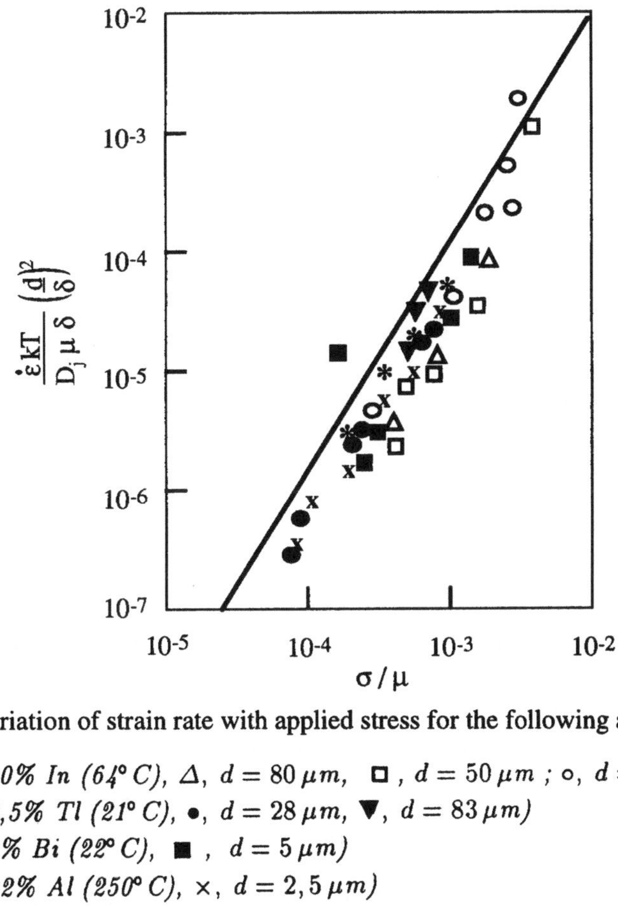

Fig. 1.38 Variation of strain rate with applied stress for the following alloys.

- Pb - 40% In ($64°C$), Δ, $d = 80\,\mu m$, □ , $d = 50\,\mu m$; o, $d = 200\,\mu m$)
- Pb - $2{,}5\%$ Tl ($21°C$), •, $d = 28\,\mu m$, ▼, $d = 83\,\mu m$)
- Sn - 1% Bi ($22°C$), ■ , $d = 5\,\mu m$)
- Zn - 22% Al ($250°C$), ×, $d = 2{,}5\,\mu m$)
- Pb - 38% Sn ($20°C$), ∗, $d = 2{,}46\,\mu m$)

In some cases (high-resistance aluminium alloys) superplastic microstructure is associated with a very special grain structure, characterised not only by small size but also by the presence of a precipitate-free zone around the boundaries, which is much softer than the material of the body of the grains. Using this, it has been found possible to make superplastic aluminium alloys by controlling the intergranular diffusion of gallium and forming a low-melting solid solution of Al-Ga in the neighbourhood of the boundaries.

After superplastic deformation the grains remain equiaxial; sometimes their size increases slightly, and often there is no deformation at all within their body. Examination of the surface shows that the grains slide with respect to one another, and that this intergranular sliding can account for over half of the total deformation. This agrees with the data given in Fig. 1.27, which show that sliding at grain boundaries becomes important at low stress.

(d) *Physical models*

Most models take as starting point the following observations:

- the value of the rate-sensitivity coefficient m increases with increasing rate, up to some critical rate which itself decreases with increasing grain size,

- intergranular sliding plays a dominant role. If two grains slide with respect to each other along their common boundary, cavitities should form at the triple points (see Chapter 2, on Damage); but in fact such cavities appear only after some delay, or not at all. This leads one to suppose that grain boundary sliding must be accompanied by mechanisms such as migration of the boundaries, deformation of neighbouring grains (both of these to only a limited extent) or diffusion of matter either in bulk or along the boundaries.

Of the models that attribute superplastic deformation to boundary slip accommodated by diffusional creep in the grains the most elaborate is that due to Ashby and Verral (1973). These authors have shown how the deformation can occur in uniform flow without the need for any deformation of the grains but as a result of sliding along the boundaries only (see Fig. 1.39) Taking as the elementary event the change of shape associated with a set of four grains, they showed that the intermediate position for which the area of the boundary is maximum is a col. The grains slide over one another and this sliding is accommodated by diffusional creep which brings about transport of material either through the volume (diffusion coefficient D_v) or along the boundaries (D_j). Assuming that the grain boundaries are perfect sources and sinks of vacancies, they have established the following relations:

$$\dot{\varepsilon}^{v} = 100\frac{\Omega}{kTd^{2}}\left(\sigma - \frac{0.72\gamma_{j}}{d}\right)D_{v}\left(1+\frac{3.3\delta_{j}D_{j}}{dD_{v}}\right) \tag{1.82}$$

where γ_j is the energy of the grain boundaries.

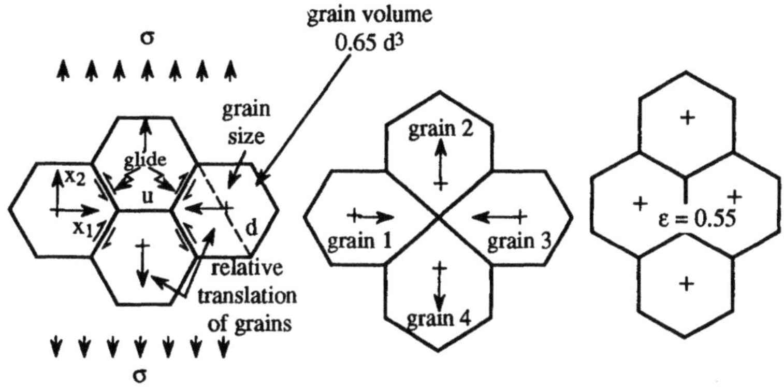

Fig. 1.39 Elementary mechanism of deformation in a superplastic polycrystalline material.

As would be expected, this has the same form as (1.79). However, it involves a threshold stress in the term γ_j/d, which could explain why in a logarithmic plot of stress against deformation rate there is often a change of slope in the region of small values of stress.

1.4 MECHANICAL MODELS OF MACROSCOPIC VISCOPLASTICITY

The rheological formulation of viscoplastic behaviour can be modelled on that of plastic behaviour, taking the same basic elements. As in the case of plasticity, in the present state of knowledge and of the computational methods available it would be unreasonable to expect to pass completely from the microscopic scale to that of mechanically-significant volumes, because of the many intermediate stages to be gone through and the heterogeneities encountered on the way. In general therefore we cannot hope that models developed on the scale of dislocations will give quantitative information concerning materials having a complex microstructure: their role is rather to enlighten the user on microscopic-scale mechanisms and to define the limits within which mechanical models will be valid. It is these latter that will be used as a last resort in structural computations; in general they rest on a broad *phenomenological* base, but the structure of the most powerful of them is suggested by the local deformation mechanisms and they involve critical variables suggested by these.

Thus it is a question again of determining:
- the form of the elastic region at the instant considered,
- the expression for viscoplastic flow,

- the evolution of the hardening variables.

The elastic region is defined with the aid of a scalar-valued *load function f*. This depends on the stress tensor and the work-hardening variables in such a way that it provides a means for representing the initial form of this region and its possible changes during the loading. The viscoplastic flow is defined by reference to this function in the classic case of associated viscoplasticity; and similarly for the work-hardening if the model is *standard*; a new element enters when the mechanisms of recovery over time or of ageing have to be taken into account. The results obtained in this chapter relate to isothermal conditions; for a simple treatment of anisothermal creep the reader is referred to Exercise 8.

1.4.1 Viscoplastic potential for a single crystal

Our study of the microscopic-scale mechanisms has shown that viscoplastic flow can be a consequence of slip or of diffusion; the second, which implies transport of material, does not come within the framework of classical models. The treatment of deformation by slip that we shall now give will be limited to displacements in the plane without any consideration of the mechanisms of climb, which involve stress components normal to the slip plane and therefore violate the assumption of normality. The success of models of this type rests on the fact that although slip is not always the dominant mechanism it is often responsible for the major part of the deformation. This is certainly true at low temperatures, but for dislocation creep it holds also at high temperatures to the extent that the climb process displaces dislocations only over short distances compared to the "long" distances over which they are displaced by slips.

Under the assumption that the material obeys Schmid's law the rate of slip $\dot{\gamma}_r$ in a given system will depend only on the resolved shear $\tau_r = \vec{m}^r . \underline{\underline{\sigma}} . \vec{n}^r$, the work-hardening variables, here represented conventionally by a vector of components Y_m, and possibly the temperature T. This allows us to define the rate of viscoplastic deformation, in the case when several slip systems are active, by a function Φ, non-decreasing with τ_r:

$$\dot{\gamma}_r = \Phi(\tau_r, Y_m, T), \quad \underline{\underline{\dot{\varepsilon}}}^v = \sum_r \dot{\gamma}_r (\vec{m} \otimes \vec{n})^r \tag{1.83-4}$$

Applying a virtual variation $d\underline{\underline{\sigma}}$ to the existing stress state represented by the tensor $\underline{\underline{\sigma}}$ we have

$$\underline{\underline{\dot{\varepsilon}}}^v : d\underline{\underline{\sigma}} = \sum_r \Phi(\tau_r, Y_m, T) d\tau_r \tag{1.85}$$

The viscoplastic potential is then defined as a scalar-valued function of $\underline{\underline{\sigma}}$, Y_m and T by

$$\Omega(\underline{\underline{\sigma}},Y_m,T)=\sum_r\int\Phi(\tau_r,Y_m,T)d\tau_r \tag{1.86}$$

This function enables us to calculate the macroscopic viscoplastic strain rate corresponding to any given state of stress and hardening; the rate vector is orthogonal to the equipotential surface through the point corresponding to that state:

$$\dot{\underline{\underline{\varepsilon}}}^v=\partial\Omega/\partial\underline{\underline{\sigma}} \tag{1.87}$$

While no system is active Ω remains constant. Without loss of generality it can be given the value zero inside and on the boundary of the elastic region, and the surfaces Ω = *constant* are nested inside one another and are convex. It is easy to show that Ω is a convex function of $\underline{\underline{\sigma}}$; the proof is based on an inequality relating to the slip systems, into which the stress and viscoplastic strain tensors are substituted, thus:

$$\left(\Phi(\tau_r,Y_m,T)-\Phi(\tau_r^0,Y_m,T)\right)(\tau_r-\tau_r^0)\geq 0 \tag{1.88}$$

(expressing the condition that dissipated energy must be positive), giving

$$\left(\frac{\partial\Omega}{\partial\underline{\underline{\sigma}}}(\underline{\underline{\sigma}})-\frac{\partial\Omega}{\partial\underline{\underline{\sigma}}}(\underline{\underline{\sigma}}^0)\right):(\underline{\underline{\sigma}}-\underline{\underline{\sigma}}^0)\geq 0 \tag{1.89}$$

from which, between two points $\underline{\underline{\sigma}}^0$, $\underline{\underline{\sigma}}^1$,

$$\Omega(\underline{\underline{\sigma}}^1)-\Omega(\underline{\underline{\sigma}}^0)\geq(\underline{\underline{\sigma}}^1-\underline{\underline{\sigma}}^0):\frac{\partial\Omega}{\partial\underline{\underline{\sigma}}}(\underline{\underline{\sigma}}^0) \tag{1.90}$$

The convexity is not strict, since the resolved shear depends only on the deviatoric part of the stress tensor; but if Φ is a strictly increasing function of τ_r, Ω is a strictly convex function of the deviator of $\underline{\underline{\sigma}}$. On the other hand, it is found that an equipotential surface is defined piecewise. If only one system is active the surface consists of hyperplanes, since, as for the elastic region in the case of time-independent plasticity, the constancy of Ω implies that of τ_r also. Between two adjacent hyperplanes, for example if two systems r, s are active, it is defined, for given temperature and work-hardening state, by

$$\int\Phi(\tau_r,Y_m,T)d\tau_r+\int\Phi(\tau_s,Y_m,T)d\tau_s=\text{constant} \tag{1.91}$$

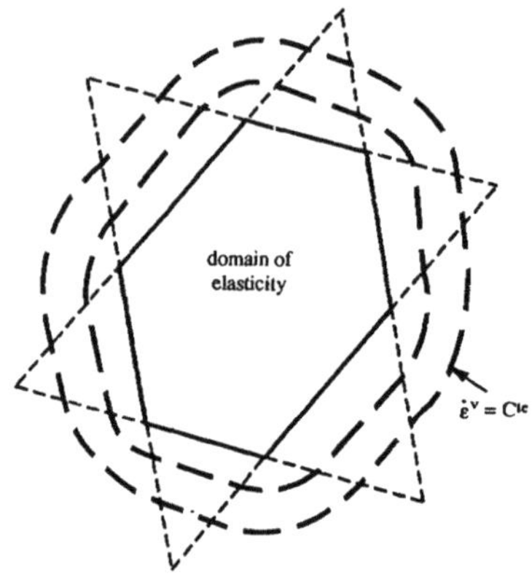

Fig. 1.40 Viscoplastic equipotentials for a single crystal.

1.4.2 Viscoplastic potential for a polycrystal

Treating a polycrystal as an assembly of single crystals gives a "macroscopic" viscoplastic potential which is the sum of the single-crystal equipotentials for the separate grains, weighted by their volume fraction. Thus for N grains

$$\Omega(\underline{\underline{\sigma}}, Y_m^1, \dots Y_m^N, \sigma_{loc}^1, \dots \sigma_{loc}^N, T) = \sum_{i=1}^{N} \frac{V_i}{V} \Omega_i \tag{1.92}$$

where each potential depends on the local stress $\sigma_{loc}{}^i$ and the work-hardening variables for the individual grain $Y_m{}^i$, as well as on the mean stress on the aggregate. This shows that the convexity properties continue to hold, and that equipotentials can again be defined and in turn define by their normals the direction of viscoplastic flow.

The phenomenological approach takes this route but reduces the number of variables used to define the state of the material: the local stresses are not evaluated explicitly and for the work-hardening only global variables relating to the "average" material are taken. The previous formalism is retained and, following once again the example of plasticity, generalised standard viscoplastic materials are defined, using the convex potentials for each of the variables $\underline{\underline{\sigma}}$ and Y_m and putting

$$\dot{\underline{\underline{\varepsilon}}}^v = \frac{\partial \Omega}{\partial \underline{\underline{\sigma}}} \quad \text{and} \quad \dot{\alpha}_m = \frac{\partial \Omega}{\partial Y_m} \tag{1.93}$$

For all models of this type the value of the mechanical dissipation for any given state of stress and work-hardening exceed that of Ω; this ensures that it is positive, and therefore that the second law of thermodynamics is not violated.

If recovery is to be taken into account this requires a rather different viscoplastic potential, with two parts. The first, Ω_v, corresponds to what we have just developed; the second, Ω_r, to the effects of recovery on the internal variables, and can exist even if the representative point for the stress lies inside the elastic region. In general, this second potential depends only on the internal variables, thus:

$$\Omega = \Omega_v(\underline{\underline{\sigma}}, Y_m, T) + \Omega_r(Y_m, T) \tag{1.94}$$

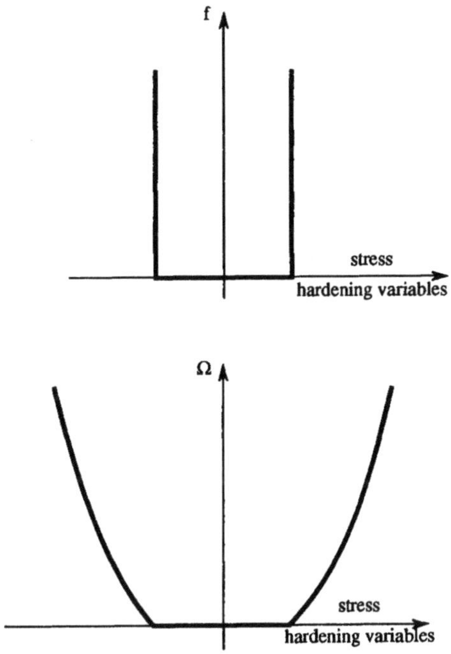

Fig. 1.41 Plastic and viscoplastic potentials. (a) plastic; f is the indicator function for the elastic region (b) viscoplastic: Ω is the viscoplastic potential

1.4.3 Time-independent plasticity and viscoplasticity compared

Time-independent plasticity can be studied by starting with a viscoplasticity law and assuming that the equipotentials defined by the function Ω are very closely packed. The inelastic strain rate will then vary very quickly with the stress, and this will tend to give instantaneous inelastic deformations. By going to a limit, Ω can be replaced by f, the *indicator function* for the elastic region, represented by a cylinder; this is shown in Fig. 1.41a, for comparison with the viscoplastic potential in Fig. 1.41b. In this case it is clear that whilst the direction of the flow is still specified, the same is not true for the magnitude, all the sections being projected on to the surface $f = 0$. Thus the deformation rate is specified to within a scalar factor, the plastic factor $\dot{\lambda}$, whose value is determined by the condition that the working point must remain on the surface $f = 0$:

$$\dot{\underline{\underline{\varepsilon}}}^p = \dot{\lambda}\partial f / \partial \underline{\underline{\sigma}} \tag{1.95}$$

Thus the essential difference between plastic and viscoplastic theory is that the latter allows freedom of choice for the rate of inelastic deformation, whilst in the former a further equation has to be satisfied, with the result that this rate is determined once the shape of the elastic region has been chosen.

1.4.4 Specific constitutive equations

The study of physical models shows that viscoplastic behaviour occurs in very diverse conditions - which explains why these models can have very different forms. They are all characterised by an expression that gives the inelastic strain rate as a function of the work-hardening and the current stress; this relation forbids instantaneous plastic strain and separates the strain into an elastic and an inelastic part. However, by passage to a limit models can be constructed for which the strain rate is practically independent of time, and also rigid-viscoplastic models. In the following paragraphs we first give models that do not involve a threshold and then models that do, finishing with some comments on the use of viscoplastic models for numerical work.

1.4.4.1 Laws not involving work-hardening

Norton's law (1.1c) is arrived at by choosing for the visco-elastic potential a power law in von Mises's equivalent stress $\overline{\sigma}$, with two coefficients K, M that characterise the material:

$$\Omega = \frac{K}{M+1} < \frac{\overline{\sigma}}{K} >^{M+1} \tag{1.96}$$

where $<H> = \max(0,H)$ and $\overline{\sigma} = \sqrt{\frac{3}{2}\underline{\underline{s}}:\underline{\underline{s}}}$ (1.97)

The viscoplastic flow is then given by

$$\dot{\underline{\underline{\varepsilon}}}^v = \frac{\partial \Omega}{\partial \underline{\underline{\sigma}}} = \frac{3}{2}\left(\frac{\overline{\sigma}}{K}\right)^M \frac{\underline{\underline{s}}}{\overline{\sigma}} \tag{1.98}$$

Under uniaxial loading, when the only non-zero component of the stress tensor is $\sigma_{11} = \sigma$, this simplifies to

$$\dot{\varepsilon}^v_{11} = (\overline{\sigma} / K)^M, \quad \dot{\varepsilon}^v_{22} = \dot{\varepsilon}^v_{33} = -\tfrac{1}{2}\dot{\varepsilon}^v_{11} \tag{1.99-100}$$

There is no work-hardening and the rate of viscoplastic deformation is determined fully by the secondary creep. M is the *Norton exponent* of 1.1c, equal to the slope of the line in the log-log plot of strain rate against stress, as in Fig. 1.3.

Other threshold-free laws can be envisaged, in particular when the power law is a poor approximation at high stresses. Provided that the equipotentials are given in terms of a von Mises equivalent, all these laws give an expression of the type

$$\underline{\underline{\dot{\varepsilon}}}^{v} = \frac{3}{2}\dot{\bar{\varepsilon}}^{v}\frac{\underline{\underline{s}}}{\bar{\sigma}} \tag{1.101}$$

where $\dot{\bar{\varepsilon}}^{v} = \sqrt{\frac{2}{3}\underline{\underline{\dot{\varepsilon}}}^{v}:\underline{\underline{\dot{\varepsilon}}}^{v}}$ is the absolute rate of viscoplastic strain rate under uniaxial loading. The most currently-used formulae involve products of exponentials and powers, sometimes a hyperbolic sine. Temperature may enter explicitly through activation terms in exp(-Q/kT):

$$\dot{\bar{\varepsilon}}^{v} = \dot{\varepsilon}_0(\bar{\sigma}/K)^q \exp(F(\bar{\sigma},T)) \tag{1.102}$$

Thus the Sellars-Teggart law is

$$\dot{\bar{\varepsilon}}^{v} = \dot{\varepsilon}_0(\sinh K\bar{\sigma})^q \exp(-Q/kT) \tag{1.103}$$

1.4.4.2 Laws with multiplicative work-hardening

All the preceding laws state that under any non-zero constant loading the flow rate of the material will be non-zero and constant: they reduce the elastic region to a single point. Hardening of the material can be represented while still retaining this property by manipulating the spacing of the equipotentials and thus introducing a work-hardening variable into the expression for the deformation rate. Although not mandatory, it has become standard to use for this the cumulative viscoplastic deformation v, defined as

$$v(t) = \int_0^t \dot{\bar{\varepsilon}}^{v}(\tau)d\tau \tag{1.104}$$

In expressions of the type of (1.96) the constant K, which has the dimensions of a stress, now depends on v and is called *drag stress*, expressing the resistance to deformation due to the rate term. It has become standard also to represent the hardening by a power law in v, $K = K_0 v^n$; this leads to a new expression for v which can be inverted to give $\bar{\sigma}$:

$$\dot{v} = \langle \bar{\sigma}/K_0 \rangle^{1/m} v^{-n/m}, \quad \bar{\sigma} = K_0\dot{v}^m v^n \tag{1.105a,b}$$

with $m, n > 0$. (1.105a) gives the hardening by deformation and shows that the strain rate is infinite at the start of the loading and tends to zero as the strain increases. For the case of creep in simple tension under a uniaxial stress σ_0 the relation can be integrated to give

$$\varepsilon_v = \left[\frac{n+m}{m} \left(\frac{\sigma_0}{K_0} \right)^{1/m} t \right]^{\frac{m}{n+m}} \tag{1.106}$$

1.4.4.3 Laws with isotropic additive work-hardening

There are several ways in which a threshold might be introduced:

- through the existence of an elastic region of constant size throughout the course of the deformation. There is no work-hardening and the relation is of the type of (1.6), in which the critical stress remains always equal to the initial elastic limit σ_0:

$$\dot{v} = \left(\frac{\sigma - \sigma_0}{K} \right)^M \tag{1.107}$$

- through the existence of an elastic region whose size increases in the course of the deformation. As in plasticity, there is isotropic additive work-hardening and the work-hardening variable depends on the cumulative viscoplastic deformation, for example exponentially:

$$\dot{v} = \left(\frac{\sigma - R - \sigma_0}{K} \right)^M, \quad R = Q\,[1 - \exp(-bv)] \tag{1.108-9}$$

R and Q have the dimensions of stress. R represents a friction stress, whose rheological image is a friction element; it corresponds to the applied stress necessary to produce inelastic deformation. To the extent that it is independent of the direction of the load its existence is normally associated, on the microscopic scale, with local obstacles to the movement of the dislocations, such as interactions between dislocations and particles or between dislocations and precipitates when the latter are sheared.

1.4.4.4 Laws with kinematic additive work-hardening

Kinematic work-hardening displaces the centre of the elastic region in the stress space, corresponding to a change of macroscopic "neutral state" for the material. As we discussed in connexion with physical models (§1.3.2), this is related to heterogeneities on the microscopic scale. These give rise to non-uniform local stress fields over "long" distances (long range stresses), so that the shear force on the dislocations can be determined from the local "effective" stress, that is, the difference between the applied stress and the "back stress", which exerts a counter-acting force tending to produce a deformation in the sense opposite to that of the load. Under a uniaxial load, with a constant threshold and a back stress σ_i, the critical stress of (1.6) is

$$\sigma_s = \sigma_i + \sigma_0 \text{ (simple tension)}, \quad \sigma_s = \sigma_i - \sigma_0 \text{ (simple compression)} \tag{1.110 a, b}$$

In both metallic alloys and composite materials there are many sources of heterogeneities: phases, grains, precipitates, dislocation cells, ... and to each there corresponds a procedure for localising the stresses and the deformations. These redistributions are very complex and are still not fully characterised, but in general they are all governed by local macroscopic deformation. Thus the corresponding phenomenological laws also are governed by the deformation; the simplest law for work-hardening, that due to Prager, is

$$\dot{\underline{\underline{X}}} = \frac{2}{3} C \dot{\underline{\underline{\varepsilon}}}^{v} \tag{1.111a}$$

which gives an elastic region and a viscoplastic potential depending on both $\underline{\underline{\sigma}}$ and $\underline{\underline{X}}$:

$$f = \overline{(\underline{\underline{\sigma}} - \underline{\underline{X}})} - \sigma_0 \tag{1.111b}$$

In viscoplasticity as in plasticity more elaborate work-hardening laws can of course be considered. For non-linear kinematic work-hardening in particular a fading memory term can be included in the expression for strain rate, whose direction can be given either by the present value of the back stress or by the instantaneous viscoplastic strain rate, thus:

$$\begin{aligned} \dot{\underline{\underline{X}}} &= \frac{2}{3} C \dot{\underline{\underline{\varepsilon}}}^{v} - \gamma \underline{\underline{X}} \dot{v} \\ \dot{\underline{\underline{X}}} &= \frac{2}{3} C \dot{\underline{\underline{\varepsilon}}}^{v} - \gamma \left(\underline{\underline{X}} : \frac{\partial f}{\partial \underline{\underline{\sigma}}} \right) \dot{\varepsilon}^{v} \end{aligned} \tag{1.112 a, b}$$

Equations (1.111) and (1.112a) were discussed in detail, in both uniaxial and multiaxial formulation, in I-§3.4.2.4. (1.112b) differs from (1.112a) only in multiaxial loading, when it has different properties concerning the ratchet effect.

1.4.4.5 Laws with recovery

Recovery, which results in a softening of the material, is often considered as a removal of internal stresses; it is thus represented by the addition to the viscoplastic potential of a term originating in a recovery potential Ω_r, as in (1.94), modifying the definition (1.111a) of the variable $\underline{\underline{X}}$ to

$$\dot{\underline{\underline{X}}} = \frac{2}{3} C \dot{\underline{\underline{\varepsilon}}}^{v} - \left(\frac{Q X_{II}}{L} \right)^{l} \frac{\underline{\underline{X}}}{X_{II}} \quad \text{where } X_{II} = \sqrt{\frac{2}{3} \underline{\underline{X}} : \underline{\underline{X}}} \tag{1.113}$$

If l is given the value 1 this corresponds exactly to the approach given in §1.3.2.1 for dislocation creep.

Other forms of recovery can be envisaged, in particular those bearing on the isotropic variable in the case of additive work-hardening.

1.4.4.6 Using the models

The work-hardening laws we have discussed should be understood as constituting an hierarchical model, in which one can take more or fewer of the elements according to the material, the temperature range and the particular application (for example, whether monotonic or cyclic loading). It should never be forgotten that the more complex the model the more tests will be needed to specify it: this is saying that having more elaborate equations, far from reducing the need for tests, increases this and requires more precautions to be taken. A fundamental requirement for characterising viscous stress is a set of test results that covers a wide range of deformation rates. Further, it is not always certain that the available models give a correct interpretation of the results of both creep tests and relaxation tests at the same time, showing that even with monotonic loading and steadily increasing viscoplastic deformation, viscosity will make the mechanical responses more complex.

Thus there are difficulties in specifying the coefficients that represent the properties of real material. Further, a viscoplastic formulation is often used in preference to a plastic formulation for reasons of computational simplicity It has been shown that the viscoplastic formulation "regularises" the plasticity, leading to more regular numerical solutions: two classical examples concern rigid-viscoplastic models and threshold-free models with power functions in which the exponents can be very large.

Rigid-viscoplastic models may or may not include a threshold term; they can express for example the model of (1.105b) in terms of total deformation, assuming that this takes place without change of volume. .At any instant $\dot{v}$ is given by $\sqrt{\frac{2}{3}\dot{\underline{\underline{\varepsilon}}}(\tau):\dot{\underline{\underline{\varepsilon}}}(\tau)}$ with $\mathrm{Tr}\,\dot{\underline{\underline{\varepsilon}}} = 0$, and v is the integral of this with respect to time. Given the data for the present state, represented by v and the deformation rate $\dot{v}$, (1.105b) enables the stress state to be determined. This formulation is particularly well adapted to finite-element calculations ($\dot{\underline{\underline{\varepsilon}}}$ given) in cases when the elastic deformation is negligible, such as metal forming. Bingham's law for viscous fluids can be obtained by the same procedure, starting from (1.107) and treating this as a limiting case of viscoplasticity.

The viscoplastic formulation leads to calculations in which the behaviour is practically independent of the deformation rate. The standard method given in the literature is to choose a Norton-type law, or multiplicative work-hardening, with a power law with very high exponent, of the order of *200*. This is the current approach for single crystals in particular. A numerical cut-off is often associated with the method, with the reintroduction of an elastic region, the approach by infinitesimal deformation rates not affecting the results:

$$\dot{\gamma} = 0 \text{ if } \tau \le \tau_c, \quad = (\tau / \tau_c)^M \text{ otherwise} \qquad (1.114\text{ a, b})$$

τ_c is a work-hardening variable, a function of the slip system to which it relates, and possibly of slips in other systems also.

1.4.5 Simultaneous treatment of plasticity and viscoplasticity

In this section we have often introduced plasticity with the aim of either comparing or contrasting the plastic formulation to the viscoplastic: the first leads to instantaneous and the second to delayed responses. Both types of response can be treated by a viscoplastic model, referred to as " unified modelling"; however, with some materials it can be useful to maintain the distinction between the two regimes, one of fast deformation and the other of slow. This is especially the case for:

- materials with considerable capacities for both work-hardening and creep, such as austenitic stainless steels, and which are inversely sensitive to the loading rate in certain temperature ranges. Here "inverse" means that the stress necessary for inelastic flow decreases as the loading rate increases,

- materials in which instantaneous deformation (plasticity) and creep (viscoplasticity) occur in distinct stress regions; for example, nickel-based alloys such as Inconel 78.

Thus in the unified models two thresholds, plastic and viscoplastic respectively, coexist and the total deformation is the sum of three parts: elastic, plastic and viscoplastic:

$$\underline{\underline{\varepsilon}} = \underline{\underline{\varepsilon}}^e + \underline{\underline{\varepsilon}}^p + \underline{\underline{\varepsilon}}^v, \text{ where } \dot{\underline{\underline{\varepsilon}}}^p = \dot{\lambda}\partial f / \partial \underline{\underline{\sigma}}, \quad \dot{\underline{\underline{\varepsilon}}}^v = \partial \Omega / \partial \underline{\underline{\sigma}} \qquad (1.115\text{ a, b, c})$$

The models most in current use are limited to a Norton law for viscoplasticity and linear isotropic or kinematic work-hardening for plasticity. Experience has shown, however, that some coupling between the plastic and viscoplastic laws must be introduced. The resulting models, which are outside the scope of this chapter, enable plasticity and viscoplasticity to be identified in the results of basic tests, which must then be supplemented by tests that show the plasticity-creep coupling. Fig. 1.42 shows the construction of the inelastic deformation for two models with kinematic and isotropic work-hardening in plasticity and viscoplasticity.

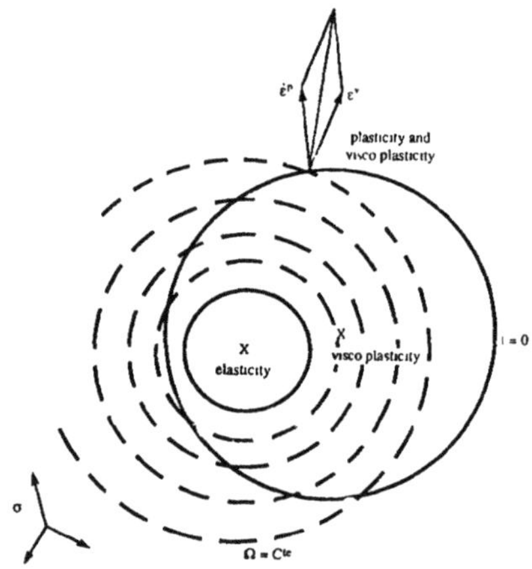

Fig. 1.42 Construction of inelastic strain rate for a model covering plasticity and viscoplasticity.

1.5 METHODS FOR REINFORCING AGAINST CREEP

We end the chapter with a few indications that may be useful in the development of materials that must resist creep under loading at high temperatures.

It will be seen from what we have shown that two approaches are possible:

- reducing the diffusional transport of material,
- blocking the slip or climb movements of the dislocations.

We study these approaches, considering first methods concerning diffusion and then two mechanisms for blocking dislocations, one by solid solution and the other by precipitation.

1.5.1 Reinforcement by diffusion

This is the appropriate choice of method when working with materials of high melting point, such as refractory metals (Ta, Mo, W, Nb) or ceramics (Al_2O_3, ZrO_2, Si_3N_4 etc.). Such materials, unfortunately, are brittle and difficult to work.

As we have seen, important contributors to creep resistance at high temperature are grain size and intergranular slip. If the load is essentially a uniaxial tension a good way to reduce slip at the boundaries is to orient them all in the direction of the load, so that no shear stress is acting on them. This can be achieved by what is called *directional solidification* (DS), used in the manufacture of gas-turbine blades by casting. Fig. 1.43 shows the principle. An even more radical solution to the problem, now being developed widely, is to use this process to eliminate the grain boundaries completely and produce single-crystal blades.

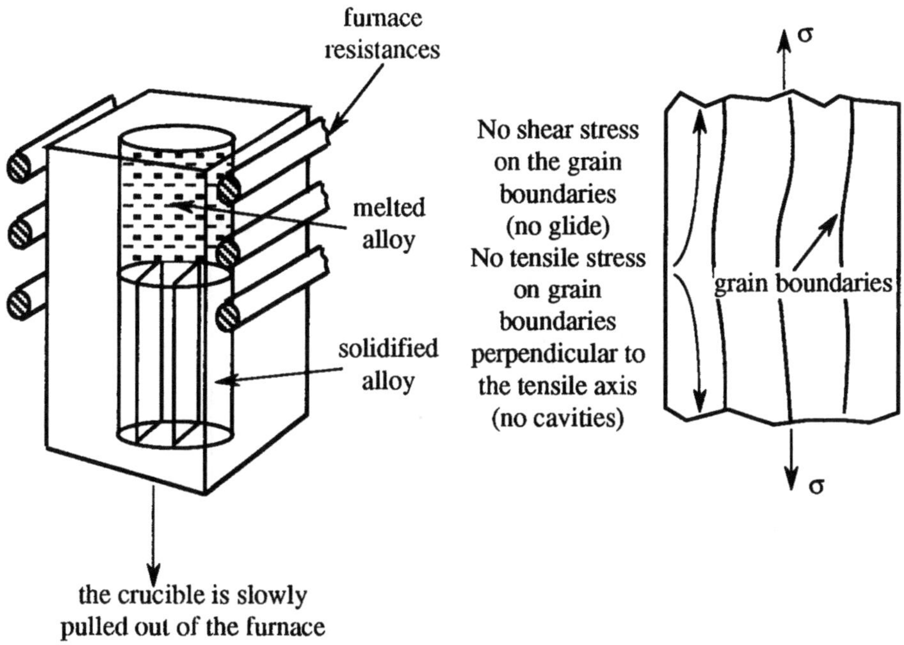

Fig. 1.43 Directional solidification of turbine blades, giving elongated grains or even single crystals (Ashby and Jones (1980) p.185).

Fig. 1.44 shows that up to 1960 turbine inlet temperatures were virtually the same as the metal temperatures. After this date methods for cooling the blades - internally by circulating air through channels running the length of the blade, or externally by creating a cool boundary layer over the surface - enabled the inlet temperature to be raised without changing the material. The figure shows also the progress made possible by the use of DS materials and replacement of nickel-based alloys by niobium alloys or composite materials; and indicates possible further improvements that may result from studies of mixtures of conventional nickel-based alloys and directional eutectics.

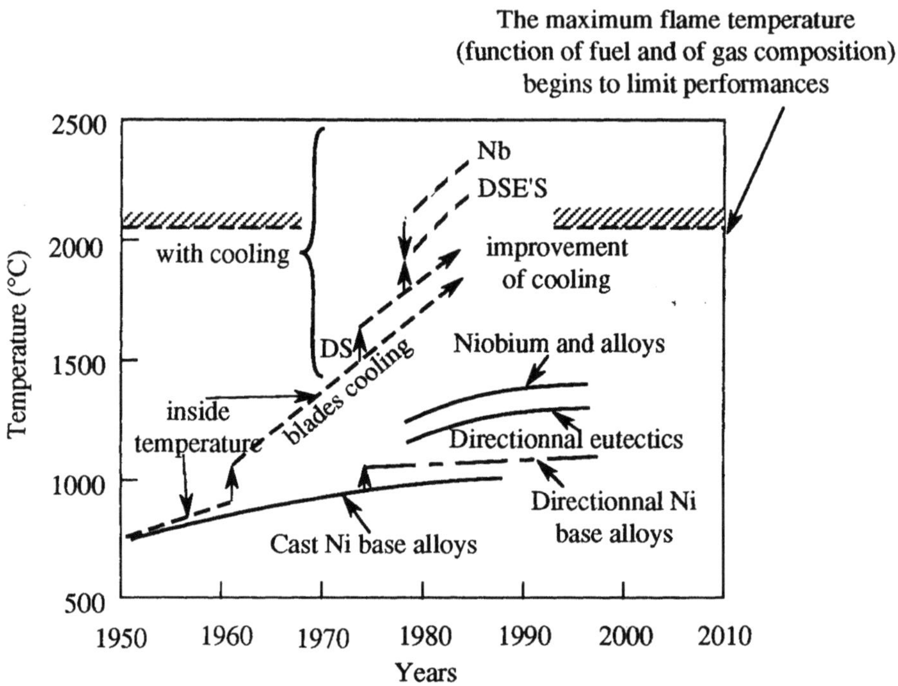

Fig. 1.44 Evolution of working temperature for turbine blades, and materials used for their manufacture. (Ashby and Jones (1980)).

1.5.2 Creep in solid solutions

Substantial reductions in the creep rate can be achieved by adding elements in solid solution. We have already mentioned the austenitic stainless steel with basic percentage composition 18Cr-12Ni, for which the rate was greatly reduced by adding *2.5%* of molybdenum. Another example, of niobium alloys, is given in Fig. 1.45; for this the reinforcement resulting from the addition of various elements in solid solution (Mo, W, Re) is expressed by a parameter S_B defined by

$$S_B = (\sigma_{alloy} - \sigma_{Nb})/\sigma_{Nb} \tag{1.116}$$

where σ_{alloy} and σ_{Nb} are the stresses that have to be applied to the alloy and to niobium respectively to give a creep rate of $10^{-5}\ s^{-1}$. It is important to realise that with the value $M = 3\text{-}5$ of the stress exponent, the values shown for S_B correspond to reductions in rate by *2* to *3* orders of magnitude.

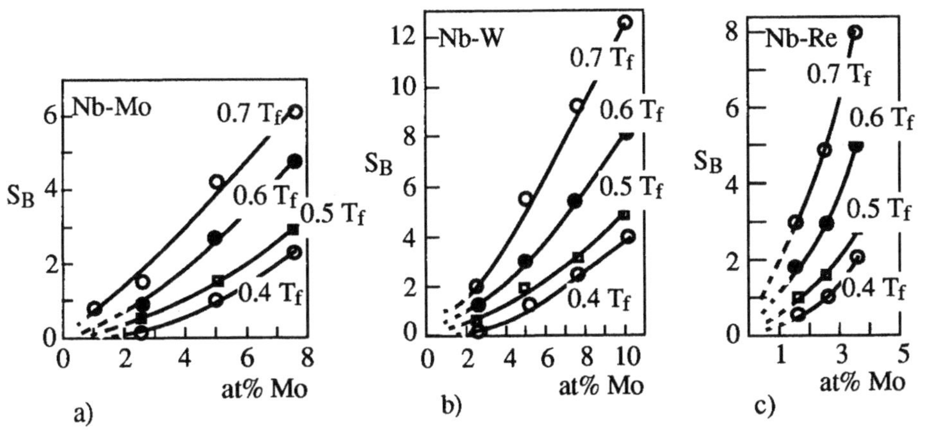

Fig. 1.45 Increased creep resistance of niobium resulting from addition of (a) Mo (b) W (c) Re.

As in low-temperature plasticity, elements added in solid solution constitute obstacles to the movement of dislocations. There are three effects to consider:

- of size,
- of modulus,
- of distortion, when the deformation field surrounding the solute atoms is not spherically symmetrical. This is especially the case for the interstitial atoms C, N, O in BCC metals (cf. I- §3.3.8.5).

Further, at high temperature the solute atoms are as mobile as those of the solvent, a situation which gives rise to what is often a rather serious difficulty in the detailed analysis of creep phenomena. It concerns essentially the choice of the diffusion coefficient which, as we have seen, appears in the various models of creep. For dilute solutions the relevant coefficient is that of the solvent itself in the solution: stated simply, it is known that adding a "fast" solute increases the self-diffusion of the basic metal, whilst a "slow" solute reduces this. The effect is greater, the higher the solute concentration, up to a limit of around *5%* by atoms. This simple rule explains, qualitatively at least, the effect of adding an atom such as molybdenum to the matrix of an austenitic stainless steel.

With concentrated solutions the choice of the diffusion coefficient raises more difficult conceptual problems.

Another aspect of high-temperature behaviour of solutions and atomic mobility is the existence of "viscous drag". The main origins of this are the elastic interaction between the dislocations and the solute atoms (Cottrell effect), the segregation of solute atoms on to stacking faults (Suzuki effect) and the destruction of short- or long-range order by the passage of a dislocation. Stated simply, this drag effect can be represented by the fact that the dislocations are subjected not only to the applied stress field but also to a force due to their interaction with the segregated atoms (see I-§3.3.8.5).

1.5.3 Creep in alloys reinforced by particles

Another, and very effective, way to block the movement of dislocations is to incorporate particles into the material. This can be done either by precipitation (as with nickel-based superalloys and creep-resistant ferritic steels) or by powder metallurgy (for example, particles of the very stable oxides of thorium or yttrium in nickel- or aluminium-based matrices, or of silicon carbide in aluminium-based matrices.)

Taking as example the nickel-based alloys used in manufacturing turbine blades, the solutions adopted are based mainly on the excellent temperature stability of compounds of the type of Ni_3Al (γ' phase). The hardening due to these γ' precipitates underlies the development of the superalloys, the most modern having over *50%* of this phase.

Fig. 1.46 shows the progress made in superalloy development since 1940; the average increase in resistance can be expressed as a gain of almost *10°C* per year over an entire *30* year period.

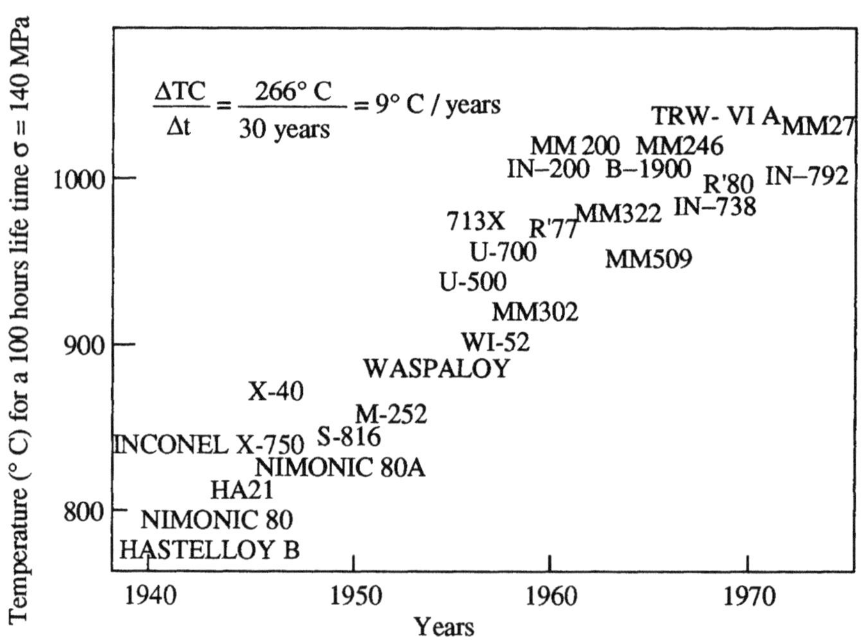

Fig. 1.46 The evolution of nickel-based superalloys.

CHAPTER 1 EXERCISES

1. Relaxation and creep

1.1 Show that relaxation under uniaxial loading in a material for which the creep law has the form $\dot{\varepsilon}^v = B\sigma^M$ is

$$\sigma = \left[(M-1)BEt + \frac{1}{\sigma_0^{M-1}} \right]^{-1/(M-1)}$$

where σ_0 is the initial stress.

Application. A steel bolt joining two plates is held at $550°C$, at which temperature creep tests have shown that in the steady state $M = 3$ and B is a constant such that for $\sigma = 275$ MPa, $\dot{\varepsilon}^v = 2.8\ 10^{-8}\ h^{-1}$. The bolt is tightened to give $\sigma_0 = 70\ MPa$. Show that at the end of 1 year this stress has fallen to $16.3\ MPa$.

1.2 Obtain the expression for the relaxation curve $\sigma(t)$ corresponding to primary creep. The form of this law can be either $\dot{\varepsilon}^v = At^{-p}\sigma^n$ or $\dot{\varepsilon}^v = B\varepsilon^{-q}\sigma^{n'}$. Show that the choice is not arbitrary, since it implies the assumption of either time hardening or strain hardening.

2. Internal stress and precipitation

Some alloys, such as ferritic steels for steam turbines and nickel-based superalloys for gas turbines, must be perfectly resistant to creep at high temperatures. They are hardened by precipitation, of carbides in the steels and of the γ' phase in the superalloys. The law for stationary creep can be written

$$\dot{\varepsilon}^v = \alpha|\sigma - \sigma_s|^4 \quad \text{where} \quad \alpha = A'\exp(-Q'/kT)$$

Some authors have suggested that the internal stress σ_s can be written

$$\sigma'_s = \beta + BR^{-p}$$

where β is the component due to the deformation (kinematic and isotropic work-hardening) and R measures the mean size of the precipitates; as a first approximation we take β to be constant. R can change during creep as a result of ageing; following Volume I Annex 1 we take

$$R = \left[R_0^3 + Ct\right]^{1/3}$$

where t is time and C a constant at any given temperature.

2.1 Putting $\beta = 0$ and assuming that σ_s can be identified with the stress necessary for crossing a precipitate by a wrapping mechanism (the Orowan stress), show that it can be expressed in the form

$$\sigma_s = \frac{\mu b\sqrt{f}}{2R(1-\sqrt{f})}$$ where f is the volume fraction of particles.

Show also that in steels, where $f \approx 0.1$, σ_s is small, about $35\ MPa$, whereas in certain superalloys where $f \approx 0.6$, σ_s is much greater, around $250\ MPa$. Note that with $p = 1$ the model shows the competition between work-hardening (if β is an increasing function of the deformation) and a recovery phenomenon.

2.2 Investigate the consequences to be expected of such a model

Show that when $t \to 0$, we can write $\left(\dot{\varepsilon}^v / \dot{\varepsilon}_0\right)^{1/4} = 1 + Kt/(\sigma - D)$

where $K = \frac{1}{3} BpCR_0^{-(p+3)}$, $D = \beta + BR_0^{-p}$, $\dot{\varepsilon}_0^v = \dot{\varepsilon}(t=0)$

Show also that as $t \to \infty$, $\left(\dot{\varepsilon}^v / \dot{\varepsilon}_0\right)^{1/4} = \left(1 - \alpha/(\sigma - \beta)\right)^{-1}$, $\alpha = BR_0^{-p}$

Investigate the evolution of the curves of the associated tertiary creep.

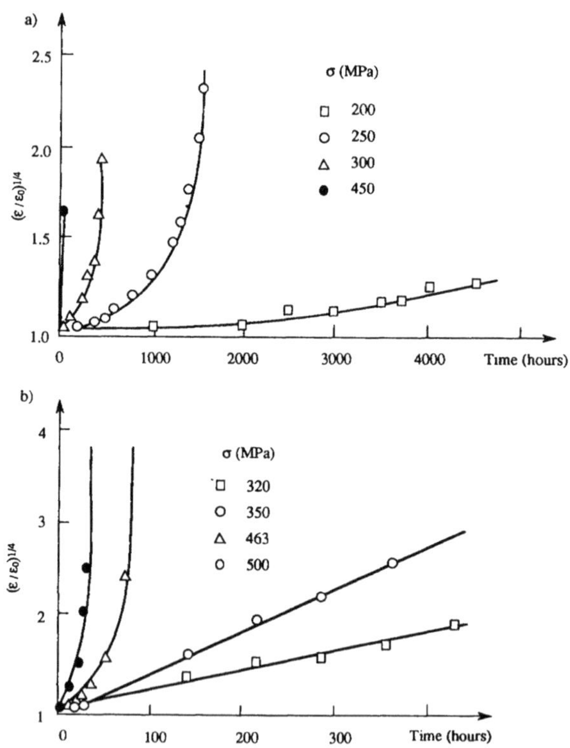

Fig. 1.47 Creep in two Ni-based superalloys at $850°C$: (a) IN 738, (b) MAR M 246

2.3 Fig. 1.47 gives the results of tests at *850°C* on two nickel-based alloys having a large volume fraction (around *50%*) of the γ' phase; these were analysed, using the representation suggested by the model of this Exercise. Show that, whilst the data demonstrate clearly the existence of tertiary creep, they are not in full accord with the model; investigate the following changes that might be made to the model, to give better agreement:

(i) taking into account the change of internal stress with deformation (the β term)
(ii) noting that in the alloys the ripening process is not isotropic, but stress-oriented (cf.I-Annex 2, §10.8)

3. A "2-bar" system
We consider a 2-phase system: a matrix A, relatively soft and viscous, with a small fraction f_B of a harder material. We assume that both have visco-elastic behaviour and have the same elastic constants: for the viscous part

$$\dot{\varepsilon}_A^v = \dot{\varepsilon}_{0A}(\sigma_A / \sigma_{0A})^{M_A}, \quad \dot{\varepsilon}_B^v = \dot{\varepsilon}_{0B}(\sigma_B / \sigma_{0B})^{M_B}$$

$$\sigma_{0B} >> \sigma_{0A}, \quad \dot{\varepsilon}_{0B}^v << \dot{\varepsilon}_{0A}^v$$

The behaviour of this composite material is modelled by a system with two parallel bars, as in Fig. 1.48.

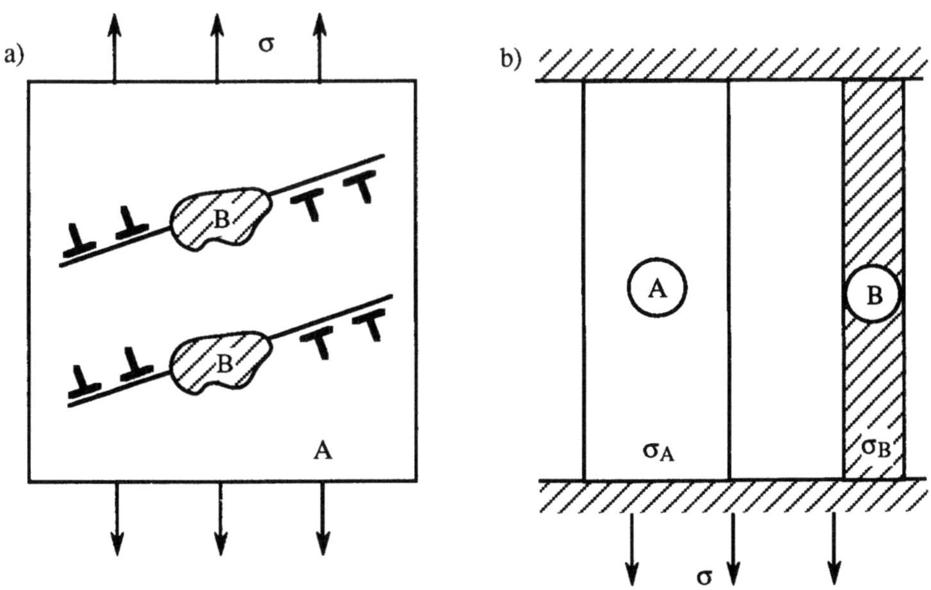

Fig. 1.48 2-phase medium, with matrix A and second phase B.
(a) Schema (b) Representation by a 2-bar model.

3.1 Show that the equations for equilibrium and for strain compatibility of the deformations are

$$\sigma = \sigma_A f_A + \sigma_B f_B$$

$$\begin{cases} d\sigma_A = E\left[d\varepsilon - \dot{\varepsilon}^v_{0A}(\sigma_A / \sigma_{0A})^{M_A} dt\right] \\ d\sigma_B = E\left[d\varepsilon - \dot{\varepsilon}^v_{0B}(\sigma_b / \sigma_{0B})^{M_B} dt\right] \end{cases}$$

Deduce that the strain rate is

$$\dot{\varepsilon}^v = \dot{\sigma} / E + f_A \dot{\varepsilon}_{0A}(\sigma_A / \sigma_{0A})^{M_A} + f_B \dot{\varepsilon}_{0B}(\sigma_B / \sigma_{0B})^{M_B}$$

3.2 The stress supported by the phase B - more precisely, the stress $\sigma_i = f_B\ \sigma_B$ - can be considered as an internal stress opposing the viscous deformation of the phase A. Show that if $f_B << f_A$ (so that f_A is approximately *1*) it follows from the equilibrium condition that $\sigma_A = \sigma - \sigma_i$

3.3 Find the behaviour of the model when $\dot{\sigma} = 0$

Start in the steady state with

$$d\sigma_A = d\sigma_B = 0,\ \dot{\varepsilon}^v = \dot{\varepsilon}^v_{SS},\ \sigma_i = \sigma_i{*}.$$

Show that

$$\dot{\varepsilon}^v = \dot{\varepsilon}^v_{SS}\left[f_A\left(\frac{\sigma - \sigma_i}{\sigma - \sigma_i{*}}\right)^{M_A} + f_B\left(\frac{\sigma_i}{\sigma_i{*}}\right)^{M_B}\right]$$

with the law of evolution for σ_i:

$$d\sigma_i / dt = f_B E\left[\dot{\varepsilon}^v - \dot{\varepsilon}^v_{0B}\left(\frac{\sigma_i}{f_B \sigma_{0B}}\right)^{M_B}\right]$$

which can be written:

$$d\sigma_i / dt = f_B E\left[\dot{\varepsilon}^v - \dot{\varepsilon}^v_{ss}\left(\frac{\sigma_i}{\sigma_i{*}}\right)^{M_B}\right]$$

showing the presence of a hardening term and a recovery term.

Investigate the variation of σ_A and of σ_i with time. To simplify the problem, consider the realistic case in which the recovery term is small compared to the hardening term, so that we can write

$$d\sigma_i / dt = Ef_B\dot{\varepsilon}^v \text{ with } \dot{\varepsilon}^v = \dot{\varepsilon}^v_{ss}\left(\frac{\sigma - \sigma_i}{\sigma - \sigma_i *}\right)^{M_A}, \quad \dot{\varepsilon}^v_{ss} = \dot{\varepsilon}^v_{0B}\left(\frac{\sigma_i *}{f_B\sigma_{0B}}\right)^{M_B}$$

4. Activation energy needed for dislocations to bypass precipitates

In nickel-based alloys hardened by precipitates of the coherent and ordered phase of the type $Ni_3(Ti, Al)$ (pseudo-FCC structure, Ll_2), it is found that a dislocation will shear small particles but will wrap around those above a certain critical size. - the Orowan mechanism, (see Volume I §3.3.8.5b and Annex 2). The exercise here is to determine the activation energies for these two modes of bypassing precipitates.

4.1 . *Shearing*

Consider two particles a distance L apart in the slip plane, on which latter there is a shear stress τ_A (Fig. 1.49). The dislocation penetrates the particles, taking the position AMB; the antiphase energy γ_A which holds the dislocation line is responsible for the curvature.

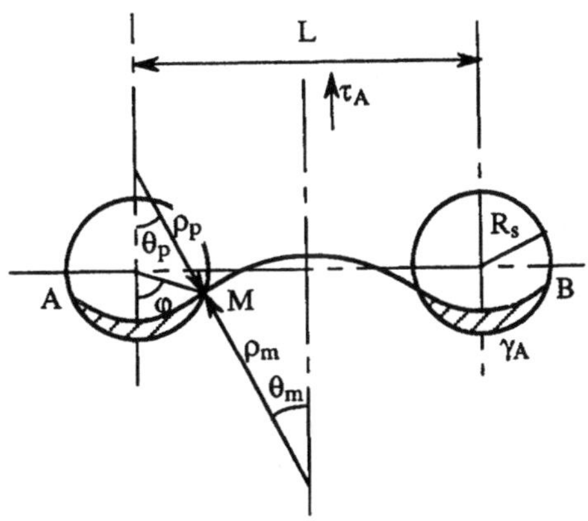

Fig. 1.49 Shearing of two precipitates.

4.1.1 By drawing the unit cell of the Ll2 structure (I, Annex 2), explain how this antiphase energy originates. Show that order can be restored by the passage of a second dislocation that has the same Burgers vector as the first.

4.1.2 Show that if R_0 is the mean radius of particles in the bulk of the material, those in the slip plane will have a mean radius $R_s = (\pi/4)R_0$

4.1.3 Show that the equilibrium equations for the arc of the dislocation are

$$\text{(i) } \rho_p = t_p / (\gamma_A - b\tau_A) \quad \text{(ii) } \rho_m = t_m / b\tau_A$$

and that the following geometrical relations hold

$$\text{(iii) } L/2 = \rho_m \sin\theta_m + \rho_p \sin\theta_p \quad \text{(iv) } \rho_p \sin\theta_p = R_s \sin\varphi$$

where t_m, t_p are the line tensions of the dislocation in the matrix and precipitates respectively. If these are equal then the junction is tangential and $\theta_m = \theta_p = \theta$.

Using these relations, show that for any given value of τ_A there are two possible values φ and π-φ that satisfy the equilibrium conditions. Show these on the energy diagram, as in Fig. 1.50b.

Note that in general, with $t_m \neq t_p$, equations (i)-(iv) are not sufficient to define the equilibrium: the further condition $d(\Delta E)/d\varphi = 0$ must be added, where

$$\text{(v) } \Delta E = (t_m \Delta l_m + t_p \Delta l_p) - b\tau_A \Delta S_m - (b\tau_A - \gamma_A)$$

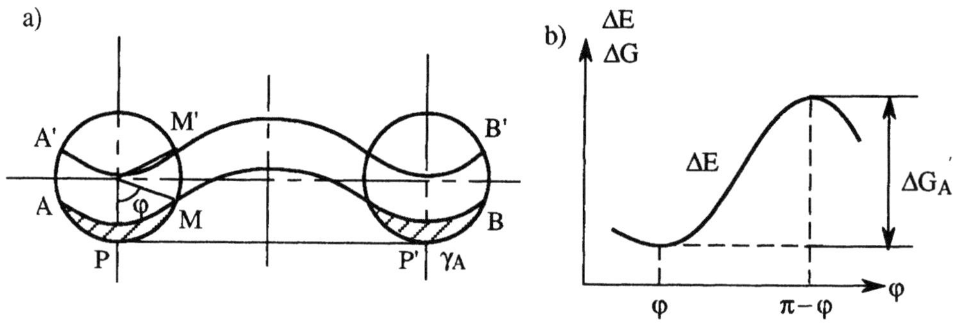

Fig. 1.50 Shearing of two particles by a dislocation:
(a) stable and metastable positions of the dislocation
(b) variation of energy ΔE; ΔG_A is the activation energy.

4.1.4 Compute the energy variation ΔE during the crossing; then, starting from some reference position, for example PP' in Fig. 1.50, find

- the increases in length Δl_m, Δl_p of the dislocation in the matrix and the precipitates respectively

- the areas ΔS_m, ΔS_p swept by the dislocation in the matrix and the precipitates

Substitute these results in (v) and, with $d(\Delta E)/d\varphi = 0$, derive the relation

$$\text{(vi) } 2[t_p \cos\theta_p - t_m \cos\theta_m] + [2R_s\gamma_A \sin\varphi - bL\tau_A]\tan\varphi = 0$$

Show that equations (i)-(iv) and (vi) together enable the complete solution of the problem to be found.

4.1.5 Assume throughout that $t_m = t_p$. Show that

$$\text{(vii)}\quad \sin\varphi = bL\tau_A / 2\gamma_A R_s \qquad \text{(viii)}\quad \tau_A = \tau_A^C(\varphi = \pi/2) = 2\gamma_A R_s / bL$$

where τ_A^C corresponds to the stress needed for crossing at absolute zero, when there is complete absence of thermal activation.

Show that the activation energy ΔG_A can be expressed in the simple form

$$\text{(ix)}\quad \Delta G_A = \gamma_A R_s^2 [\pi - 2\varphi - \sin 2\varphi]$$

and the activation area A^*

$$\text{(x)}\quad A^* = 2LR_s \cos\varphi = 2LR_s \left[1 - \frac{bL\tau_A}{2\gamma_A R_s}\right]^{1/2}$$

4.1.6 In the neighbourhood of $0°$ K, the critical shear stress to cut the precipitates is given approximately by (viii). Show that here the expression for ΔG_A can be simplified to

$$\text{(xi)}\quad \Delta G_A = \pi\gamma_A R_s^2 \left[1 - \frac{2bL\tau_A}{\pi\gamma_A R_s}\right]$$

4.1.7 A numerical application

Evaluate ΔG_A for $\gamma_A = 200.10^{-3}\ Jm^{-2}$, $R_s = 22.10^{-10}\ m$ and compare the value with the thermal activation energy at ambient temperature. Is it likely that shearing by thermal activation will be possible?

4.2 *Orowan bypassing*

Use the same reasoning and the same methods for attacking the problems as before. Fig. 1.51, corresponding to Fig. 1.49, gives the geometry; note the two equilibrium positions for the dislocation

4.2.1 At $0°$ K, with no thermal activation, $\varphi = \pi/2$ and the two equilibrium positions coincide. Deduce that the Orowan stress is

$$\text{(xii)}\quad \tau_A^C = \frac{2t}{b(L - 2R_s)}$$

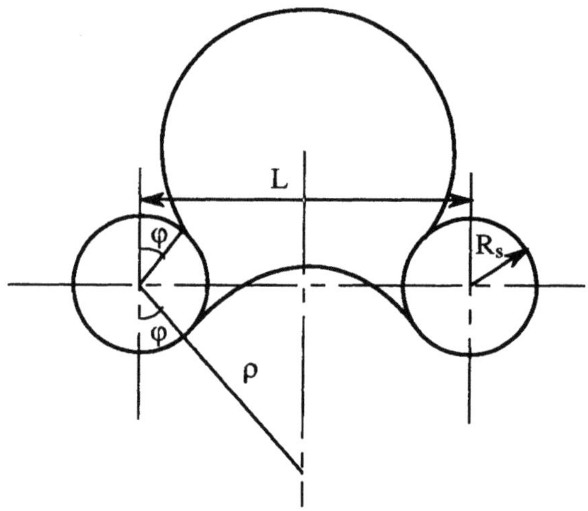

Fig. 1.51 Bypassing of two precipitates by Orowan mechanism.

4.2.2 Proceeding as in 4.1.6, show that

$$\text{(xiii)}\quad \Delta G_A = \frac{(t + bR_s\tau_A)^2}{b\tau_A}[\pi - 2\varphi - \sin 2\varphi]$$

4.2.3 Show high energies are needed for this mode

5 Diffusional creep; Coble model

This model applies to materials that creep under low stress at moderate temperatures (cf. Coble (1963)), with the creep rate determined not by bulk diffusion as in the Herring-Nabarro mode, but by diffusion along the grain boundaries. It is based on simple physical arguments.

We assume that the grains are spherical and subjected to a uniaxial tension (see Fig. 1.52), and that diffusion occurs only at the surface, along the boundary; by symmetry, we need consider only a hemisphere. We aim first to determine the concentration gradient along the boundary and then, using Fick's equation, the strain rate.

5.1 Describe in outline the paths of the atoms and the vacancies.

5.2 If C_0 is the concentration of vacancies, show that the excess at A in Fig. 1.52 is

(i) $\Delta C = (\sigma\Omega/kT)C_0$, where $\Omega \approx b^3$ is the atomic volume.

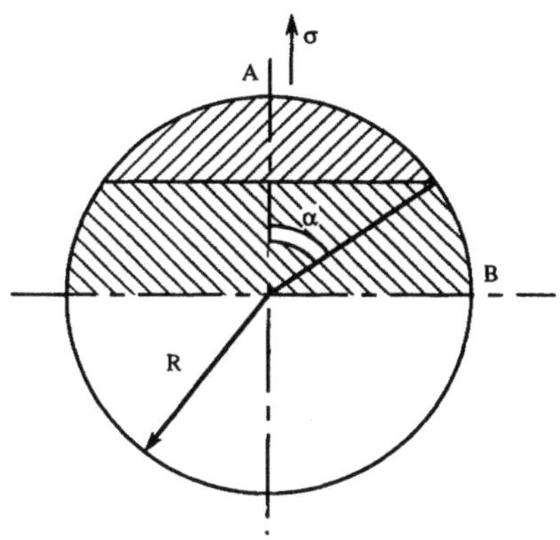

Fig. 1.52 Source and sink of vacancies in a spherical grain.

5.3 For simplicity, we divide the hemisphere into a source part and a sink part, separated by a plane parallel to the diametral plane, with the concentration uniform in each. Show that the location of the separating plane is such that $\alpha = \pi/3$ and that the concentration gradient, maximum at $\alpha = \pi/3$, is given by

$$\text{(ii)} \quad \frac{1}{\mathrm{R}}\left(\frac{dC}{d\alpha}\right)_{\alpha=\pi/3} = \gamma \frac{\Delta C}{\pi R/2}$$

where γ is a numerical constant, whose value is found to be 2.15. Deduce that the flux of vacancies is

$$\text{(iii)} \quad \Phi = \gamma D_{\mathrm{L}} \frac{\Delta C}{\pi R/2} \delta_j 2\pi R \frac{\sqrt{3}}{2} = 7.45 D_L \delta_j \Delta C \qquad (\gamma = 2.15)$$

where δ_j is the width of the boundary

5.4 To evaluate the corresponding strain rate we have

$$\text{(iv)} \quad \dot{\varepsilon}^v = \frac{1}{R}\frac{dR}{dt} = \frac{\Phi\Omega}{\pi R^3}$$

since the source, of area πR^2 and assumed uniform, generates a volume change given by $\Phi\Omega = \pi R^2\, dR/dt = 7.45\delta_j \Delta C \Omega$. Show that

$$\text{(v)} \quad \dot{\varepsilon}^v = \frac{7.45 D_j \delta_j \sigma \Omega}{\pi R^3 kT}$$

It follows from this, by taking R equal to the radius of the grains, of diameter d, that

$$\dot{\varepsilon}^{v} = B'\frac{D_j\delta_j\sigma\Omega}{d^3kT}, \quad \text{where } B' \approx 20$$

6. Diffusional creep: experimental demonstration

Crossland et al (1973) studied Coble creep in a number of metals, using a simple but ingenious experimental technique [J. Phys. (D): Appl. Phys. **6** (1973) 1040)]. A sample of the metal (Ag, Cd, Cu, Mg, Ni) was made into the form of a helical spring, suspended from one end and allowed to creep under its own weight. Thus each turn of the spiral was subjected to a shear (at least for small deformation) which increased linearly with the distance from the free end. The experiment is easily carried out with a spring made with the common Pb-Sn alloy used for solder, when it will be found that the separation of the turns increases linearly with time and with the distance from the free end (se Fig. 1.53).

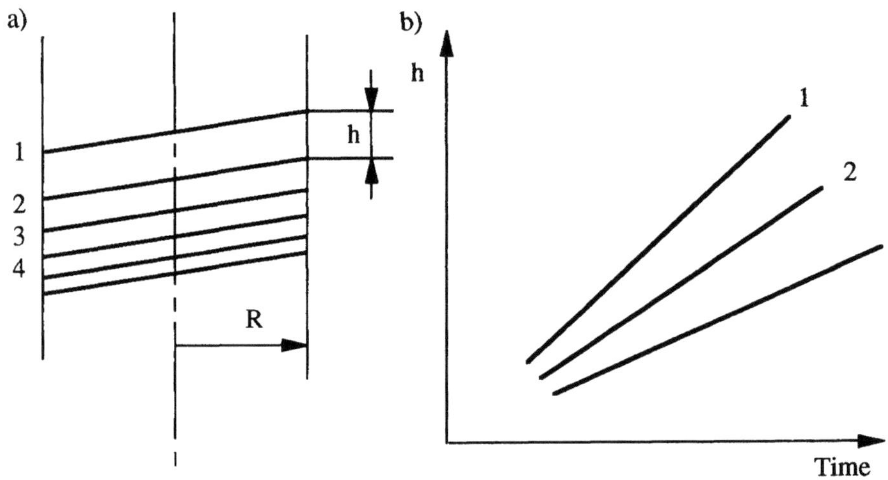

Fig. 1.53 Diffusional creep. (a) Spring creeping under its own weight; h is the distance between successive turns. (b) Time-variation of h for various turns.

6.1 Show that the behaviour of this "mechanical clock" is characteristic of diffusional creep.

6.2 Show that the shear stress and shear strain at the surface of the wire are

$$\tau = 2FR/\pi r^3, \ \gamma = hr/2\pi r^2$$

where F is the applied load (which varies along the length of the spring), h the distance between successive turns, r the radius of the wire, R that of the helix.

7. Creep of a beam in bending: importance of diffusional mechanisms

We study the creep of a beam under bending, making the standard assumptions of small deformations, plane sections remaining plane, and uniaxial stress. The length of the beam is L, its axis is parallel to the x-axis and the y-axis is vertical; the loads we shall consider generate bending moments about the z-axis (Fig. 1.54).

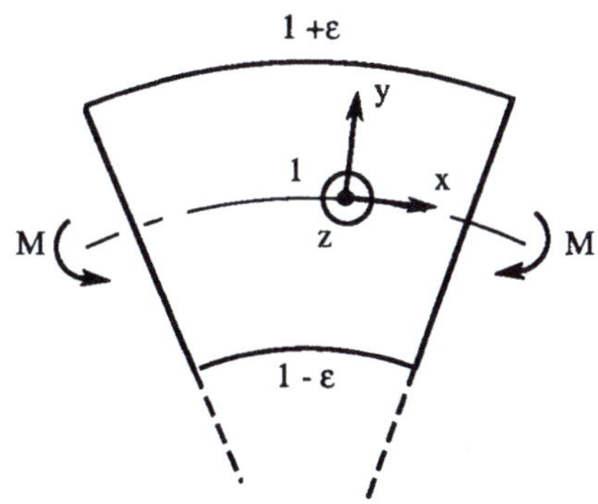

Fig. 1.54 Section of a beam undergoing creep.

σ and ε are the stress and deformation respectively in the x-direction. We use a simple Norton law for the viscoelastic behaviour:

$$|\dot{\varepsilon}^v| = \left(\frac{|\sigma|}{K}\right)^M$$

7.1 *Beams of arbitrary section*

7.1.1 Show that at any instant t in the course of the creep the variation of stress in the y-direction can be written

$$|\sigma| = \sigma_{\max}\left|\frac{2y}{h}\right|^{1/M}$$

where σ is the stress at that section and that instant, and σ_{max} the maximum stress in the beam. What simplifying assumptions have been made in deriving this result? Recall that assuming that plane sections remain plane is equivalent to putting

$$\varepsilon = \varepsilon_{\max}(2y/h) \quad \text{or} \quad \dot{\varepsilon}^v = \dot{\varepsilon}^v_{\max}|2y/h|$$

7.1.2 Express the maximum stress as a function of the bending moment M

7.1.3 If I is the moment of inertia of the section as used in elasticity ($I = bh^3/12$ for a rectangular section of breadth b and depth h), show that the stress can be written

$$|\sigma| = (Mh/2I)\, f(y, m)$$

where for $M = 1$ f is linear in y and has the value 1 at $y = h/2$. Give the qualitative form of f as a function of y for $M = 1$ and $M = 12$.

7.1.4 Defining an "equivalent moment of inertia" I_C for creep by $\sigma_{max} = Mh/2I_C$, express the curvature d^2y/dx^2 as a function of bending moment and time. Express I_C as a function of I.

7.2 *Beams of circular section*

7.2.1 What are the mechanisms of creep when $M = 1$?

7.2.2 Creep of a tube under its own weight. The photograph of Fig. 1.55 shows the present form of lead pipework installed *20* years ago. Anchored at points A, B one metre apart, it has crept under the action of its own weight; the distance δ is about *40 mm*.

Fig. 1.55 Lead piping fixed at A and B, having crept under it own weight (Ashby and Jones (1980) p.159)

The stresses here are small, σ_{max} about *1 MPa*. Lead has a density of *11.35 g/cm³* and a melting point of *327° C (600° K)*; its deformation chart is given in Fig. 1.56.

7.2.1 Using the deformation chart, show that Coble creep is very likely here: recall that $1/K = 14\pi\Omega\delta_j D_j/kTd^3$ (d is the grain size, other quantities are defined in §1.3.2.2). Accepting that this is the case (Coble creep), show, again with the help of the diagram, that at $25°\ C$ the value of $1/K$ is of the order of 10^{-15}, with σ in Pa and $\dot{\varepsilon}^v$ in sec^{-1}.

7.2.2 With this type of creep ($M = 1$), the sag can be determined by transposing the solution already obtained into elasticity and expressing it in terms of rate, replacing E by K. For a thin-walled tube of radius R and mass per unit volume ρ

$$\dot{\delta} = \frac{10}{384}\frac{\rho g l^4}{KR^2}$$

Taking $1/K = 10^{-15}$, $R = 2\ cm$, calculate the sag after 20 years; compare the result with what is shown in Fig. 1.55. Give an approximate value for the grain size of the material from which the tube was made: is this value reasonable?

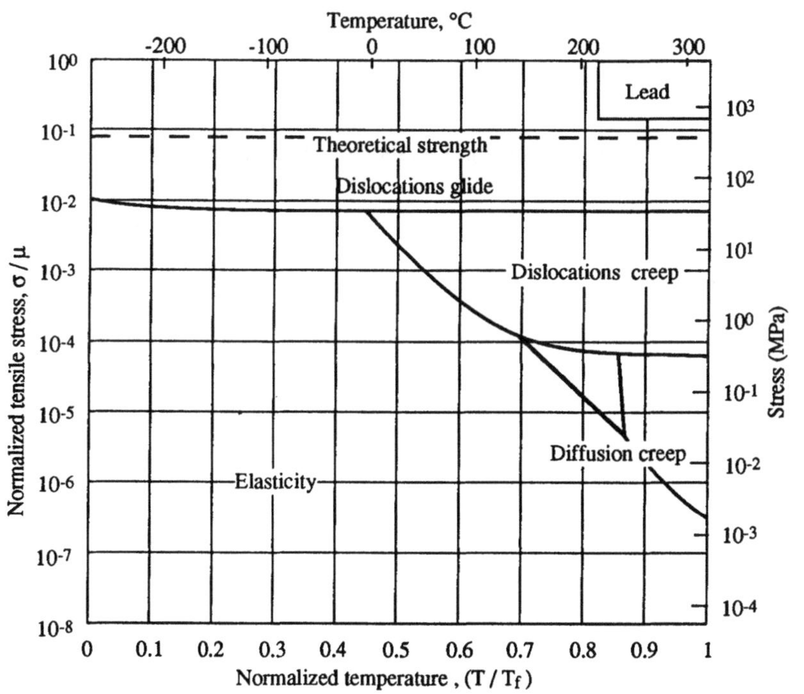

Fig. 1.56 Deformation chart for lead.

8. Anisothermal creep, residual stresses

Fig. 1.57 shows a bar M fixed at its ends to two rigid, immovable walls and subjected to a temperature cycle. The problem is to find the residual stresses in the bar at the end of the cycle. We assume that the material of the bar is elasto-viscoplastic (no plasticity) with Young's modulus E and coefficient of dilatation α independent of temperature, and that the creep law can be written

$$\dot{\varepsilon}^{v} = B|\sigma|^{M} \operatorname{sgn}(\sigma) \quad \text{where } B = f(T)$$

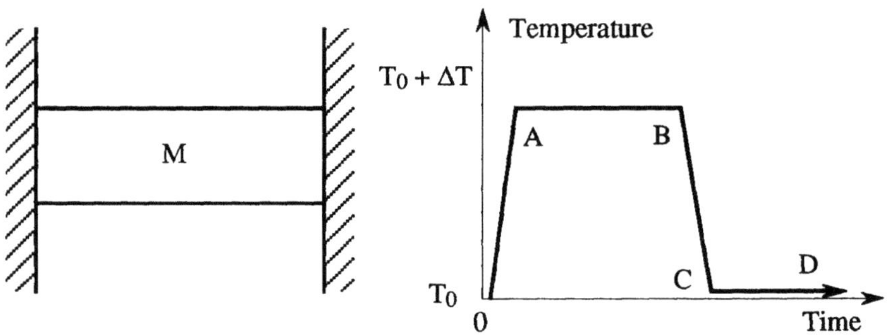

Fig. 1.57 Bar fixed between rigid walls and subjected to a temperature cycle.

8.1 Show that the deformation of the bar can be written

$$\varepsilon = \varepsilon_e + \varepsilon^v + \varepsilon_{th}, \text{ with } \varepsilon_e = \sigma / E,\ d\varepsilon_{th} = \alpha dT,\ \dot{\varepsilon}^v = B|\sigma|^M$$

8.2 Show that if $\varepsilon_m = \varepsilon_e + \varepsilon^v$, the behaviour of the bar during the temperature cycle OABCD can be represented qualitatively in the σ–ε_m plane by the path OABCD of Fig. 1.58. Discuss the relation of the amplitude of the relaxation over the part CD to that over AB.

8.3 Compute the behaviour throughout the cycle; show that

$$\sigma_A = -\alpha E \Delta T,\ \sigma_B = \left[\frac{1}{(\alpha E \Delta T)^{M-1}} + (M-1)BEt\right]^{1/(M-1)},$$

$$\sigma_C = \sigma_B + \alpha E \Delta T$$

8.4 The material of the bar was tested for creep under tension, with $\Delta T = 50° C$; it was found that at $T_0+\Delta T$, $\dot{\varepsilon}^v = B\sigma^3 = 2.8\ 10^{-8}$ h^{-1} for $\sigma = 27.5$ MPa.
For this material $E = 200{,}000\ MPa$, $\alpha = 1.5\ 10^{-5}$ per degree C.

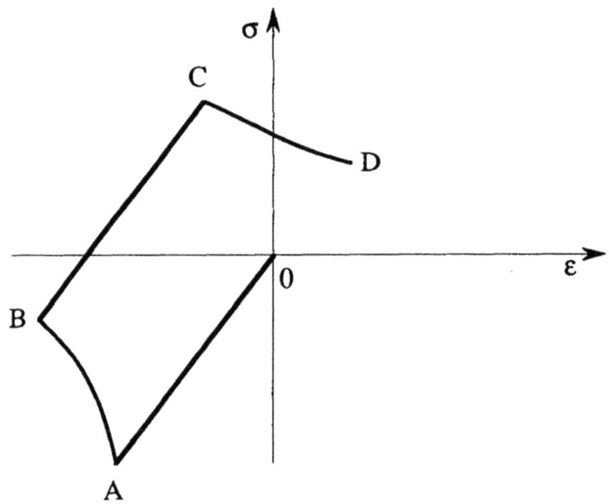

Fig. 1.58 σ–ε diagram for the bar of Fig. 1.57

(a) Show that $\sigma_A = 150\ MPa$, and that at the end of 24 hours $\sigma_B = 132\ MPa$.

(b) To what creep mechanism does the value $m = 3$ correspond?

8.5 Recognise in this exercise the following simple rule for residual stress of thermal origin: when the temperature returns to ambient, the parts of the structure that were hottest will be in tension and those that were coolest will be in compression.

CHAPTER 2 DAMAGE

2.1 THE VARIOUS TYPES OF DAMAGE

2.1.1 Generalities

In Chapter 1 of Volume I we surveyed the various ways in which materials can be deformed, and the corresponding constitutive equations; this enabled us to calculate the deformations and stresses experienced by a sample subjected to external loads of various types. The question now arises of determining the limits that must not be exceeded if the material is not to suffer damage to its structure, possibly leading to rupture. If we are considering a volume of the material that is large compared to the size of the damage, the latter must be integrated into the behaviour law; and for a rigourous treatment the coupling between the mechanisms of deformation and rupture must be taken into account. If the microfissures, the microcavities resulting from the

Fracture type	*Volume damage*	*Mixed damage*	*Surface damage*
Immediate	Cavitation Cleavage (trans/intergranular)		Embrittlement by liquid metals (Hg, Cd, Ga)
Delayed	Embrittlement by impurities		Oxydation corrosion
	Embritlement by hydrogen		Stress corrosion
			Fatigue corosion
	High-temperature creep	Fatigue creep	Fatigue
		Creep in concrete	
	Embrittlement by irradiation		

↓

formation of macroscopic cracks

Table 2-1. Modes of damage

growth or the coalescence of defects, are large enough this approach to the problem is no longer valid; recourse must then be had to fracture mechanics, which is the subject of the next chapter. Here we shall describe various modes of damage and try to explain how they develop and the influence of the main parameters: internal, which include temperature and rate of loading; external, which include grain size and volume fraction of inclusions.

Table 2-1 summarises the main modes. These can be understood by recalling the elementary, atomic-scale mechanisms that create new surfaces by cleavage, slip and cavitation, shown diagrammatically in Fig. 2.1. Cleavage results from the breaking of interatomic bonds in a direction perpendicular to the cleavage plane (Mode I in fracture mechanics). Slip, from breaking in a direction parallel to this plane (Modes II and III): this was studied in the chapter on plasticity. High temperature cavitation is the result of vacancies coalescence, which can happen if there is local supersaturation of vacancies and if they can move; mobility requires the temperature to be high enough, so this is essentially a high-temperature creep phenomenon

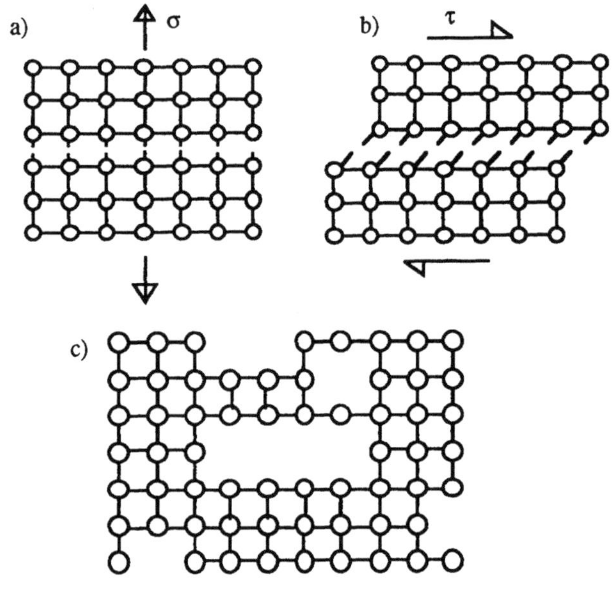

Fig. 2.1 The three elementary atomic-scale mechanisms of damage :
(a) cleavage (b) slip (c) cavitation.

2.1.2 Cleavage and related fractures

Cleavage mechanisms operate in crystalline solids, metals, ceramics and covalent materials. (Fig. 2.2). Analogously, fracture occurs also in glasses, polymers and concrete, but here it cannot be in a crystallographic direction since no such thing exists in these materials. In polymers the cleavage planes that result from failure of the weak Van der Waals bonds are linked by filaments consisting of molecular chains. These microfissures, whose presence is revealed by the material's whitening or becoming opaque, are known as *crazes* (Fig. 2.3); the filaments, forming bridges between the two faces,

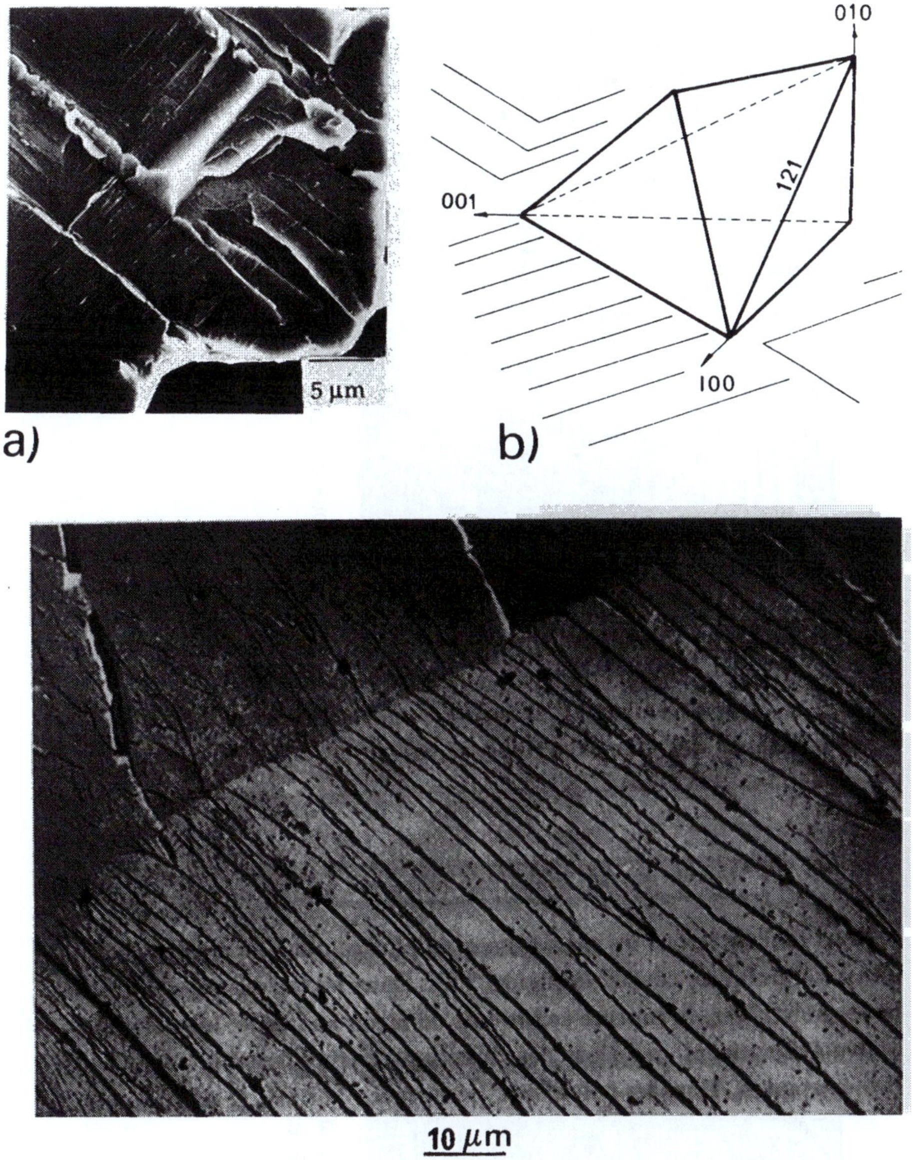

Fig. 2.2 (a) Scanning electron micrograph showing a tongue on the surface formed by cleavage of the ferrite in a cast steel having a duplex ($\alpha+\gamma$) structure. The faces correspond to *(001)* cleavage planes diverted by the intersection of the principal plane (shaded in (b)) with a mechanical twin (i.e resulting from stress). (c) Formation of "rivers" in a polycrystal of zinc, ruptured by cleavage; note the occurrence of rivers across the grain and their "direction of flow", which shows the direction of local propagation of the crack.

Fig. 2.3 Crazing in a polycarbonate. Note the interference fringes, analogous to Newton's rings and corresponding to contours of equal separation of the crazes

are extracted from the matrix as the craze opens. Fracture can also occur along interfaces: along grain boundaries, when it is known as *intergranular rupture*, or along inclusions, fibres in the case of composites or granulates in the case of concrete.

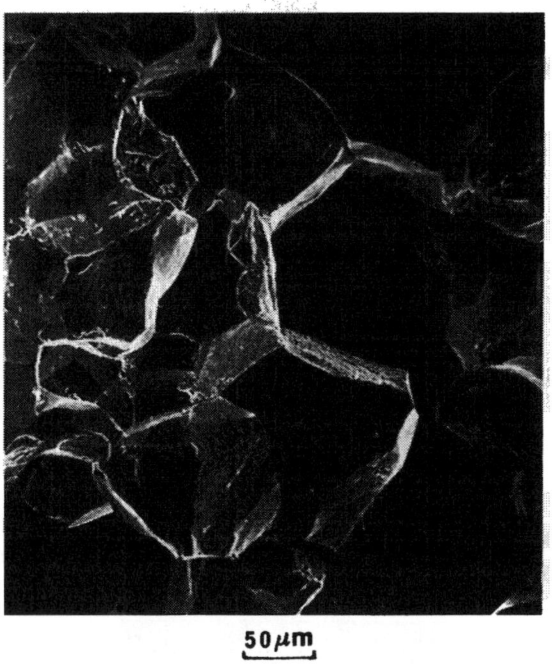

Fig. 2.4 Intergranular rupture following a tensile test carried out at *-196°C* on a 16 MND 5 steel subjected to an embrittlement treatment which has resulted in atoms of phosphorous being segregated on the grain boundaries. The treatment was: *2 hrs*. at *630°C* - slow cooling at *33°/hr*. to *530°* - held there *24 hr*. -cooling at *1°/hr*. to *470°* held there *72 hr*.

2.1.3 Ductile fracture

It can happen that a *slip* starting in one particular plane in a single crystal continues until the crystal separates completely into two parts: this can occur, for example, in CPH zinc, in which the *(0001)* plane is of easy slip (Fig. 2.5a). More generally, in a single crystal several slip systems can be active at the same time, leading to a thinning of the crystal, necking (Fig. 2.5b) and finally to fracture (Exercise 1)

In ductile polycrystals , and also in polymers, plastic deformation can lead to instability through necking (Fig. 2.5c), this resulting from the increase in stress due to the reduction of the cross-section not being compensated for by work-hardening (see Vol. I Ch.3, §2.10) However, not only the work-hardening but also the strain rate plays a part here. In general, as the latter increases it produces a hardening which has a stabilising effect (the Hart condition, Exercise 2); if the material is very sensitive to strain rate the necking does not develop, we have *superplasticity* (see Chapter 1) and deformations of around *1000%* can be reached before rupture occurs - as with chewing gum. A very few materials show the opposite behaviour of softening as the deformation rate increases, and for these the instability can be spectacular. Finally there is the temperature effect. It can happen that thermal dissipation of the plastic deformation raises the temperature locally, softening the material and so leading to instability: this is the *adiabatic shearing*, seen when the rate of deformation is too great for conduction to give uniform temperature, and in metals at very low temperatures, when they conduct heat badly (Fig. 2.6).

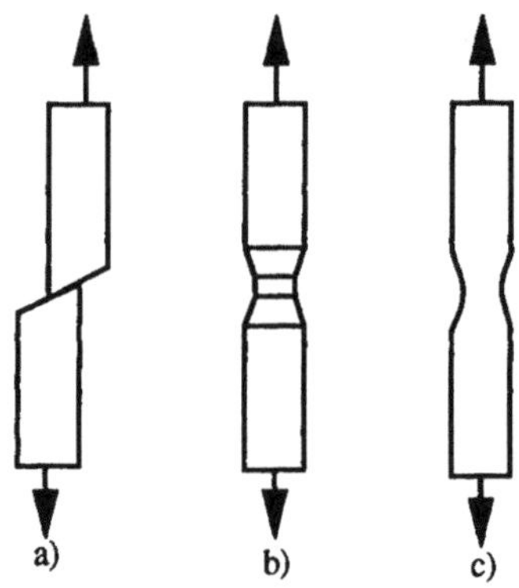

Fig. 2.5 Fracture by slip: (a) single crystal of zinc, slipping along the base plane (b) single crystal of copper, with multiple slips (c) a polycrystal.

Fig. 2.6 Adiabatic shearing in an austenitic stainless steel at *4°K*; the shear band crosses several grains.

It is very rare to find fracture resulting solely from necking, alhough this can occur under very high hydrostatic pressure.

Fracture normaly results from the coalescence of cavities; in general these originate on the inclusions and grow as a result of the plastic deformation - not by diffusion of vacancies. Their coalescence is an expression of local plastic instability and examination of the rupture surface shows a covering of ***dimples***. (Fig. 2.7))

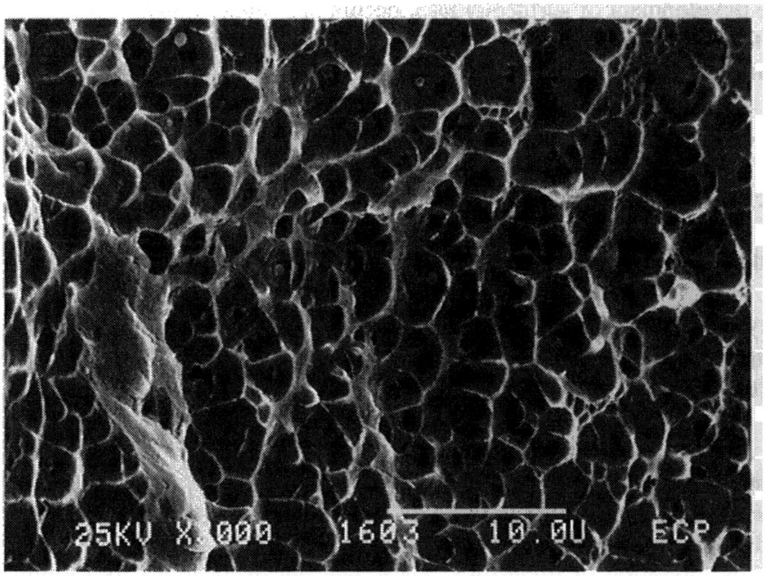

Fig. 2.7 Ductile fracture of austenitic stainless steel Z 02 CND 17 13, showing dimples. Note the presence of particles at the bases.

2.1.4 Fatigue

In Volume I, §3.3.8.2 (h), we saw that under cyclic loading the deformation becomes concentrated in heavy slip bands. If the deformation were perfectly reversible, the steps formed on the surface would vanish when the load reversed; but it is not, so that intrusions and extrusions form (Fig. 2.8)and, gradually, cracks that follow the slip lines. This is Stage I of *fatigue*, which thus appears as a manifestation of surface damage. Sometimes, but only in very special circumstances, these microcracks form below the surface; this is the case in contact fatigue, because it is there that the stress is maximum (see Chapter 4)

2.1.5 Creep

At temperatures at which creep occurs the grain boundaries are the sites at which cavities are most likely to appear. These can result, for example, from the blocking of slip lines by the boundaries, or, and most commonly, from the slipping of the boundaries themselves, which will occur at high temperatures. These microcavities grow under the effect of the plastic deformation and also of the diffusion of vacancies. The final fracture is intergranular (Fig. 2.9).

Under cyclic loading at high temperature there is an interaction between fatigue and creep-induced damage: this is *creep fatigue.*

2.1.6 Stress corrosion and corrosion fatigue

Damage is often made worse by the environment. *Stress corrosion cracking,* resulting from the development of surface cracks, occurs only under conditions of static loading in a hostile environment, which can be of very varied nature: high atmospheric humidity is sufficient in the case of some materials.

Fatigue corrosion occurs under cyclic loading; very often this is not just the developmen of (cyclic) stress corrosion cracks, but a synergetic effect accelerates the rate of damage. For many materials pure fatigue is seen only under vacuum.

It should not be forgotten that the environment can play a part in creep fatigue also.

One very special environment is that of the liquid metals mercury and gallium; these can brings about a dramatic embrittlement in certain alloys - *liquid metal embrittlement*, often following the grain boundaries.

2.1.7 Embrittlement

We conclude this survey of the modes of damage with a review of *embrittlement.* This is a modification of a material's resistance to fracture by a series of structural changes, especially by migration of impurities. These can pin dislocations by forming *atmospheres* (see Vol.I §3.3.8.5δ), or collecting together in the grain boundaries. One particular impurity is hydrogen and *hydrogen embrittlement* has some special features, mostly related to the high mobility of this atom in alloys, for example that it can precipitate either as a gas or in the form of hydrides.

Finally, neutron bombardment produces *irradiation embrittlement* by creating a supersaturation of vacancies and interstitial atoms which are eliminated at different rates. The effect is to increase the hardness and, if the intensity of the radiation is high enough, to form cavities in which rare gases may collect under pressure.

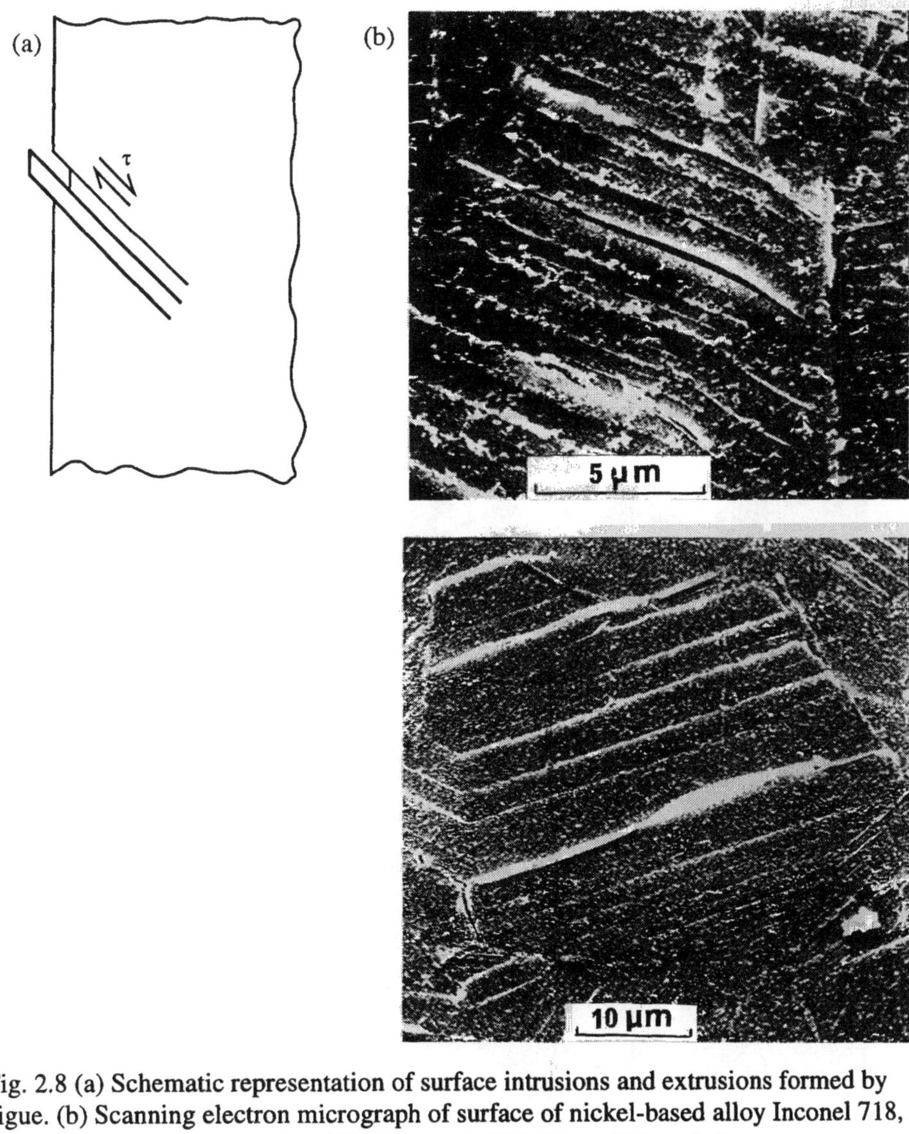

Fig. 2.8 (a) Schematic representation of surface intrusions and extrusions formed by fatigue. (b) Scanning electron micrograph of surface of nickel-based alloy Inconel 718, fatigue-tested at *25°C* ($\Delta\varepsilon_p = 0.62\%$, $N_R = 924$ *cycles*), showing intrusions and extrusions.

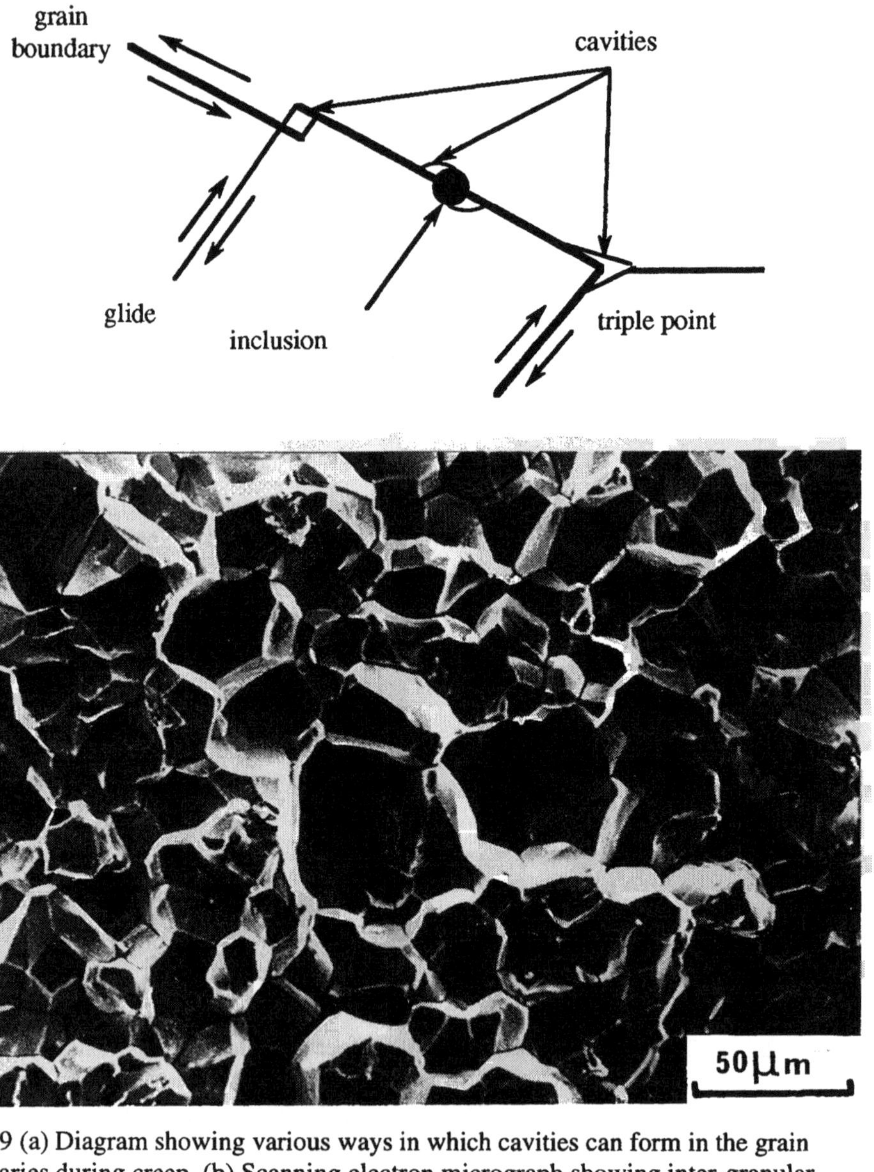

Fig.2.9 (a) Diagram showing various ways in which cavities can form in the grain boundaries during creep. (b) Scanning electron micrograph showing inter-granular rupture of nickel-based alloy Inconel 718 in creep-fatigue test, held under maximum load at *650°C*.

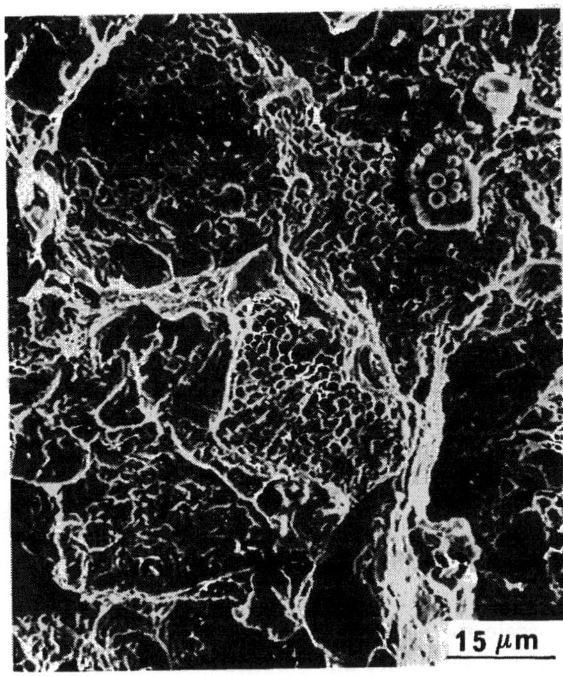

Fig. 29 (c) Scanning electron micrograph showing intergranular facet resulting from fracture in steel 1Cr-1Mo-0.25V at *540°C*.

2.1.8 Sudden and delayed fracture

Rupture can occur suddenly when the load is increasing steadily, by cleavage, by crazing or by ductile cavitation. But in many cases it can be delayed, as in fatigue, stress corrosion cracking and combinations of these, and in the various forms of embrittlement; and these impose greater restrictions in practice.

We now look in greater detail into the various damage and fracture phenomena.

2.2 FRACTURE BY CLEAVAGE

2.2.1 Cleavage

Cleavage occurs preferentially over dense atomic planes. Table 2.2 lists some cleavage planes that have been determined by experiment.

Structure	Cleavage plane	Some materials
BCC	{*100*}	ferritic steels. Mo, W, Ta
FCC	{*111*}	very rarely seen
CPH	{*0001*}	Be, Mg, Zn, Ti, graphites
diamond	{*111*}	diamond, Si, Ge
NaCl	{*100*}	NaCl, LiF, MgO, AgCl
ZnS	{*110*}	ZnS, BeO
CaF_2	{*111*}	CaF_2, UO_2, ThO_2

Table 2.2 Cleavage planes

The orientation of a cleavage plane will change when it crosses a sub-boundary, a twin boundary or a grain boundary, and steps appear on the fracture surface to compensate for the disorientation. In the case of twins these take the form of a type of indentation called a *tongue* (see Fig. 2.2a,b), in that of boundaries or sub-boundaries the steps exert a restraining force on the cleavage front. To maintain the equilibrium of the front the nearest steps join together (Exercise 3), forming a single step of greater height; a micrograph will show the "rivers" that characterise this process, which seem to run in the direction of propagation of the cleavage (Fig. 2.2c).

On the macroscopic scale the surfaces of the cleavage are normal to the maximum principal stress. In fracture mechanics this is called Mode I.

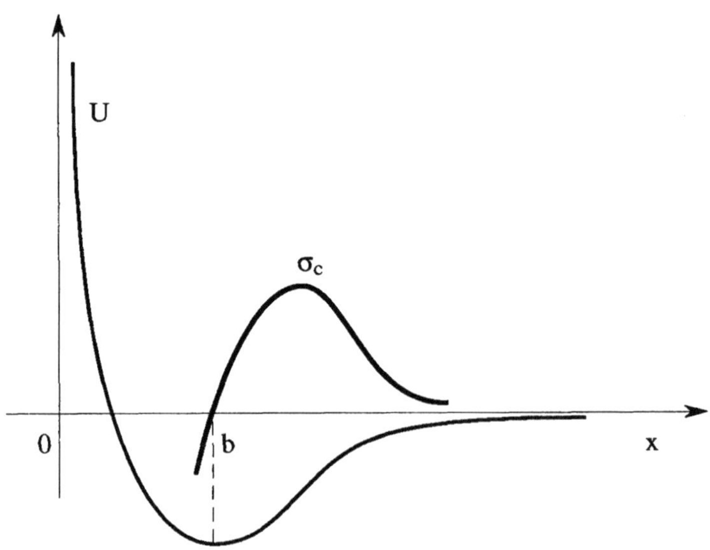

Fig. 2.10 Variation of binding energy *U* and corresponding stress with interatomic distance.

2.2.2 Theoretical cleavage stress

In the treatment of plasticity in Volume I §3.3.2 we calculated the shear stress theoretically needed to make a crystal slip along a crystallographic plane. We can now make an analogous calculation of the normal stress σ_C theoretically needed to fracture the crystal by cleavage (Fig. 2.10). The force required to separate the planes of the cleavage is the derivative of the binding energy with respect to distance; as the distance increases the stress, which is zero at the distance b of interatomic equilibrium, goes through a maximum which is the value σ_C that we are seeking. Hooke's law holds for very small displacements. Cleavage requires an expenditure of energy corresponding to the creation of a pair of new surfaces, that is, $2\gamma_s$ per unit area broken, where γ_s is the surface energy. Taking the variation of the stress with displacement of the crystallographic planes to be sinusoidal, it is easily shown (Exercise 4) that

$$\sigma_C = (E\gamma_s/b)^{1/2} \tag{2.1}$$

where E is Young's modulus.

With the typical values $E = 2.10^{11}\ Nm^{-2}$, $b = 3.10^{-10}\ m$, $\gamma_s = 1\ Jm^{-2}$ this gives $\sigma_C = 2.6\ 10^{10} \approx E/10$. Alternatively, $\gamma_s \approx 0.1\mu b$ where μ is the shear modulus, giving $\sigma_C \approx 0.2E$.

This theoretical value is much higher than what is usually found, about $10^3\ MPa$ for steels. However, for whiskers the observed values are of the same order as the theoretical, showing that in their case the calculation is valid.

2.2.3 Representation of a crack by dislocations

How are we to explain this difference between the theoretical value of cleavage stress and what is found in practice? Movement of dislocations provided the reason in the case of shear; the analogue for cleavage would be the climb of an edge dislocation. Actually, a cleavage break does not happen instantaneously but grows as it were from a seed and is propagated by the successive breaking of the atomic bonds at its edges.

It is not difficult to calculate the stress needed to increase the diameter of a ring-shaped edge dislocation by climb: if the radius is a the result is (Exercise 5)

$$\sigma = \frac{\mu b}{2\pi(1-\nu)}\frac{1}{a}\left(1 + \log\frac{a}{b}\right) \tag{2.2}$$

which tends to zero as the radius increases. This climb can be taken to represent the propagation of a cleavage provided that the energy needed to break the atomic bonds is taken into account, which means adding a term γ_s/b to (2.2). We recall that the definition of a dislocation involves a cut, the edges of which are displaced with respect to each other as defined by the Burgers vector; a crack differs only in that this displacement is not constant over the whole of its surface, and therefore can be represented by a continuous distribution of dislocations (see Fig. 2.11)

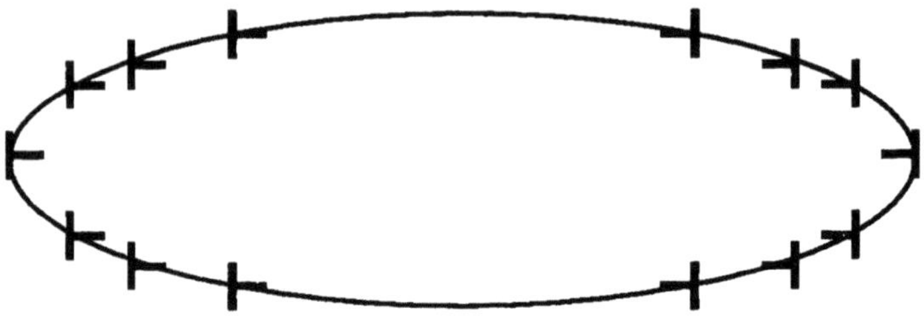

Fig. 2.11 Representation of a crack by a distribution of edge dislocations.

2.2.4 Cleavage or intergranular fracture?

We saw in Chapter 3 of Volume I, in connection with plastic deformation, that the grain boundaries, surfaces separating two differently-oriented systems, have a surface energy γ_j which is a function of this difference in orientation. Therefore if a crack forms along such a boundary the breaking energy $2\gamma_s$ of the atomic bonds is reduced by this amount, and hence it must be easier to cause intergranular ruptures than transgranular cleavages. However, the fact that the surface energy γ_s is anisotropic must be taken into account: this will clearly be minimal for a cleavage plane, and will always be less than that for an intergranular surface: a factor of *1.2* seems to be appropriate. The preference for intergranular rupture over cleavage is thus measured by the ratio $\gamma_j/2\gamma_s$; if we define the parameter R_{CI} (with an obvious notation) by

$$R_{CI} = \frac{2\gamma_s^{in} - \gamma_j}{2\gamma_s^{cl}} = 1.2 - \frac{\gamma_j}{2\gamma_s^{cl}} \tag{2.3}$$

Intergranular fracture will be favoured if $R_{CI} < 1$

Cottrell considers γ_j to depend mainly on the shear modulus μ, whilst γ_s depends on the bulk modulus k; given this, R_{CI} becomes a function of the ratio μ/k:

$$R_{CI} = 1.2 - \alpha\mu/k \tag{2.4}$$

A detailed calculation for the case of nickel gave the value $\alpha = 0.95$. Table 2.3 gives the values of R_{CI} for a number of metals.

	Au	Ag	Cu	Pt	Ni	Rh	Ir	Nb	Ta	V	Fe	Mo	W	Cr
μ/k	0.11	.19	.22	.24	.34	.52	.52	.25	.31	.32	.33	.48	.52	.82
R_{CI}	1.09	1.02	.99	.97	.87	.71	.70	.97	.91	.89	.88	.75	.71	.42
$(\mu/k)_{CD}$	0.36	.43	.57	.38	.49	.39	.32	.59	.55	.65	.56	.35	.45	.68

Table 2.3 Transition parameters for fracture: CI for cleavage/intergranular fracture, CD for cleavage/ductile fracture.

Allowing for the approximations, it seems certain that Rh, Ir, Mo, W and Cr will show intergranular brittleness.

2.2.5 Initiation of cleavages

According to the calculation of the climb of a dislocation loop, a stress of the order of γ_s/b is necessary for the propagation of a large cleavage crack, where b is the width of the crack; whilst the mechanics of the fracture shows that in reality the variable separation of the edges releases elastic energy which compensates for the added term. Nevertheless, a nucleus is needed to start the cleavage, for which an extra stress $\mu/2\pi(1-\nu)$, or $E/4\pi(1-\nu^2)$, must be provided, which is of the order of the theoretical stress σ_C (Exercice 5).

Thus for a cleavage to start there has to be a concentration of stress. This can result from geometrical defects, in particular from small surface indentations; we know, for example (see Fig. 2.12), that around an elliptical notch in a material in which there is a uniform stress field σ the stress concentration factor is

$$K_T = \sigma_{max}/\sigma_{nom} = 1 + 2\sqrt{(a/\rho)} \tag{2.5}$$

where $2a$ is the major axis of the ellipse and ρ is the radius of curvature at its extremity.

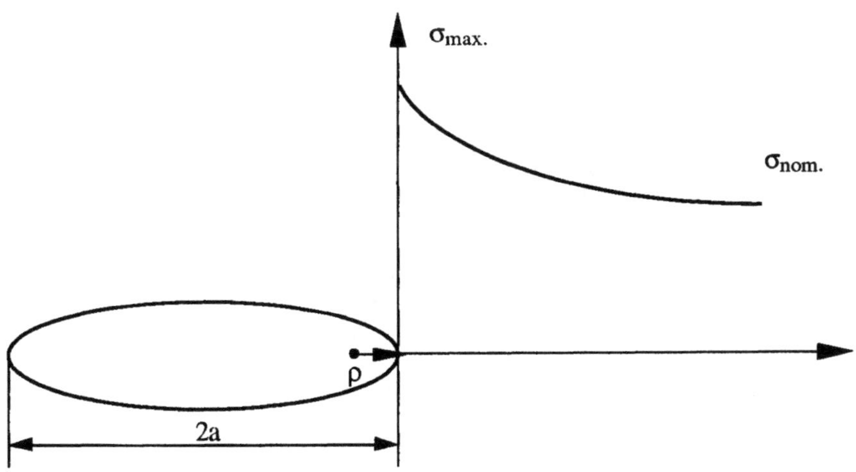

Fig. 2.12 Stress concentration around an elliptical notch.

This shows that it is only necessary for the ratio a/ρ to be high enough for the theoretical stress σ_C to be reached locally; the breaking stress for the piece of material containing this indentation will then be

$$\sigma_f \approx \frac{1}{2}\sigma_C\sqrt{\rho / a} = \frac{1}{2}\sqrt{\frac{E\gamma_s\rho}{ab}} \tag{2.6}$$

It has been found by experiment that for cleavage to start this stress must exceed a threshold value, and therefore ρ has to be less than a critical radius ρ_s, related to the microstructure of the material. This is because it is not sufficient that the stress reaches the critical value σ_C at the tip of the indentation, but it must exceed this value over a certain minimum distance, for example the length of a grain.

Heterogeneities in the deformation - heterogeneous slips and twins -form another source of stress concentrations. There will be such a concentration at the head of a pile-up of dislocations, n in number, say, blocked by a grain boundary (Fig. 2.13) or by an inclusion (see I-§3.3.8.4). For the pile-up, let the distance between the source and the head be L; it can be equated to a crack of length *2L* in an infinite mass of the material, subjected to an effective shear stress $\tau - \tau_i$ (Mode II); we shall show in Chapter 3 that for such a crack the normal stress in the neighbourhood of its end is

$$\sigma = (\tau - \tau_i)\sqrt{L / 2r}\; f(\theta) \tag{2.7}$$

where r, θ are cylindrical coordinates relative to the head of the pile-up.

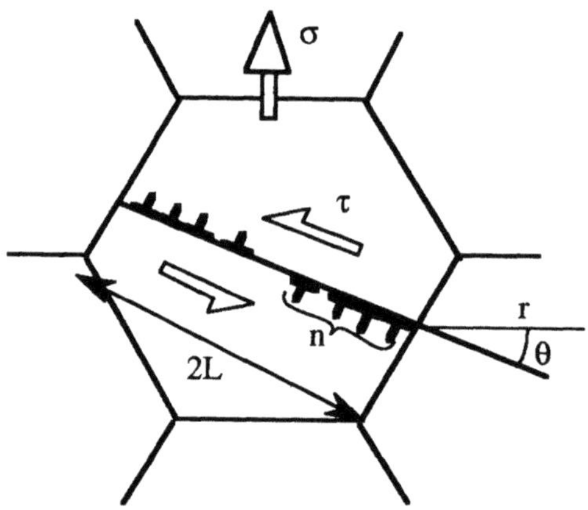

Fig. 2.13 Concentration of stress at the head of a pile-up of dislocations, capable of generating a cleavage.

Cleavage can start if σ reaches the theoretical critical value σ_C over a sufficiently long distance, say *3b*; it then follows from (2.7) that

$$\tau_f - \tau_i = \sigma_C (6b / L)^{1/2} / f(\theta) = \left(6E\gamma_s / L\right)^{1/2} / f(\theta) \tag{2.8}$$

Since the length L that characterises the size of the dislocation pile-up is a linear function of d, a dimension characterising the grain size, this shows that the stress necessary for the initiation of a cleavage varies like $1/\sqrt{d}$.

Many variants have been proposed to this elementary mechanism, which is due to Zener (1949). Cottrell (1956), for example, suggested that self-blocking by two {110} slips in a BCC material could bring about a cleavage in a {100} plane (Fig. 2.14). In hexagonal materials it is often seen that a cleavage starts on the base planes at the intersection with tilt boundaries, sheared by slip in these planes This is the mechanism of Friedel, Stroh and Gilman, Fig. 2.15: see Friedel (1964), Stroh (1954), Gilman (1954), also McClintock and Argon (1966) Chapter 17. Finally, twins can give rise to heterogeneities in the deformation, which can initiate cleavage (Fig. 2.16).

An important fact is that these mechanisms do not operate in FCC materials, the reason that in these there are very many possibilities for slip. Further, any stress concentrations are very easily relaxed.

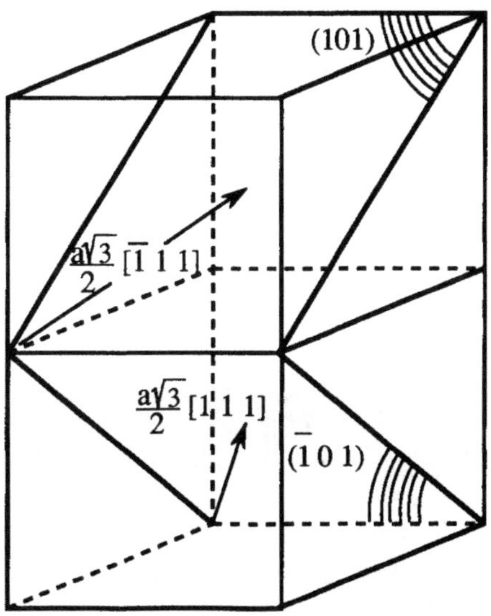

Fig. 2.14 Cottrell's mechanism: {*100*} cleavage initiated in BCC material by self-blocking of two {*110*} slips.

The above calculation of the stress required to start a cleavage has not addressed the question of whether the process is possible on energy grounds. When a cleavage crack of length $2a$ appears, the dislocations in the two pile-ups climb rapidly up the crack which can thus be regarded as a dislocation with Bürgers vector $nb\sqrt{2}$ and a core whose size, according to Cottrell (1956), is a. At the same time two surfaces of area $2a \times 1$ are created and potential energy is released; fracture mechanics shows that the amount of this is $-\pi\sigma^2 a^2(1-\nu)/4\mu$

Thus the change in energy is the sum of the energy of the giant dislocation representing the crack, the surface energy and elastic energy released:

$$E_C = \frac{\mu(nb\sqrt{2})^2}{4\pi(1-\nu)}\log\frac{R}{a} + 4\gamma_s a - \frac{\pi\sigma^2 a^2(1-\nu)}{4\mu} \tag{2.9}$$

The condition for instability of the cleavage is $\partial E_C/\partial a \leq 0$, a second-degree inequality in a. If the applied stress σ is too small the energy E_C will first decrease as a increases and then go through a stable minimum (Exercise 6). The cleavage cannot propagate until the stress is great enough to ensure that E_C is always decreasing: this critical value is

$$\sigma = \sigma_f = 4\gamma_s/nb \tag{2.10}$$

In (Volume I, Ch.3) we showed that the number n of dislocations that can be piled up is given by

$$nb = \pi(1 - \nu)L(\tau - \tau_i)/\mu \tag{2.11}$$

and therefore

$$\sigma_f(\tau - \tau_i) = 4\gamma_s\mu/\pi(1 - \nu)L \tag{2.12}$$

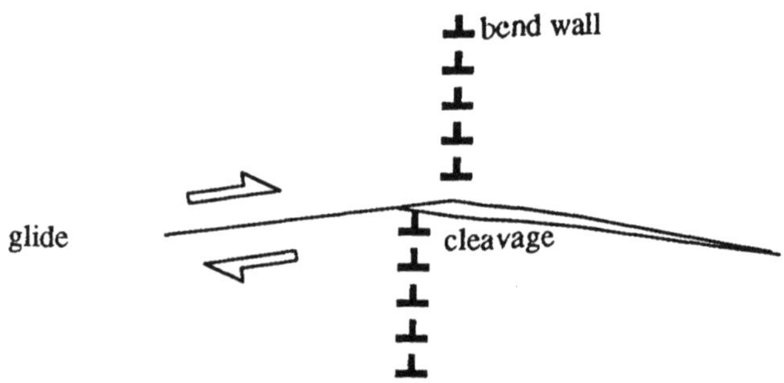

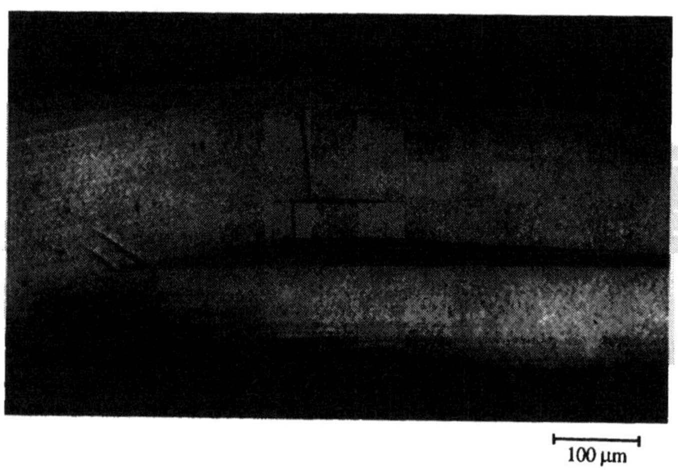

Fig. 2.15 (a) Gilman's mechanism: cleavage initiated by a slip crossing a tilt boundary (b) Formation of such cleavages in zinc (Gilman (1954)).

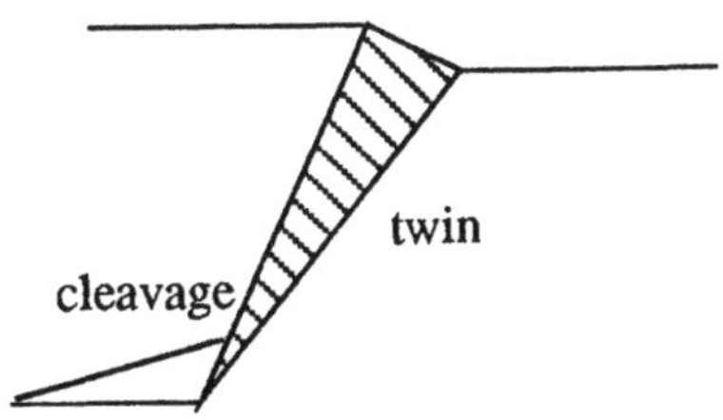

Fig. 2.16 (a) Cleavage starting in the neighbourhood of a twin (b) Scanning electron micrograph showing cleavage microcrack at the intersection of two mechanical twins B,C; the planes of the twins and of the cleavage are indicated. Duplex (austenitic- ferritic) steel.

Note that the critical length for the nucleus for a cleavage, corresponding to the critical condition 2.10, is very small, of the order of 10^{-7} *m*.

Since cleavage is a consequence of plastic deformation, the effective stress acting on the dislocations is given as a function of grain size by Petch (1953):

$$\tau - \tau_i = k_y d^{-1/2} \tag{2.13}$$

and since the size L of the pile-ups is proportional to d, say $L = \alpha d$, we have finally

$$\sigma_f = \frac{4\gamma_s \mu}{\alpha\pi(1-\nu)k_y d^{1/2}} \approx \frac{\mu\gamma_s}{k_y d^{1/2}} \tag{2.14}$$

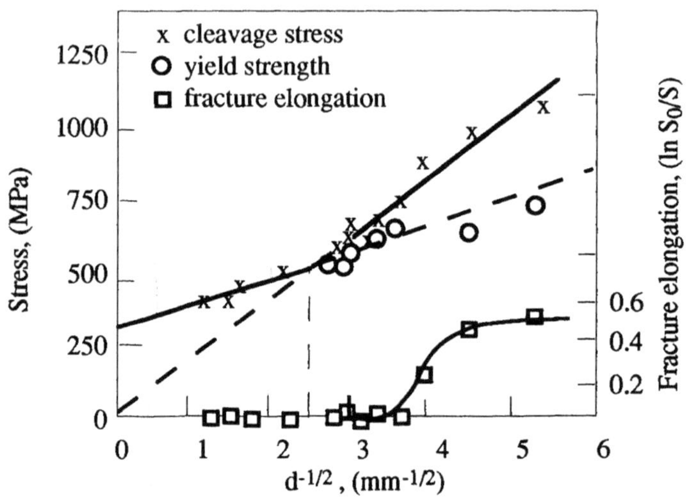

Fig. 2.17 Variation of cleavage stress, yield strength and fracture elongation with grain size, for a mild steel.

Experiments show that cleavage stresses σ_f are indeed proportional to $d^{-1/2}$ (Fig. 2.17) but the constants of proportionality that are found lead to energy values much greater than $2\gamma_s$. We shall show that these large values are due to dissipative mechanisms which add to the work ($2\gamma_s$) required to break the atomic bonds.

2.2.6 Blunting of cleavage

In crystalline solids cleavage cracks can be *blunted* by the emission of dislocations - a dissipative mechanism. Rice and Thomson (1974) investigated the conditions under which this can operate and derived a criterion for the transition between pure and blunted cleavage.

The emission of a dislocation loop (Fig.2.18) corresponds to an energy change, u_D say, and the mechanism will be active if this is negative. It consists of:

(a) the energy of formation of a dislocation loop of radius r in a slip plane making an angle φ with the cleavage plane,

(b) the energy of the step created by this loop,

(c) the elastic energy released by the formation of the loop.

It is found (Exercise 8) that -

$$u_D = \frac{2-\nu}{8(1-\nu)}\mu b^2 r \log\left(\frac{8r}{e^2 r_0}\right) + 2\beta'\gamma_m b(r-r_0) - \frac{0.986\beta}{3\sqrt{1-\nu}}\sqrt{\mu\gamma_c}(r^{3/2}-r_0^{3/2})b \qquad (2.15)$$

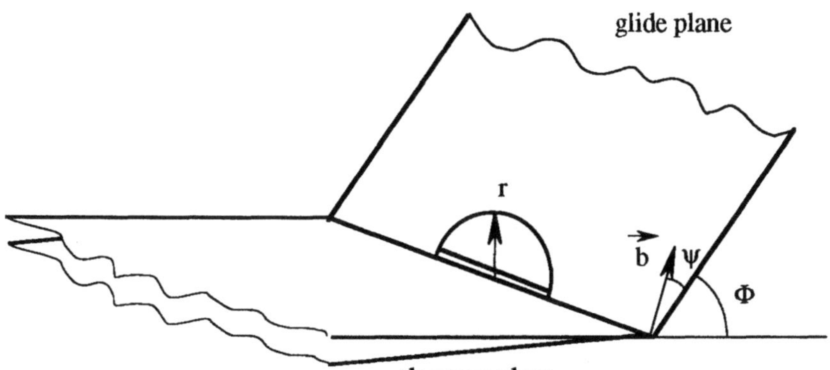

Fig. 2.18 Formation of a dislocation loop of radius r in a slip plane at the tip of a cleavage crack.

Here r is the radius of the loop, b its Burgers vector and r_0 its core radius; γ_m is the surface energy of the step and γ_C the energy of the cleavage; $\beta' = \sin\varphi \cos\psi$, $\beta = \sin\varphi \cos\varphi/2 \cos\psi$, where ψ is the angle between the Burgers vector b and the projection of the crack propagation direction on the slip plane.

The maximum of u_D as r varies is found by setting $\partial u_D/\partial r = 0$. We define the following dimensionless parameters:

$$U = \frac{8(1-\nu)}{(2-\nu)^2} \cdot \frac{\beta^2}{\beta'} \cdot \frac{u_{D\max}}{\mu b^3} \frac{\gamma_C}{\gamma_m}$$

$$S = \frac{16(1-\nu)}{5(2-\nu)} \beta' \frac{\gamma_m}{\mu b} \tag{2.16}$$

$$R_0 = \frac{16\beta^2}{5(2-\nu)\beta'} \cdot \frac{r_0}{b} \frac{\gamma_C}{\gamma_m}$$

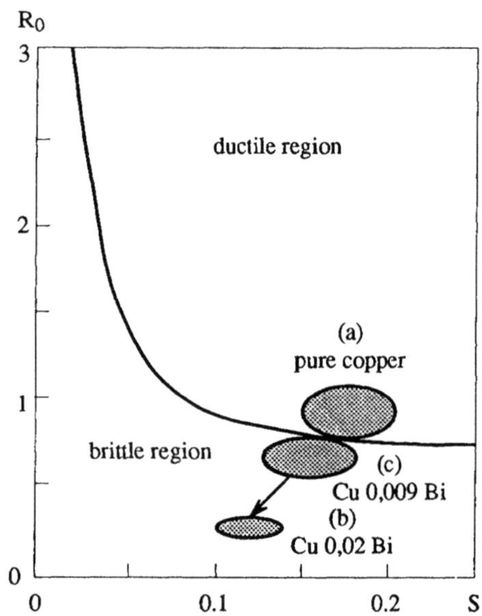

Fig. 2.19 Differentiation of the non-blunting zone for a cleavage, where the energy of formation of a dislocation loop is positive, from the blunting zone where this is negative. R_0 is the parameter for the core size and S that for the surface energy.

In the S-R_0 plane the curve $U = 0$ (Fig. 2.19) separates the region in which rupture by cleavage predominates from that in which blunted rupture predominates. It seems that this certainly is the case if $S > 0.2$, that is, for $\mu b/\gamma_m \geq 10$, to which there corresponds a critical value for the ratio μ/k given by

$$(\mu/k)_{CD} = 10\gamma_s/bk \tag{2.17}$$

Values for this are given in Table 2.III, which show that the only metals in which we should expect to find cleavage are rhodium, iridium, molybdenum, tungsten and chromium. But these are just the metals in which intergranular rupture is found, so the conclusion is that cleavage should not occur in pure metals: since this is not the case, any occurrence observed will be due to impurities having altered the surface and boundary energies and reduced the mobility of the dislocations.

2.2.7 Propagation of cleavages

Fracture mechanics shows that the stress needed to propagate a circular (penny-shaped) crack of radius a is

$$\sigma_P = \frac{1}{2}\left[\frac{\pi E \gamma_R}{(1-\nu^2)a}\right]^{1/2} \tag{2.18}$$

where E is Young's modulus, ν is Poisson's ratio and γ_R is the fracture energy.

Thus once started, a cleavage can propagate under a decreasing stress; if there is an external force acting on the solid this will lead to instability, with the excess energy that is available becoming transformed into kinetic energy. However, there are two effects which can restrain the propagation: relaxation by dislocation slip, and crossing of grain boundaries. We now consider these.

2.2.7.1 *Relaxation by dislocation slip*

At the end of a cleavage crack there will be a high concentration of stress, capable of causing the dislocations to slip - that is, of producing plastic deformations, which in turn can blunt the cleavage. This slip will absorb the deformation energy γ_P which will be added to the $2\gamma_s$ needed to break the bonds; thus the cleavage energy γ_R is equal to $\gamma_P + 2\gamma_s$.

The work γ_P done in propagating a crack at a speed v_C is equal to that needed to displace the dislocations in the neighbourhood of the tip of the cleavage, which will acquire a speed v depending on the local stress τ. Experiments have shown that the relation between speed and stress can be of the form

$$v = v_0(\tau/\tau_0)^m \tag{2.19}$$

where $m > 10$. Alternatively, exponential relations can be assumed, from which it can be deduced that the speed of movement of the dislocations is limited to that of Rayleigh waves in the material.

Fracture mechanics shows that at a distance r from the tip of a crack of length $2a$ the shear stress varies like $1/\sqrt{r}$:

$$\tau = \sigma\sqrt{a/r}\,.r^*(\theta) \tag{2.20}$$

where σ is the applied stress and $r^*(\theta)$ is an oscillatory function of θ which we shall neglect.

To determine γ_P (Exercise 7) we calculate the plastic power dissipation $dW/dt = \tau\, d\varepsilon/dt$ at the tip of the cleavage. With the stress σ equal to that needed to start the cleavage, that is, $\mu\gamma_s/k_y d^{1/2}$, and putting $v_\mu = v_0(\mu/\tau_0)^m$, the speed that the dislocations will reach under a stress μ, we find

$$\gamma_P / 2\gamma_s = \frac{\pi}{m-1} \frac{\mu \gamma_s^m}{k_y^{m+1} b^{(m-3)/2}} \rho_D \frac{v_\mu}{v_C} \tag{2.21}$$

were ρ_D is the density of mobile dislocations, a complex function of the work-hardening. The result shows that the energy of a cleavage should be inversely proportional to its speed. It has been found by experiment that the breaking energy can be up to a thousand times that of the surface energy.

It follows from (2.21) that γ_P, whilst much greater than γ_s, depends very strongly on the latter, through the exponent m; thus any change in the surface energy will be automatically reflected, and greatly amplified, in the total breaking energy.

2.2.7.2 *Interaction with grain boundaries*

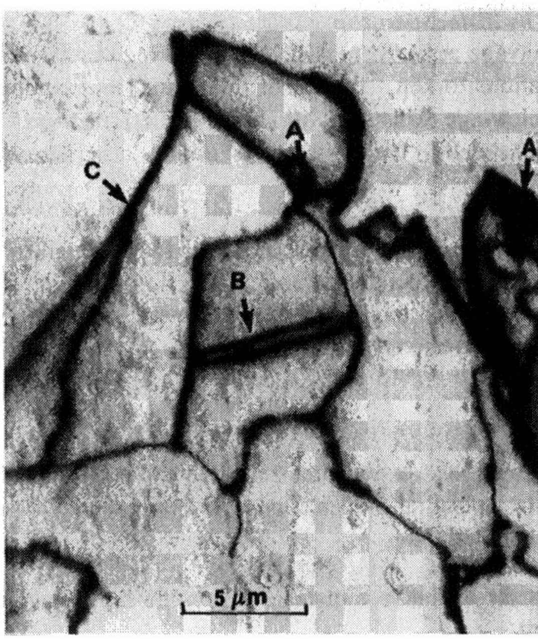

Fig. 2.20 Cleavage crack (B) stopped by grain boundaries in a steel of *0.04%* carbon content; the low value explains the small amount of pearlite (A). Observation by optical micrography of a transverse cut across a fracture surface (C) which had been covered by electrolytically deposited nickel.

As we have seen, the change in orientation of a cleavage plane when passing from one grain to another results in the appearence of many steps on the broken surface. Further, the stress normal to the cleavage plane will be less in the new grain, for otherwise the cleavage would have started in the latter. Finally, the arrival of a crack in the neighbourhood of a new grain can activate the sources of dislocations and cause new modes of relaxation. For all these reasons the crossing of a grain boundary constitutes an obstacle, expressed by an increase in the cleavage energy which we may estimate as a factor of 2 at least.

From (2.18), giving the stress needed for the propagation of a cleavage we find that that needed for the crossing of a boundary is

$$\sigma_G = \left(\frac{\pi}{1-\nu^2}\right)^{1/2}\left(\frac{E\gamma_R}{d}\right)^{1/2} \qquad (2.22)$$

This can exceed the value $\mu\gamma_s/k_y d^{1/2}$ needed to start the cleavage, in which case the cleavages become blocked at the boundaries, as can be seen in optical micrographs (Fig. 2.20)

2.2.8 Fracture of carbides in mild steels

In mild steels we can treat cleavages as originating in very brittle platelets of cementite. The necessary stress concentrations will have been generated by dislocation pile-ups in the ferrite, and it is in this that the cleavages will be propagated. According to Smith (1966) the corresponding necessary condition is expressed by

$$(c_0/d)\sigma^2 + (\tau-\tau_i)^2\left[1+(c_0/d)^{1/2}4\tau_i/\pi(\tau-\tau_i)\right]^2 \geq 4E\gamma_P/\pi(1-\nu^2)d \qquad (2.23)$$

where c_0 is the thickness of the platelets and d the size of the ferrite grains. The terms on the left relate to the applied stress and express on the one hand the direct effect of this on a crack of length c_0 (first term), on the other the indirect effect of the stress concentration resulting from the pile-up (second term). The term on the right represents the resistance of the ferrite to the propagation of the cleavage.

If we are at the elastic limit then the Petch relation $\tau - \tau_i = k_y d^{-1/2}$ holds. Smith's relation then gives

$$c_0\sigma_f^2 + k_y^2\left[1+(4\tau_i/\pi k_y)\sqrt{c_0}\right]^2 = 4E\gamma_P/\pi(1-\nu^2) \qquad (2.24)$$

The ferrite grain size (d) does not appear in this relation, whilst experiments with mild steel show that the cleavage stress varies approximately as $d^{-1/2}$ (Fig. 2.21); the discrepancy is explained by the fact that there is a correlation between the platelet thickness c_0 and the grain size.

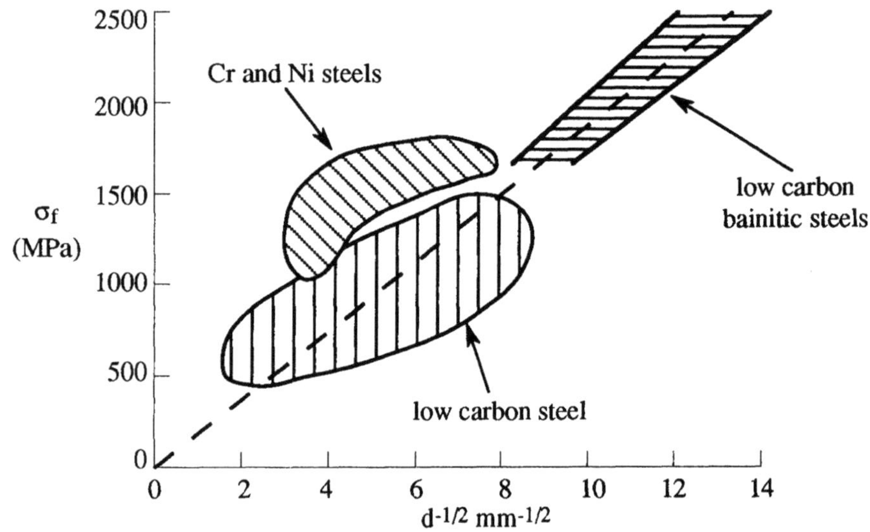

Fig. 2.21 Variation with grain size of cleavage stress in steel (Knott (1977)).

2.2.9 Parameters of the cleavage stress

The external parameters of temperature and rate of loading can affect the cleavage energy through the intermediary of the fracture energy. Raising the temperature will increase the mobility of the dislocations, increasing v_μ and consequently, by (2.21), γ_P also. Fig. 2.22 shows that there is indeed this effect in manganese steels, but that for mild steels the stress remains roughly constant as the temperature increases. The loading rate can affect the propagation speed v_C; one would expect the stress to fall as the rate increases, and again Fig. 2.22 shows that such an effect is very marked in the case of manganese steels but very weak for mild steels. The conclusion is that in general both temperature and loading rate have negligible effect on cleavage stress.

The crucial metallurgical parameter is the grain size: reducing this is the most effective way to increase the cleavage stress. Dislocation mobility and density of mobile dislocations are also important; hardening by solid solution and especially by dislocation pinning by atmospheres have a marked effect on the cleavage stress, which is reduced accordingly. This is the reason for the "blue brittleness" of steels, so-called because it occurs after work-hardening and holding at *300-400°C*, at which temperature the blue oxide FeO forms. The Cottrell atmospheres (I-§3.3.8.5δ) appear in this temperature range.

To conclude, we can say that the criterion for cleavage that is generally adopted is a critical value for the maximum principal stress; this is independent of temperature and is inversely proportional to the square root of the grain size.

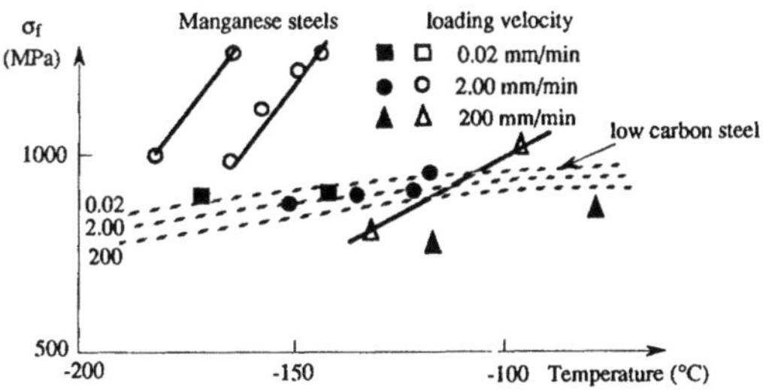

Fig. 2.22 Variation of cleavage stress with temperature for different rates of loading. The dotted lines are the predictions of the Smith model (Eqn. 2.23)

2.3 BRITTLE FRACTURE OF CERAMICS

2.3.1 Nature of the fracture

Ceramics are covalent and ionic materials; at low temperatures dislocations in them are firmly anchored in the Peierls valleys, so that plastic deformation is not possible and they are very fragile. Observed breaking stresses are, broadly, less than theoretical because of the presence of impurities, particularly at the grain boundaries; these give rise to very high concentrations of stress, even if they are of small extent.

How can these disadvantages be reduced? The first point to note is that rupture here is intergranular, and therefore it is important to examine the structure of the grain boundaries carefully. These materials, which have very high melting points, are generally manufactured by sintering, when any impurities on the surface of the powder move into the boundaries. Further, additives are often introduced, to activate the sintering by increasing the surface energy; the driving force for the reaction being provided by the difference between this energy and the lower energy of the grain boundaries (sintering is the inverse of fissuring). Thus for example sintering of silicon carbide at above *2000°C* is made possible by the addition of boron or carbon. The impurities tend to gather in the grain boundaries where they form amorphous films (see Vol.I, Fig. 1.36) which are fractured more easily than the crystalline grains. As will be readily appreciated, the sintering will not be complete: pores will be left, which will act as generators of stress concentrations.

2.3.2 Improvements resulting from actions on the boundaries

One way to increase the breaking strength of a ceramic is to reduce the grain size; this enables the size of the defects to be limited, with consequent reduction of the stress concentrations.

Another is to improve the strength of the boundaries by post-sintering heat treatment; this can produce its effect by:

(i) putting into solution in the grains the elements which form the inter-granular vitreous phase,

(ii) forming new crystalline phases,

(iii) causing the vitreous phase to react with the grains,

(iv) causing certain elements, metals in particular, to migrate to the surface by inter-granular diffusion.

Thus for the ceramic SiAlON, of which the main phase is a solid solution of alumina in silicon nitride, annealing by holding for *300* hours at *430°C* under nitrogen results in inter-granular precipitation of alumina with an increase of *50%* in the fracture load at ambient temperature.

2.3.3 Improvement by incorporating metastable zirconia

Zirconia has a reversible allotropic transformation from monoclinic form, stable at ambient temperature, to tetragonal at *1100°C*; it is accompanied by a reduction of volume of *3%*. On cooling, the particles of zirconia, finely dispersed in a matrix which is not transformed, remain in the tetragonal form, and therefore are metastable. If a crack starts to form, the stress around the zirconia particles will be relaxed, allowing the reverse transform to the monoclinic form to take place with the accompanying increase in volume, which will oppose the opening of the crack. By this means "partially stabilised zirconia" (PSZ) materials can be produced, with breaking loads of over *800 MPa* and fracture toughness K_{IC} of over $8\ Mpa\sqrt{m}$. The same phenomenon makes it possible to improve the mechanical properties of alumina greatly by adding *15%* by volume of zirconia.

2.3.4 Improvement by microcracking

A great drawback of a very brittle material is that it can break without warning - it lacks the ability to *yield*, a property provided by plasticity. One way to compensate for this lack is to allow a stable system of microcracks to develop; this will reduce the rigidity, and the proliferation of the cracks will increase the fracturing energy because this depends on their total area. The effect can be achieved by adding particles, around which the microcracks can develop; this is what happens with zirconia particles after the tetragonal-monoclinic transformation has occurred, which is the case if they are large enough (Exercise 9).

2.3.5 Sensitivity to thermal shock

In addition to being fragile, ceramics are very poor conductors of heat and consequently are very sensitive to thermal shock. Rapid changes of temperature result in high thermal gradients developing stresses which are all the higher, the greater the coefficient of dilatation.

2.3.6 Delayed fracture of ceramics

Observation shows that, when loaded, ceramics break after a certain interval; this is known as *static fatigue*. It is attributed to the progressive development of surface microcracks under the effect of the environment, especially of humidity, and in particular on the vitreous phase of the grain boundaries. The phenomenon is related to stress corrosion; we shall meet it again in connection with damage to glass.

2.4 FRACTURE OF GLASS

2.4.1 Surface microcracks

The great fragility of glass is attributed to the presence of microcracks on the surface. The difference between the observed and the theoretical fracture stress can be explained by assuming these to have a diameter of 5 microns and a radius of curvature at the tip of the order of the interatomic distance - see Equation 2.6, with $\rho \approx b$. The small size of the opening makes it impossible to see them, but they can be detected by etching.

Such cracks are produced by a variety of effects in the course of handling the glass - contact with abrasives or with dust, impingement of pointed objects etc. Further, glass is much more resistant immediately after manufacture than after some time in use: this deterioration can be made very evident by comparing the ease with which a bottle can be broken after a few pebbles have been shaken in it with that for a new, unused bottle. Another source is chemical reactions with dust.

The situation can be improved by surface treatment, either by coating with a protective material, for example a polymer, or by generating residual compressive stresses there, which can be done in two ways. If a newly-cast sheet of glass is cooled quickly the surface will contract more than the interior. Initially the latter will be viscous enough to relax the thermal stresses but later the contraction of this will result in the cooled, and now rigid, surface being in compression. This is the reason for the good resistance of quenched glass; but if a crack begins to form the glass immediately shatters into a thousand fragments.

The second way is to introduce elements into the surface which will dilate the glass by ion exchange, for example replacing small sodium ions by larger potassium. The resulting distribution of stress across the thickness is very different from that in quenched glass, the compression zone being much narrower than that produced by the thermal gradient. The interior tension, which must balance the surface compression, is therefore much smaller, and the sheet will not shatter as does a quenched sheet.

2.4.2 Reinforcement by metallic particles

Metallic particles incorporated into glass can improve its resistance to fracture because of their plastic deformation energy. However, the elastic moduli and dilatation coefficients must be matched very carefully if a satisfactory result is to be obtained. If the particles contract less than the glass when this is cooled after manufacture, tensile stresses will be generated and will make the glass fragile; in the converse case, the interfaces break and the inclusions then behave as cavities; in a way, they can somewhat decrease the sharpness of the cracks. The best results are obtained if the thermal stresses vanish and cracks are propagated by crossing the particles. This can be achieved for example for glass M (percentage composition 50 SiO_2, 37 Na_2O, 8 Al_2O_3, 5 Li_2O) whose dilatation coefficient is $16.10^{-6}/°C$ and Young's modulus *68.9 GPa*, with particles of aluminium having the same values. Further, a surface coating of aluminium gives good adhesion at the glass/particle interface. The result is that the addition of *20%* by volume of aluminium raises the fracture energy of this glass from $12\ Jm^{-2}$ to $600\ Jm^{-2}$.

2.4.3 Delayed fracture of glass

As with ceramics, the resistance of glass under load falls with the passage of time, especially in the presence of humidity or water vapour.

Several mechanisms can cause surface microcracks to grow, a major cause being the interaction with water molecules. The vigour of the attack depends on the composition of the glass, the pH of the water and the temperature; it is faster, the higher the temperature and the greater the number of hydroxyl ions (the higher the pH, that is). The first reaction to be considered is the exchange between hydroxyl ions and hydrogen atoms. Because of the very small size of the latter, the surface is in tension. If the pH is greater than *9* the hydroxyls react with the silica. However, the mechanisms of attack are made complex by the kinetics: the fluid must first reach the ends of the fissures, and this requires it to flow in a very narrow channel. After this its composition at the crack tip changes as a result of the chemical reactions.

2.5 FRACTURE OF CONCRETE

2.5.1 Fracture mechanisms

Concrete is made by mixing granulates ("aggregate") with cement. The cement paste consists mainly of hydrated carbonates and silicates, abbreviated to CSH, and is porous; the hydration reaction does not take place immediately but develops slowly over a long time. The mechanical properties change with time as the concrete hardens, but do so increasingly slowly; it is usual to record these at *28* days. It is important not to forget that water remains in the pores, adsorbed on to their surface as liquid or as vapour; it can migrate within the mix and there are continuing exchanges with the external medium. Thus for a proper appreciation of the properties of a concrete it is important to know the conditions in which it has been stored.

The aggregate is of very varied size, shape and composition, according to the sites from which it is quarried and the structures to be built. The interface with the cement is often a weak zone, where the composition is different from that in the body; it could be called a transition envelope. The changes in volume that takes place during setting give rise to stresses, even cracks, more easily at the lower part of the aggregates because of the low resistance resulting from the water meniscus formed during casting. Thus concrete could be regarded as a multi-fissured material, with cracks ranging in scale from the finest CSH platelets (a few nanometres) to the largest pebbles (a few centimetres).

For each of these cracks there is a corresponding critical propagation stress. It is not an easy matter to construct criteria for fracture on this scale, because of the different orientations of the fissures and the interactions between these and the aggregate. Since in use concrete is most often in compression, relative movement of the edges of a crack is restrained by friction. As a load is applied it is the cracks that are closest to the critical condition which will begin to propagate; under stress along the normal they would become unstable if they did not encounter aggregate, which in fact forces them to divert and follow the interfaces: whilst cement paste is very weak, concrete is much less so because of the stabilising role of the aggregate, which allows multi-fissuring to develop. This effect plays a minor role in tension but a major one in compression, when the critical cracks propagate under the action of shear stress (Modes II and III) and divert, tending to move into Mode I, that is, normal to the direction of the maximum normal stress (Fig. 2.23).

In a pure compression test these diversions orient the cracks parallel to the axis of the compression, in which direction they are no longer subjected to the load and are therefore stabilised. However, in concrete the stress field is not uniform on the scale of the aggregate and there can be local tensile stresses, averaging to zero over the test piece as a whole but sufficient to cause cracks to propagate, running parallel to the compression axis until they meet and are stopped by aggregate. Fracturing, resulting from a very diffuse fissuring, is progressive and absorbs much more energy than it would under tension, when rapid localisation of stress would generate serious instability. This is why concrete is much stronger in compression than in tension. The final fracturing of a

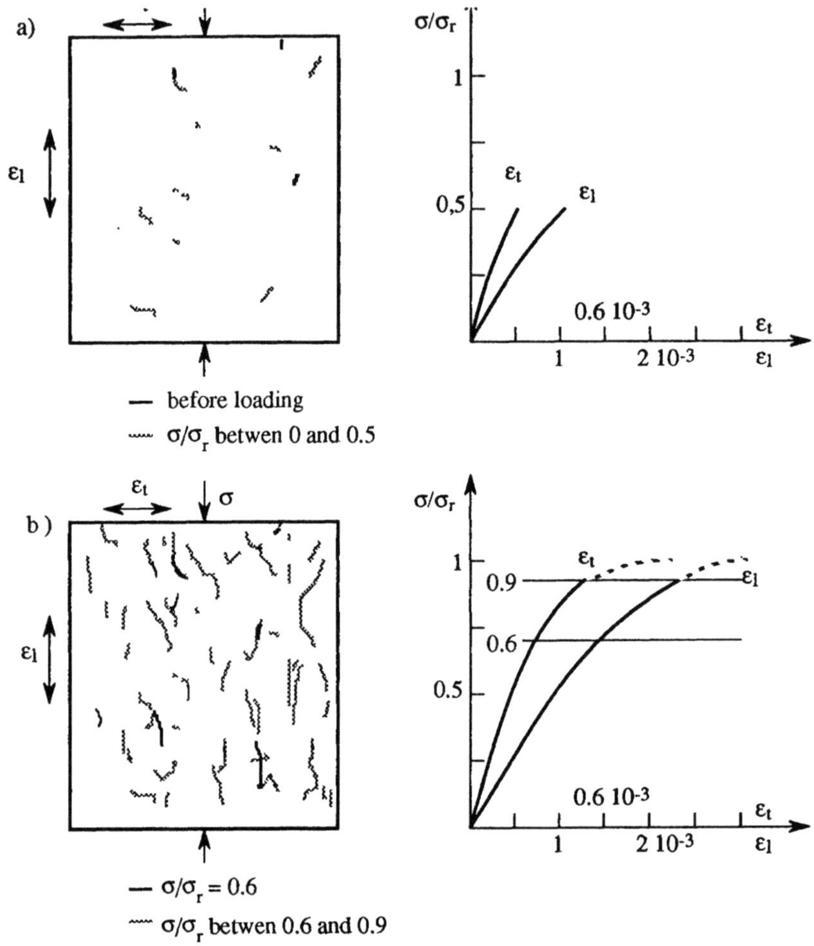

Fig. 2.23 Development of microcracks in concrete under compression. (a) at half (b) at *90%* breaking load (J. Mazars)

test piece under compression is due to fragmentation into small columns parallel to the compression axis.

2.5.2 Criteria for fracture

On the microscopic scale the fracture criterion cannot be expressed in the form of a critical stress since, in a compression test, cracks can open under zero macroscopic-stress. However, a criterion for maximum deformation can be useful since in such a test the Poisson effect gives positive lateral deformations. Multi-axial stress tests show that the criterion for fracture can be taken to be a *maximum positive deformation* of the order of 10^{-4}. Fig. 2.24 gives contours of the elastic limit and breaking strength of concrete under bi-axial stresses, and for comparison the contour derived theoretically from the deformation criterion proposed by Mazars (1984).

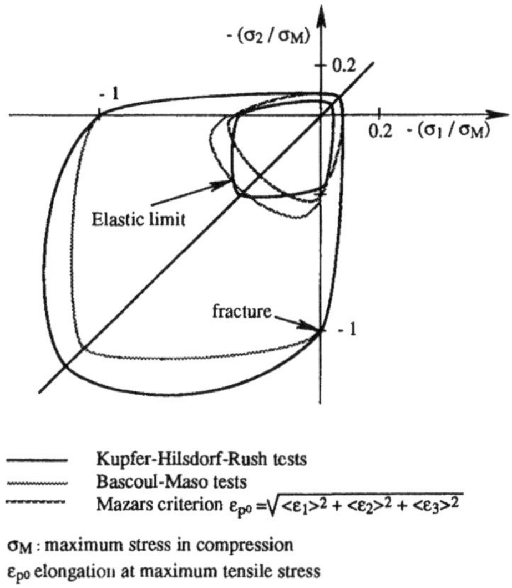

Fig. 2.24 Experimentally-determined contours of elastic limit and breaking strength of concrete.

A tensile test on concrete gives a load-displacement curve which begins to depart from linearity shortly before the breaking load. If what is set is the load, the fracture is sudden and the deformation very small. If the test is controlled by monitoring the deformation, the part of the curve in which the load decreases with increasing displacement can be recorded and the behaviour analysed after the event. It is then found that under successive loading and unloading the stiffness of the material falls steadily (Fig. 2.25), and that irreversible deformations occur. This can be interpreted as the development of cracks in the test piece, which increase its capacity for deformation; the irreversibility could be due to debris in the cracks preventing them from closing completely when the load is reduced, and also to relaxation of the internal hydration stresses.

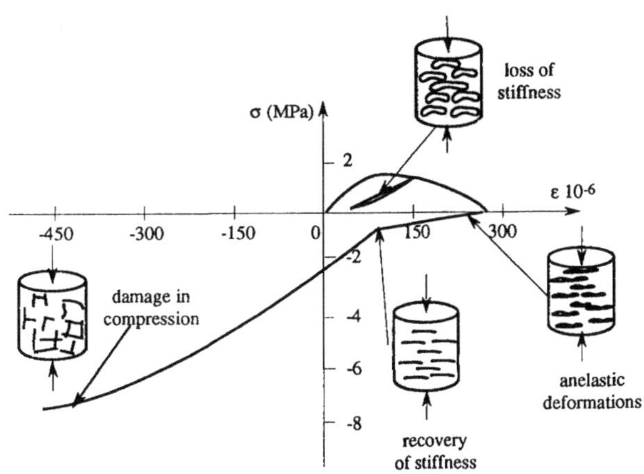

Fig. 2.25 Stress-deformation curve for concrete under uniaxial load of tension followed by compression.

2.5.3 Damage mechanics

The study of this was started by Kachanov (1958) and Rabotnov (1969) and has been developed in the framework of the thermodynamics of irreversible processes by Lemaître and Chaboche (1990) in particular.

Here we shall give only the essentials of the subject. In this treatment, damage is characterised by a *damage parameter D*, with values between *0* and *1*. Fracture occurs with certainty when $D = 1$ (total damage) but can occur, and generally does, when a somewhat lower critical value is reached.

In a sample of concrete an arbitrary section of area S will include an area S^* of voids, so that the stress on S is in fact distributed over an area $S - S^*$; this suggests defining the damage parameter as S^*/S.

In a uniaxial test, if F is the load then the nominal stress σ over the section is F/S and the effective stress over the void-free part is

$$\sigma_{eff} = F/(S - S^*) = \sigma/(1 - D) \tag{2.25-6}$$

If the behaviour is linear the modulus is reduced by the damage to a value E_D such that

$$\varepsilon = \sigma/E_D = \sigma_{eff}/E \tag{2.27}$$

and therefore

$$E_D = E(1 - D) \tag{2.28}$$

Thus the evolution of the damage parameter can be determined experimentally by measuring that of the elastic modulus.

The study in Volume 1, Ch.2 (Exercise 8) of the effect of porosities enables a relation to be derived between the elastic moduli and the volume fraction f_v of uniformly and isotropically distributed voids.

If k, μ and ν are the incompressibility modulus, shear modulus and Poisson ratio respectively for the material, the self-consistent model gives

$$k_D = k\left[1 - \frac{3(1-\nu)}{2(1-2\nu)} f_v\right], \quad \mu_D = \mu\left[1 - \frac{15(1-\nu)}{7(1-2\nu)} f_v\right] \tag{2.29}$$

Since, statistically,

$$f_v = S^*/S = D \tag{2.30}$$

This gives a relation between the effective moduli and the damage parameter which is more soundly-based than (2.28), since σ_{eff} is not in fact strictly equal to $F/(S - S^*)$; but in practice (2.28) is more commonly used, since E_D is more easily found by experiment.

However, in real materials the voids and cracks are not distributed isotropically, so the representation of damage by a scalar parameter is only an approximation: convenient on the macroscopic scale but very imperfect on the microscopic scale. We therefore introduce an assumption of *equivalence of deformations*, or rather of a *volumetric density of complementary potential* ϖ. For the equivalent material -

$$\varpi = \frac{1}{2}\underline{\underline{\sigma}}_{eff} S \underline{\underline{\sigma}}_{eff} \tag{2.31}$$

whilst for the damaged material

$$\varpi = \frac{1}{2}\underline{\underline{\sigma}} S_D \underline{\underline{\sigma}} \tag{2.32}$$

The equality of the complementary potentials enables the *compliance matrix* S_D for the damaged material to be defined: thus

$$\underline{\underline{\varepsilon}} = \frac{\partial \varpi}{\partial \underline{\underline{\sigma}}} = S_D \underline{\underline{\sigma}} \tag{2.33}$$

Damage is now represented by a fourth-order tensor. This is greatly simplified if we assume that the principal axes of the tensor, the stress and the deformation coincide; we then have, for biaxial loading,

$$S_{D1111} = \frac{1}{E(1-D_{11})^2}, \quad S_{D2222} = \frac{1}{E(1-D_{22})^2}$$
$$S_{D1212} = S_{D2121} = \frac{-\nu}{E(1-D_{11})(1-D_{22})} \tag{2.34}$$

With the damage parameters thus defined we must now find how they change during loading. For concrete, from what we know about the process of fracture it seems natural to relate this evolution to the deformation; essentially, the relation is based on experimental observations and we have, for example

$$D_{11} = \left[A + B\sigma_{22}/\sigma_{11}\right]\left(\varepsilon_{11} - \varepsilon_{11}^S\right)^C, \quad D_{22} = \left[A + B\sigma_{22}/\sigma_{11}\right]\left(\varepsilon_{22} - \varepsilon_{22}^S\right)^C \tag{2.35}$$

with $A = 150 \pm 10$, $B = 40 \pm 5$, $C = 0.8 \pm 0.02$.

ε_{11}^S, ε_{22}^S are the stress thresholds above which damage develops, and depend on the particular concrete.

Thus

$$\varepsilon_{11}^{S} = 79.5(\pm 6) + 43.4 f_G - 406.1 f_G^2 + 707.2 f_G^3 - 1150 f_G^4 + 763 f_G^5$$

where f_G is the volume fraction of aggregate.

The treatment can be simplified further by adopting a scalar as damage parameter; we have

$$\underline{\underline{\varepsilon}} = \frac{1}{E(1-D)}\left[(1+\nu)\underline{\underline{\sigma}} - \nu\left(tr\,\underline{\underline{\sigma}}\right)\underline{\underline{I}}\right] \tag{2.36}$$

with $D = f\left(\left|\underline{\underline{\varepsilon}}\right|\right)$. $\left|\underline{\underline{\varepsilon}}\right|$ is a scalar, the norm of the deformation tensor.

With cyclic loading it is necessary to distinguish between the damage due to tension and to compression respectively, say D_t and D_c. The behaviour of the damaged material is now given by

$$\underline{\underline{\varepsilon}} = \frac{1}{E(1-D_t)}\left[(1+\nu)\underline{\underline{\sigma}}^{+} - \nu(tr\,\underline{\underline{\sigma}}^{+})\underline{\underline{I}}\right] + \frac{1}{E(1-D_c)}\left[(1+\nu)\underline{\underline{\sigma}}^{-} - \nu(tr\,\underline{\underline{\sigma}}^{-})\underline{\underline{I}}\right] \tag{2.37}$$

where $\underline{\underline{\sigma}}^{+}$, $\underline{\underline{\sigma}}^{-}$ are the positive and negative eigen-parts of the stress tensor and E, ν are the Young's modulus and Poisson's ratio for the undamaged concrete.

We have now to determine how these parameters evolve with the deformation. For this it is useful to resort to the thermodynamic damage force Y, analogous to the rate of release of energy in the mechanics of fracture and related to D by the Legendre transformation:

$$\begin{aligned} Y_t &= \frac{1}{6E(1-D_t)^2}\left\{(1+\nu)\left[3\sum_i(\sigma_i^{+})^2 - (tr\,\underline{\underline{\sigma}}^{+})^2\right] - (1+2\nu)(tr\,\underline{\underline{\sigma}}^{+})^2\right\} \\ Y_c &= \frac{1}{6E(1-D_c)^2}\left\{(1+\nu)\left[3\sum_i(\sigma_i^{-})^2 - (tr\,\underline{\underline{\sigma}}^{-})^2\right] - (1+2\nu)(tr\,\underline{\underline{\sigma}}^{-})^2\right\} \end{aligned} \tag{2.38}$$

It follows that the evolution of the damage parameters is given by:

$$\frac{\dot{D}_t}{\dot{Y}_t} = \frac{\sqrt{Y_t^0}\,(1-a_t)}{2Y_t^{3/2}} + \frac{a_t b_t}{2\sqrt{Y_t}\exp\left[b_t\left(\sqrt{Y_t} - \sqrt{Y_t^0}\right)\right]}$$

$$\frac{\dot{D}_c}{\dot{Y}_c} = \frac{\sqrt{Y_c^0}\,(1-a_c)}{2Y_c^{3/2}} + \frac{a_c b_c}{2\sqrt{Y_c}\exp\left[b_c\left(\sqrt{Y_c}-\sqrt{Y_c^0}\right)\right]} \tag{2.39}$$

where Y_t^0, Y_c^0 are thresholds and a_t, b_t, a_c, b_c are parameters that depend on the concrete being considered (Exercise 10).

The damage parameters do not change during unloading.

A model of this type makes it possible to retain the memory of successive damages, in particular during cyclic loading.

2.6 STATISTICAL ASPECTS OF BRITTLE FRACTURE

2.6.1 Weakest-link theory

Cleavage fractures and fractures in ceramics and in glass initiate in microstructural elements - grains, indentations - that are distributed statistically through the material. The same holds for the fracture stress; in the case of cleavage, for example, it is the largest and most favourably oriented grain that initiates the fracture, and in general it is the least resistant element that is responsible for brittle fracture. Thus statistical theories based on weakest links can be used in studying these phenomena.

These theories rest on a few basic assumptions. The solid is considered as a juxtaposition of elements which, so far as fracture is concerned, are completely independent, and the failure of the weakest element entails the fracture of the whole solid. They do not apply to the situation in which cleavages are blocked at grain boundaries, since then the microcracking can alter the stress distribution and so favour the fracture of neighbouring grains. Although in these circumstances the elements can no longer be considered as independent; in some cases it is possible to make a change of scale and recover independence for sufficiently large elements. Clearly it is important always to consider very carefully if the assumption of independence is justifiable.

Taking the assumption of the weakest link to be valid, we assign to each element a probability of fracture under given stress (or deformation). Let the probability that a single element will fracture at a stress σ or less be $P_0(\sigma)$; the probability that it will not fracture - is then $1\text{-}P_0(\sigma)$, and if independence can be assumed the probability that none of a group of N elements will fracture is $[1 - P_0(\sigma)]^N$. Another assumption is that if at least one elements fails then the whole set fails, so the probability of failure for the set of N is

$$P_R(\sigma) = 1 - [1 - P_0(\sigma)]^N \tag{2.40}$$

which can be written

$$P_R(\sigma) = 1 - \exp\{-N \log[1/(1-P_0(\sigma))]\} \tag{2.41}$$

At this stage we do not know how P_0 depends on σ; various assumptions can be made in order to give a simple mathematical form.

Let V_0 be the volume of an element and V that of the total, so that $N = V/V_0$. Equation (2.40) shows that the probability of fracture increases with V; if we put

$$f(\sigma) = \log[1/(1 - P_0(\sigma))]$$

this becomes

$$P_R(\sigma, V) = 1 - \exp[- (V/V_0) f(\sigma)] \tag{2.42}$$

Suppose now that we are dealing with a piece of the material that is subjected to a heterogeneous stress system; we can divide it into a large number of elementary volumes ΔV, each subjected to a stress σ, so that the probability that it does not fracture is

$$1 - P_R = \Pi[1 - P_R(\sigma, \Delta V)] = \exp[-\Sigma(\Delta V/V_0) f(\sigma)] \tag{2.43}$$

and going to the limit

$$P_R = 1 - \exp[\int_V f(\sigma)\, dV/V_0] \tag{2.44}$$

We must now consider which stress σ must be used: it could be, for example, the maximum principal stress (provided that this is positive), but we shall see that there are alternative interpretations.

It seems, however, that the probability of fracture is a function of the distribution of stress in the solid and that, for example, the probability of a piece failing under bending will be less than that for one of the same volume failing under a tensile stress equal to that acting on the most extended fiber of the first.

2.6.2 Weibull's theory

Weibull (1951) has suggested the following form for $f(\sigma)$:

$$f(\sigma) = [(\sigma - \sigma_0)/\sigma_u]^m \tag{2.45}$$

where σ_0 is a threshold stress, below which the probability of fracture is zero; in most cases we can take $\sigma_0 = 0$.

The parameter m is the *Weibull exponent*; it gives an idea of the dispersion, which is greater, the smaller the value of m. Probability of fracture is often represented by a graph of loglog $[1/(1 - P_R)]$ against log σ; Weibull's law holds if the points lie on a straight line, when the slope of this gives the value of m (Fig. 2.26).

The Weibull parameter σ_u is related to the V_0 we introduced above, and in fact it is the product $V_0\sigma_u^m$ that is significant; roughly, σ_u is the mean breaking stress for an item of volume V_0. For simplicity, in what follows we shall use a form with only two parameters, m and $V_0\sigma_u^m$, which is equivalent to assuming σ_0 to be zero.

For a solid subjected to heterogeneous stresses, assuming that the principal stress is positive,

$$P_R = 1 - \exp[-\int_V (\sigma/\sigma_u)^m \, dV/V_0] \tag{2.46}$$

If σ_{max} is the maximum stress in the solid we define the *effective volume* V_{eff} by

$$V_{eff} = \int_V (\sigma/\sigma_{max})^m \, dV \tag{2.47}$$

The probability that the solid will fracture can therefore be written

$$P_R = 1 - \exp[-(\sigma_{max}/\sigma_u)^m \, (V_{eff}/V_0)] \tag{2.48}$$

For a beam of length L and rectangular section of sides B, h (moment of inertia $I = Bh^3/12$) subjected to a bending moment M, constant over the whole length, the maximum stress is

$$\sigma_{max} = Mh/2I \tag{2/49}$$

and the stress at a distance y from the neutral axis is

$$\sigma = My/I \tag{2.50}$$

Thus $$V_{eff} = \int_0^{h/2} (My/I)^m / (Mh/2I)^m BLdy = V/2(m+1) \tag{2.51}$$

Some authors prefer to use the Weibull stress σ_w rather than V_{eff}, defined by

$$\sigma_w^m = (1/V_0)\int_V \sigma^m dV \tag{2.52}$$

and leading to

$$P_R = 1 - \exp[-(\sigma_w/\sigma_u)^m] \qquad (2.53$$

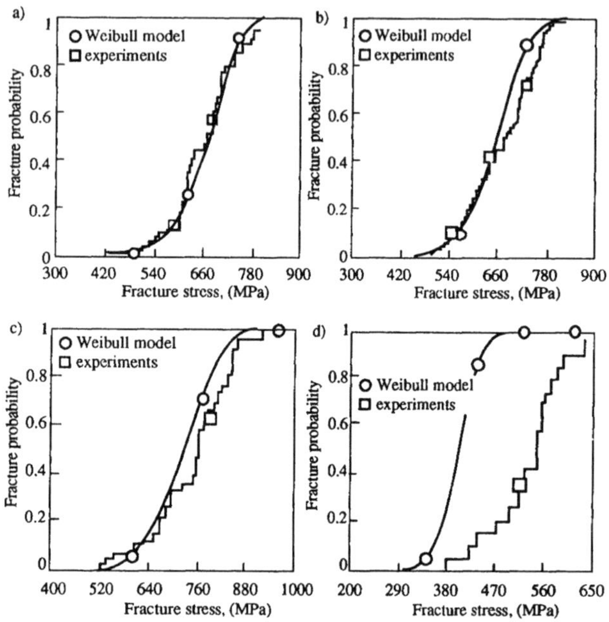

Fig. 2.26 Statistical distribution of breaking stress of silicon nitride and comparison with Weibull law. (a) 3 - point bending test (b) 4 - point bending test (c) biaxial bending (d) tensile test ($m = 7$, $\sigma_0 = 200$ *Mpa*, $\sigma_u = 570$ *MPa*). The poor agreement for tension is probably due to the use of a slightly different material.

In general, the Weibull stress and the effective volume are related by

$$\sigma_w/\sigma_{max} = (V_{eff}/V_0)^{1/m} \tag{2.54}$$

giving for the beam considered above

$$\sigma_w = (Mh/2I)[V/2(m+1)V_0]^{1/m} \tag{2.55}$$

2.6.3 Relation to the distribution of microcracks

So far we have not given any consideration to the mechanism of fracture; this is something we must do before we can deal with multiaxial stress systems.

Consider first a body subjected to a uniaxial tension σ, in which there are circular microcracks all perpendicular to the axis of the stress and far enough apart for them not to interfere with one another.

It will be shown in chapter 3 (§3.3.2.4) that the stress intensity factor K_I for such a crack of radius a in an isotropic body is

$$K_I = \frac{2}{\sqrt{\pi}}\sigma\sqrt{a} \tag{2.56}$$

Thus the stress and radius for the critical crack are related by

$$\sigma = \frac{\sqrt{\pi}}{2}\frac{K_{IC}}{\sqrt{a_c}} \tag{2.57}$$

where K_{IC} is the *fracture toughness* of the material

Let $n(\sigma)$ be the number of cracks per unit volume that will initiate fracture at a stress less than σ - that is, of radius greater than $(\pi/4)(K_{IC}/\sigma)^2$. Suppose that N items, each of unit volume, are subjected to stress σ and that N_R break; $N - N_R$ will survive. An increase $\Delta\sigma$ in the stress will cause a further $(N - N_R)\Delta n$ cracks to be propagated; the propagation of a single crack is enough to break one item, so the further number of items that will break will be

$$\Delta N_R = (N - N_R)\,\Delta n(\sigma) \tag{2.58}$$

Taking N_R/N to measure the probability $P_R(\sigma)$ of fracture, the change in this is then

$$\Delta P_R(\sigma) = \Delta N_R/N = [1 - P_R(\sigma)]\Delta n(\sigma) \tag{2.59}$$

whence on integration

$$P_R(\sigma) = 1 - \exp[-n(\sigma)] \qquad (2.60$$

and from Weibull's law $P_R(\sigma) = 1 - \exp[-(V/V_0)(\sigma/\sigma_u)^n]$ with $a_u = \frac{1}{4}\pi(K_{IC}/\sigma_u)^2$, we get

$$n(\sigma) = (1/V_0)(\sigma/\sigma_u)^m = (1/V_0)(a_u/a_c)^{m/2} \tag{2.61}$$

Thus we have found the distribution of defects that corresponds to the Weibull law. The result implies a density of microcracks that tends to infinity as the size tends to zero, which does not seem realistic; but this casts doubt only on the high probabilities of fracture.

2.6.4 Batdorf's theory (Brodt and Hasselman (1978)

In the calculation just given we have assumed for simplicity that all the microcracks are oriented normal to the axis of the traction: this will not be so in reality and Batdorf's treatment takes this fact into account. He introduces an effective stress σ_e which causes a crack to propagate as soon as it exceeds a critical value σ_c; σ_e is a function of the normal stress σ_n and shear force τ that act on the plane of the crack. Let $d\Omega$ be the solid angle containing the normals to those cracks for which σ_e lies between σ_c and $\sigma_c+d\sigma_c$; this will be a function of the stress field σ and of σ_c, and will take different values for uniaxial and biaxial traction, for example.

As before we denote by $n(\sigma)$ the number of cracks per unit volume which propagate under an effective stress σ and now assume these to be randomly oriented. If the stress increases from σ to $\sigma+d\sigma$ the number of cracks activated per unit volume will be $n(\sigma)$ $d\Omega/4\pi$, so the number of critical cracks in a volume dV is

$$dN(\sigma) = dV\int_{\sigma_c=0}^{\sigma_c=\sigma} n(\sigma)\frac{d\Omega(\sigma)}{4\pi} \tag{2.62}$$

Finally, the probability that the item fractures is

$$P_R = 1-\exp\left[-\int_V\int_{\sigma_c=0}^{\sigma_c=\sigma} n(\sigma_c)\frac{d\Omega(\sigma_c)}{4\pi}dV\right] \tag{2.63}$$

A simplifying assumption is to say that the cracks will propagate only under the action of the normal stress, and that the shear force has no effect; this is equivalent to putting $\sigma_e = \sigma_n = \vec{g}.\underline{\underline{\sigma}}.\vec{g}$, where $\vec{g}$ is the unit vector normal to the crack.

In uniaxial traction $\sigma_e = \sigma\cos^2\varphi$ where φ is the angle between the normal to the crack and the tensile axis; then

$$d\Omega = 2\pi\frac{d(\sigma_c/\sigma)}{(\sigma_c/\sigma)^{1/2}} \tag{2.64}$$

In biaxial traction $\sigma_e = \sigma\sin^2\varphi$, where φ is the angle between the normals to the crack and the plane of the traction respectively; then

$$d\Omega = 2\pi\frac{d(\sigma_c/\sigma)}{(1-\sigma_c/\sigma)^{1/2}} \tag{2.65}$$

With a Weibull distribution we find the probabilities of fracture for the two cases respectively

$$\begin{aligned} P(\sigma,0,0) &= 1-\exp\left[-\frac{V}{V_0}\left(\frac{\sigma}{\sigma_u}\right)^m\int_0^1\frac{x^m}{2}\frac{dx}{\sqrt{x}}\right] \\ P(\sigma,\sigma,0) &= 1-\exp\left[-\frac{V}{V_0}\left(\frac{\sigma}{\sigma_u}\right)^m\int_0^1\frac{x^m}{2}\frac{dx}{\sqrt{1-x}}\right] \end{aligned} \tag{2.66}$$

More generally, we need to know the effective stress σ_e as a function of σ_n and τ; the difficulty here is that propagation of the cracks is generally accompanied by bifurcation. If we suppose for example that a crack propagates when the maximum principal stress at its head reaches a critical value we find, anticipating some results to be derived in §3.2.3.2 below,

$$\sigma_e = 0.5\{\sigma_n + [\sigma_n^2 + 8\tau^2]^{1/2}\} \tag{2.67}$$

but if instead we assume that propagation takes place without bifurcation when the energy release rate reaches a critical value we find (§3.2.4.3)

$$\sigma_e = [\sigma_n^2 + \tau^2]^{1/2} \tag{2.68}$$

All this can change if the cracks have preferential orientations, but to predict the changes we need to know the distribution of the orientations.

2.6.5 Some difficulties

These theories are valid only so long as the initial assumptions hold, in particular that of the independence of the volume elements. Thus if the normal stress is negative a crack can still propagate, but in general it will bifurcate, so that the stress intensity factor decreases and the crack becomes inactive. Even so it will contribute to the progressive damaging of the material, but the "weakest link" assumption will cease to hold (Fig. 2.27).

Consider a biaxial compressive stress field σ_1, σ_2. Cracks will propagate only in Mode II, with bifurcation, in the direction of the maximum tensile stress at their extremities. Friction between the edges is to be taken into account, which involves using the normal stress. Failure will result if a sufficient number of defects coalesce. With the assumption that the cracks are randomly oriented, the probability of failure can be calculated as a function of the distribution of the size of the cracks.

Another case in which the previous theories do not apply is that in which cracks propagate but are halted on meeting obstacles, which stabilise them; this occurs in concrete and also in certain ceramics.

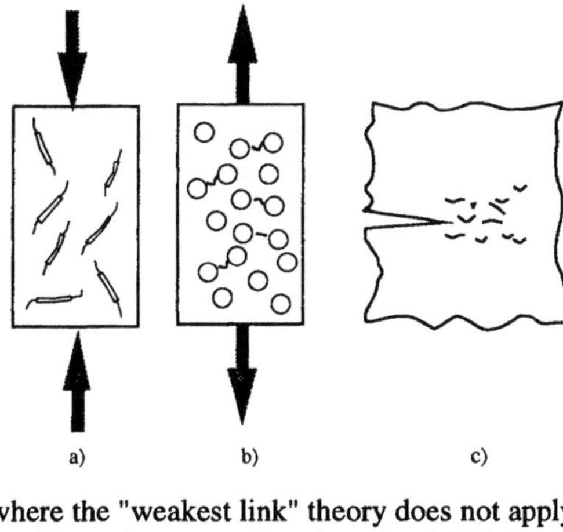

Fig. 2.27 Cases where the "weakest link" theory does not apply: (a) compressive loading (b) cracks stopped by obstacles (c) high gradients.

A further difficulty arises when the stress gradients are very high, for example in the region of the end of a crack where the stress varies like $r^{-1/2}$, r being the distance from the end. We shall show in chapter 3 (§3.2.3.2) that along the prolongation of the crack the stress in a sector of small angle θ is

$$\sigma = \frac{K_I}{\sqrt{2\pi r}} \tag{2.69}$$

where K_I is the stress intensity factor.

If we apply Weibull's theory here we find that, provided $m > 4$,

$$\log\frac{1}{1-P_R} = B\int_{-\theta}^{\theta} d\theta \int_{r_0}^{\infty}\left(\frac{K_I}{\sqrt{2\pi r}}\right)^m \frac{rdr}{V_0\sigma_u^m} = \frac{4B\theta\, K_{IC}^m}{(m-4)(2\pi)^{m/2} V_0 \sigma_u^m r_0^{(m-4)/2}} \tag{2.70}$$

From this we can find the macroscopic toughness K_{IC} which, for any given probability, is a function of the Weibull parameters, the length B of the crack front and a critical distance r_0. This last has to be introduced to avoid the second integral diverging at the lower limit; divergence would mean that the material could not support the infinite stresses that would be given by this expression, calculated assuming a linear elastic material whose fracture strength is infinite. In reality, at the head of the crack the damage shields the stress field; the cracks interfere with one another so as to make the weakest-link theory inapplicable. The problem now is how to choose a value for r_0. A possible course is to say that this distance is related to some characteristic dimension of the microstructure, the grain size, say, or the size of the damaged zone, which will be of the order of $(K_I/\sigma_u)^2$. With this second choice we get for the fracture toughness

$$K_{Ic} = \left[\frac{(m-4)(2\pi)^{m/2}}{4}\frac{V_0}{B\theta}\log\frac{1}{1-P_R}\right]^{1/4}\sigma_u \tag{2.71}$$

This procedure is admittedly approximate, but it enables the influence of several parameters on the macroscopic toughness to be evaluated; in particular, (2.71) shows the effect of scale, with the toughness falling as the length B of the crack front increases. The choice of V_0 is far from being without effect, and here we again come up against the complexity that results from the fact that the stress must remain approximately constant throughout this volume, which is large compared to microcracks. The theory is not applicable if the stress gradient is large.

To avoid these difficulties more general treatments have been developed in which random boolean functions, or random dead leaf functions, are used and in which Weibull's theory appears as a very particular case. These methods use scale effects that differ from his, and effects of shape that do not appear at all in his treatment. They require more elaborate experiments than those normally used for determining the

Weibull constants; in order to make reliable extrapolations from experimental results we should have to make measurements with test pieces that vary in both size and shape.

2.7 FRACTURE OF COMPOSITES

The very great variety of composite materials makes it impossible to give an exhaustive account of damage and fracture in these; moreover, these are far from being fully understood. We shall give only a few indications here.

2.7.1 Effect of residual stresses

It is essential to consider first the residual stresses generated by differences among the coefficients of thermal expansion of the various constituents. A fiber that contracts on cooling more than the matrix will generate compressive stresses in the latter. In the opposite case it is the fibers or the particles, and in particular the boundaries, which will be in compression, a state of affairs that opposes their cracking. Thus with *thermal cycling* there is risk of fatigue, and of the *ratchet effect*: every cycle extends the material. (Fig. 2.28). Application of the law of normality (I §3.4.2.3a) shows that at each reversal of the shear a material under constant traction accumulates a positive extension.

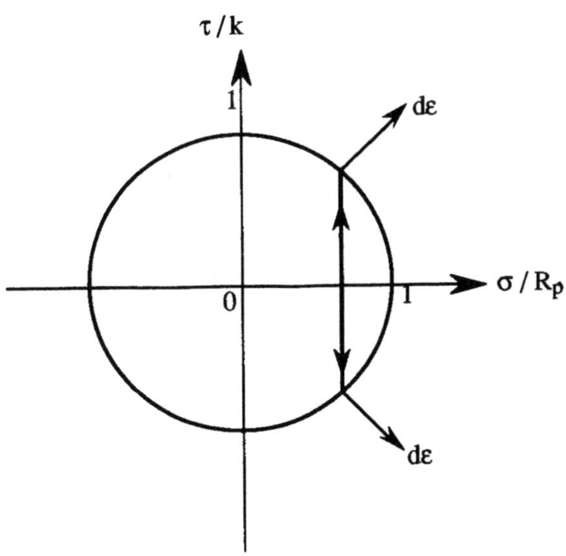

Fig. 2.28 The ratchet effect.

2.7.2 Composite with long fibers, more brittle than the matrix

A typical example is that of glass fibers in an epoxy resin. In I-§3.3.12.3 we showed that plastification of the matrix results in a tensile stress in the fibers, increasing from the ends; these will break at the point where this stress reaches the breaking value R_f. Damage in a composite in which brittle fibers of diameter d_f are embedded in a ductile medium first occurs as breaking of the fibers at distances equal to the critical length:

$$l_c = d_f R_f / R_{pm} \qquad (2.72)$$

where R_{pm} is the yield strength for the matrix material. This is in fact a mean distance, since the fracture is of a statistical nature.

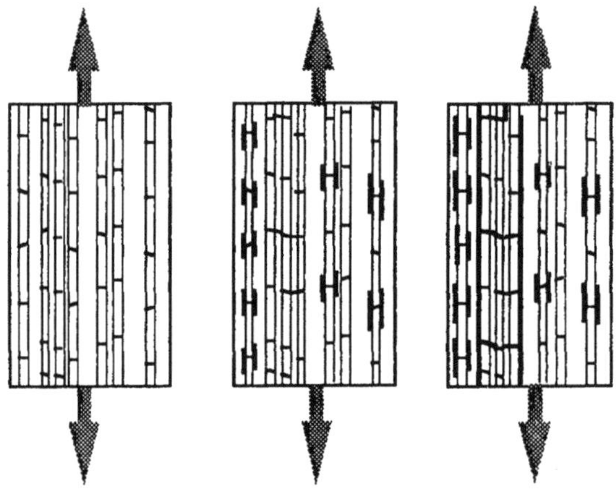

Fig. 2.29 Cracking in a composite material with long brittle fibers. (a) breaking of the fibers (b) cracks propagate in the zones where the fibers are close together (c) beyond a certain size the fissures cause delamination of the material.

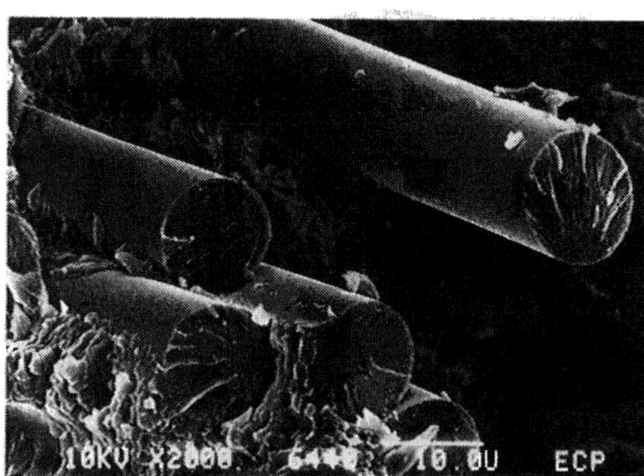

Fig. 2.30 Epoxy-glass fiber composite. In the regions where the fiber density is high, they break without pull-out; where this is low they are pulled out from the matrix.

The distribution of fibers in the composite is never uniform (Fig. 2.29). The first breaks cause cracks to appear in the matrix in the regions where the density is high; numerical calculations show that when such a crack, in which several fibers and the surrounding matrix are involved, propagates, the shear stress that it generates on the boundaries increases faster than the normal stress on the plane of the cracking. Beyond a certain size the cracks are stopped and branch, shearing the boundaries. The final rupture consists of a splitting of the composite.

Where the fibers are less dense they are loosened from the matrix. These effects are seen in the micrograph of Fig. 2.30; the reasoning for this loosening is easily seen when the high rate of shearing towards the ends of the fibers is studied. It is clear therefore that departures from homogeneity are of fundamental importance in the damaging of composites. Misalignment of the fibers also has important effects: those that are most closely aligned in the direction of the traction bear the highest stress and fail first. Finally, it is very important to take *edge effects* into account, as it is here that splitting occurs most easily.

2.7.3 Composites reinforced with short fibers or particles

A typical example is a metallic matrix reinforced with short carbon fibers, whiskers or particles of silicon carbide.

Fig. 2.31 Propagation of a crack in aluminum reinforced with whiskers of silicon carbide.

Damage results either from the growth of cavities in the matrix at the ends of the fibers or particles under traction, or to fracturing at the boundaries; it is followed by coalescence of these cavities, usually by shearing in the regions where the volume fraction of the reinforcement is high. (Fig. 2.31).

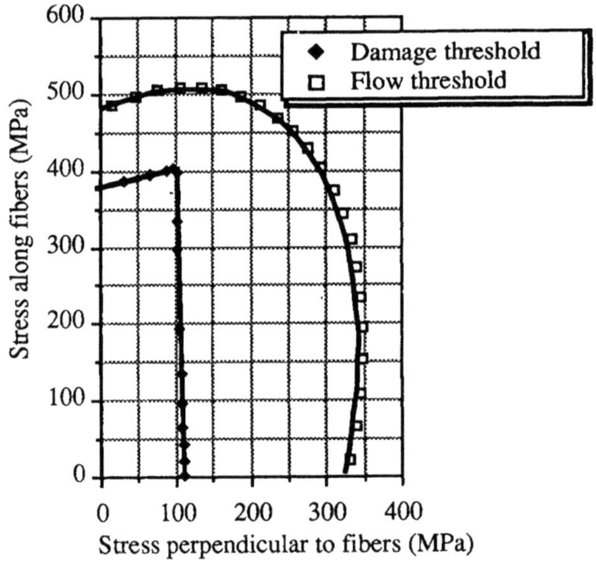

Fig. 2.32 Yield strength and damage threshold for aluminum reinforced with carbon fibers, under biaxial stress.

The threshold at which an applied stress will initiate damage can be determined from the stress on the inclusions and especially on the boundaries, calculated using Eshelby's theory, given in I-§2.7.2; it is usual to add these to the stresses resulting from differences in thermal expansion. It is often possible to assume that there is no chemical binding at the boundary, so that damage will start as soon as the applied stresses cancel the residual compressive stress there; a profile of the damage surface can then be constructed, of which Fig. 2.32 is an example.

2.7.4 Criteria for macroscopic fracture

It is seldom possible to carry through calculations on the micro-mechanics scale so as to determine, by homogenisation, a macroscopic criterion for the rupture we have been discussing. A more arbitrary macroscopic criterion might be envisaged, incorporating the effects of anisotropy and of hydrostatic stress, which latter would oppose the opening of cracks.

Taking first unidirectional composites under simple loading along the axes of anisotropy, the behaviour, and in particular the fracture load, can be approximated very simply by treating these as consisting of elements that work in parallel. It can then be assumed, for example, that if the fibers break then the matrix alone bears a tensile load in the direction of the length.

A simple criterion in this case would be to state that the stresses in the directions of the axes of the material must not exceed the fracture stresses for those directions, determined by simple tensile tests; this would give, for example, the following limits:

$\sigma_{1R}{}^{C} < \sigma_{1} < \sigma_{2R}{}^{R}$	along the fiber axis
$\sigma_{2R}{}^{C} < \sigma_{2} < \sigma_{2R}{}^{T}$	perpendicular to the fiber axis
$\tau < \tau^{C}$	in a shear test

where T, C refer to tension and compression respectively.

This gives a surface in stress space that has several sheets, with angles between these which can pose computational problems.

Tsaï has proposed a criterion derived from that of Hill for anisotropic materials (cf I-§3.4.2):

$$F(\sigma_{22}-\sigma_{33})^2 + G(\sigma_{33}-\sigma_{11})^2 + H(\sigma_{11}-\sigma_{22})^2 + 2L\sigma_{23}{}^2 + 2M\sigma_{31}{}^2 + 2N\sigma_{12}{}^2 = 1 \qquad (2.73)$$

of which the von Mises criterion is a particular case, for isotropic materials.

For loads restricted to being in the plane of the composite (2.73) becomes

$$(\sigma_1/\sigma_{1R})^2 + (\sigma_2/\sigma_{2R})^2 - (\sigma_1/\sigma_{1R})(\sigma_2/\sigma_{2R}) + (\tau/\tau_R{}^2) = 1 \qquad (2.74)$$

in which there is no distinction between tension and compression; this restriction is removed by the following criterion due to Tsaï Wu:

$$\frac{\sigma_1^2}{\sigma_{1R}^T\sigma_{1R}^C} + \frac{\sigma_2^2}{\sigma_{2R}^T\sigma_{2R}^C} + \frac{\tau^2}{\tau_C^2} + 2F_{12}\sigma_1\sigma_2 + \left(\frac{1}{\sigma_{1R}^T} - \frac{1}{\sigma_{1R}^C}\right)\sigma_1 + \left(\frac{1}{\sigma_{2R}^T} - \frac{1}{\sigma_{2R}^C}\right)\sigma_2 = 1 \qquad (2.75)$$

Here F_{12} is a coupling term which can be determined by conducting a biaxial test with $\sigma_1 = \sigma_2 = \sigma$; if this gives a fracture stress σ_R we have, from (2.75),

$$F_{12} = \frac{1}{2\sigma_R^2}\left[1 - \left(\frac{1}{\sigma_{1R}^T} - \frac{1}{\sigma_{1R}^C} + \frac{1}{\sigma_{2R}^T} - \frac{1}{\sigma_{2R}^C}\right)\sigma_R - \left(\frac{1}{\sigma_{1R}^T\sigma_{1R}^C} + \frac{1}{\sigma_{2R}^T\sigma_{2R}^C}\right)\sigma_R^2\right] \qquad (2.76)$$

Experience has shown that no serious error results from taking $F_{12} = 1/\sigma_{1R}^T\sigma_{1R}^C$, for the sake of simplicity.

2.8 DUCTILE FRACTURE OF METALS

2.8.1 Initiation of cavities

We have explained how ductile fracture of metals results from the initiation, growth and coalescence of cavities, ending with fracture showing a dimpled surface. The cavities usually start to form on inclusions, and these can often be seen at the bases of the dimples; it is however possible that they may still form in the absence of inclusions, although there is no convincing proof of this, nor has any acceptable model been proposed for such a mechanism.

Initiation on inclusions can result from fracture either of these or of the interface between the inclusions and the matrix, which can be greatly weakened by segregation of impurities (§2.13). As with all questions of fracture, two conditions must be satisfied: the local stress must at least equal the fracture stress and the energy released must at least equal the fracture energy. This second (energy) condition is expressed by equating the released energy, proportional to the cube of the size of the inclusion (and to the square of the deformation, as shown in I-§2.7.2.1) to the fracture energy, proportional to the square of this size. This gives a critical deformation which decreases with increasing size of the inclusions; it applies only to sub-microscopic inclusions.

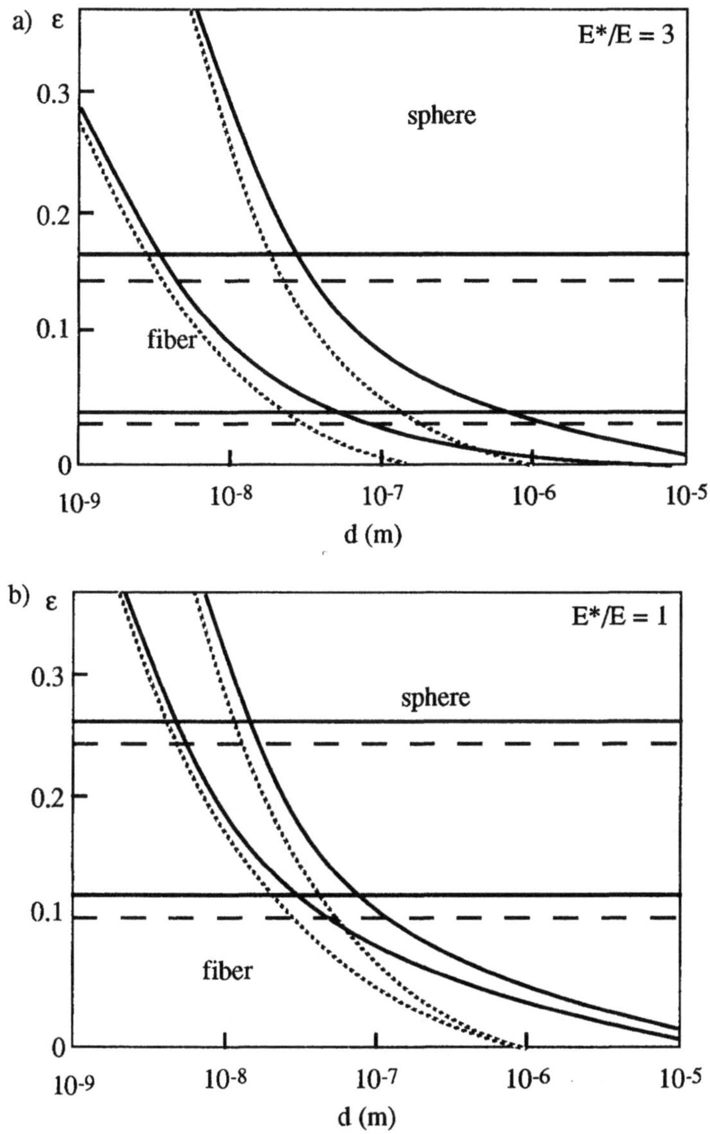

Fig. 2.33 Plastic deformation required to separate the inclusion and the matrix for (a) spherical particles (b) fibers, as a function of dimensions. Ratios E^*/E of elastic moduli of reinforcement and matrix respectively are (a) *3* (b) *1*. Curves are for energy criterion, horizontal lines for critical stress; full curves for $\sigma_{33}^A = 0$, dotted curves for $\sigma_{33}^A = E/100$. (Tanaka et al (1971)).

We shall confine our treatment mainly to the first (stress) criterion, expressed by equating the stress in the inclusion, or at the interface, to the corresponding fracture stress. In I-§2.7 we showed how the fracture stress can be calculated, using Eshelby's theory. From (I - Eq. 2.92a), considering a constitutive equation for the matrix such as (I - 3.23) with linear strain-hardening ($\varphi'(\sigma) = 1/\varepsilon^p$), and neglecting the elastic deformation , it follows that the volume remains constant for a rigid inclusion. Then

$$\underline{\underline{\sigma}} = \underline{\underline{\Sigma}} + \frac{2}{3} E_p \left(S^{-1} - I\right) : \underline{\underline{\varepsilon}}^p \tag{2.77}$$

where $\underline{\underline{\sigma}}$ is the stress in the inclusion, $\underline{\underline{\Sigma}}$ the macroscopic stress, E_p the tangent modulus for the plastified matrix (assuming linear work-hardening), S the Eshelby tensor (dependent on the shape of the inclusion) and ε^p the plastic deformation. For a uniaxial applied stress Σ, and more generally for the maximum principal stress, this becomes

$$\sigma = \Sigma + (2/3)\alpha E_p \varepsilon_{eq}^p \tag{2.78}$$

where σ is the stress in the inclusion or at the interface and α is a shape factor. $E_p\varepsilon_{eq}^p$ in this expression can be replaced by Σ_{eq} -R_p, with Σ_{eq} the von Mises equivalent applied stress.

With $\sigma = \sigma_d$, the initiation stress, the condition that a cavity starts to form is

$$\Sigma_R = \sigma_d - \lambda(\Sigma_{eq} - R) \tag{2.79}$$

where λ is a shape factor, some values for which are given in Table 2.4. This gives the following initiation stresses, all in *MPa*: *1120±60* for longitudinal cleavage fracture of sulfides in steel and *810* for decohesion at the interface (transverse stress); *80* for decohesion of graphite nodules in GS cast iron; *1700* for cementite; *1000* for particles of CuCr in copper; *1780* for particles of TiC in a maraging steel.

In steels, manganese sulfide particles, which are plastic at the temperatures of hot forming, stretch in the same way as the surrounding matrix. They thus take the form of elongated ellipsoids, for which, from Table 2.IV, the shape factor is greater in the longitudinal direction than in the transverse; although the fracture stress for cleavage is greater than that for decohesion at the interface, they break by cleavage when loaded longitudinally. To reduce λ in order to increase the initiation stress Σ_R the inclusions must be made spherical; this can be achieved with manganese sulfides by adding rare earths, of cerium in particular, which will harden them sufficiently in hot casting. Alternatively they can be given a coating of lead.

The size of the common initiation stresses is such that they would produce deformations that are small compared to those at fracture.

	sphere	ellipsoid (long)	particular case	disk	particular case
	$a = b = c$	$a >> b = c$ $K = c/a$ $L = \log [(2-K)/K]$ $\gamma=(3/2)K^2(2L-1)$	$K = 0.17$	$a = b >> c$ $K = c/a$	$K= 0.17$
λ longitudinal	1	$K^2/2\gamma - 2/3 + 2/3\gamma$	3.2	$20/9\pi K-2/3$	3.3
λ transverse	1	$K^2/12\gamma + 1/2 + 1/6\gamma$	1.5	$8/9\pi K-2/3$	0.9

Table 2.4. Shape factor λ for calculating stresses in inclusions.

2.8.2 Growth of cavities

2.8.2.1 *Isolated cavity*

Cavity growth results from plastic deformation of the matrix; in its full generality it presents a complex problem, in which the shape of the cavity has to be taken into account. Here we shall study only volume changes.

One of the first models to be suggested is that of McClintock (McClintock and Argon (1966)), who investigated the growth of a cylindrical cavity of radius R in a perfectly plastic material stretched in the direction of the axis of the cylinder (Oz) and subjected to a radial stress Σ_r at infinity. This gives (Exercise 11)

$$\frac{dR}{R} = dE_z\left[\frac{\sqrt{3}}{2}\sinh\left(\sqrt{3}\frac{\Sigma_r}{\Sigma_z - \Sigma_r}\right) - \frac{1}{2}\right] = dE_{eq}\left[\frac{\sqrt{3}}{2}\sinh\sqrt{3}\left(\frac{\Sigma_m}{\Sigma_{eq}} - \frac{1}{3}\right) - \frac{1}{2}\right] \qquad (2.80)$$

The same calculation for a viscous material that obeys a Norton law $\sigma = \sigma_0\dot{\varepsilon}^n$ gives a result which represents also the behaviour of a cavity in a plastic material with work-hardening exponent n.

This model is clearly very approximative, but it does make clear the very important influence of the degree of *stress triaxiality* Σ_m/Σ_{eq}, the ratio of the hydrostatic stress to the equivalent stress.

A more realistic model has been proposed by Rice & Tracey (1969), who consider spherical cavities of radius R in an infinite medium whose behaviour is rigid-perfectly plastic, loaded axisymmetrically. Their result for the growth of cavities is

$$\frac{dR}{R} = A.\chi.dE_{eq}\exp\left(\frac{3}{2}.\chi.\frac{\Sigma_m}{\Sigma_{eq}}\right) - B.\chi.dE_{eq}\exp\left(-\frac{3}{2}.\chi.\frac{\Sigma_m}{\Sigma_{eq}}\right) \tag{2.81}$$

where $\chi = +1$ if the axial stress is greater than the radial and -1 conversely. First estimates for the constants A, B were $A = 0.283$, $B = 0.275$; later work has shown that A may be greater than this, and further, that it may increase with increasing stress triaxiality Σ_m/Σ_{eq}. In general only the first term in (2.81) will be taken, since we are interested primarily in high degrees of triaxiality of positive stresses with axial extensions. The second term must however be included when the stress is negative, for example in sintering under load.

For high degrees of triaxiality McClintock and Argon's solution becomes essentially $dR/R = 0.243 \exp(1.73\Sigma_m/\Sigma_{eq})$, which differs little from that of Rice & Tracey.

Budiansky, Hutchinson & Slutsky have generalised Rice & Tracey's model to deal with a Norton-Hoff viscosity law $\sigma = \sigma_0\dot{\varepsilon}^m$, applying equally to a work-hardenable material. Their result is

$$\frac{\dot{R}}{R} = \frac{1}{2}(\operatorname{sgn}\Sigma_m)\dot{E}_{eq}\left\{\frac{3}{2}m\frac{|\Sigma_m|}{\Sigma_{eq}} + (1-m)\left[1 + m(0.4175 + 0.0144\chi(\operatorname{sgn}\Sigma_m))\right]\right\}^{1/m} \tag{2.82}$$

For a Newtonian material $m = 1$ and the exact result is

$$\dot{R}/R = (3/4)(\Sigma_m/\Sigma_{eq})\dot{E}_{eq} \tag{2.62}$$

The Rice & Tracey model corresponds to $m = 0$.

This model shows that the rate of growth of the cavity falls as the work-hardening exponent, the analogue of m, increases, even though calculating its influence is not easy.

2.8.2.2 *Interaction between cavities*

The models we have been discussing all concern isolated cavities, not interacting with their neighbours. For a spherical cavity of volume V enclosed in a sphere of whose volume it occupies a fraction f and which is subjected to a positive hydrostatic load, the exact solution for the rate of growth is

$$\dot{V}/V = (3/2)(m/\sigma_0)^{1/m}(3\Sigma_m/2)^{1/m}(1-f^m)^{+1/m} \tag{2.84}$$

expansion of which in the region of $f = 0$ gives

$$\dot{V}/V = (\dot{V}/V)_0[1 + (1/m)f^m + \ldots] \tag{2.85}$$

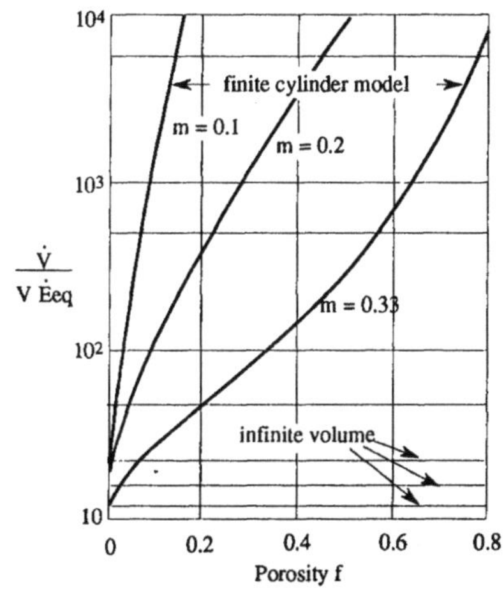

Fig 2.34 Effect of porosity *f* on rate of growth of circular cylindrical cavities, axisymmetrically loaded, in a viscoplastic material for which the stress/strain-rate relation is $\sigma = \sigma_0 \dot{\varepsilon}^m$, for various values of *m*.

Equation 2.85 shows that for small values of m the rate increases very rapidly with *f*: if, for example, $m = 1/3$ a volume fraction of *1/1000* increases the growth rate by *30%*. Thus it is very important to take porosity into account, as is confirmed by experimental observations. Fig. 2.34 gives values calculated by Licht & Suquet (1988) for cylindrical cavities in a viscoplastic material, a problem analogous to that of McClintock. These rates, normalised by dividing by the equivalent strain rates, show clearly the rapid increase with porosity.

2.8.2.3 *Behaviour of porous materials*

The problem of the behaviour of porous materials can be attacked by methods involving micro-mechanical studies and homogenisation; since such a material will be compressible the constitutive equation will involve the hydrostatic stress.

For a perfectly porous rigid-plastic material, a frequently-used equation for the yield surface is that due to Gurson (1977):

$$\Psi(\Sigma) = \left(\Sigma_{eq}^2 / \sigma_0^2\right) + 2f \cosh\left(3\Sigma_m / 2\sigma_0\right) - 1 - f^2 = 0 \qquad (2.86)$$

This reduces to the von Mises surface in the case of a dense ($f = 0$), and therefore incompressible, material. Tvergaard (1981, 1982) has modified the criterion by scaling up the porosity by a factor *1.5*, which gives better agreement with observations.

Rousselier (1987), using thermodynamic considerations, has proposed a yield surface depending on the square of the stress triaxiality; and it can be shown that the macroscopic potential of a porous material is equal to the mean of the microscopic potential of the matrix $\varphi(\underline{\underline{\sigma}})$. In fact,

$$\frac{\partial}{\partial \Sigma_{ij}} < \varphi(\sigma_{ij}) > = < \frac{\partial \varphi(\sigma_{ij})}{\partial \sigma_{kl}} \frac{\partial \sigma_{kl}}{\partial \Sigma_{ij}} > = < \dot{\varepsilon}_{kl} \frac{\partial \sigma_{kl}}{\partial \Sigma_{ij}} > \tag{2.87}$$

Now $\Sigma_{kl} E_{kl} = < \sigma_{kl} \dot{\varepsilon}_{kl} >$ and the derivative of this with respect to Σ_{ij} gives

$$< \dot{\varepsilon}_{kl} \frac{\partial \sigma_{kl}}{\partial \Sigma_{ij}} > = \frac{\partial}{\partial \Sigma_{ij}} \Psi(\sigma_{ij}) \tag{2.89}$$

It follows that if the microscopic potential is homogeneous of degree p then the same is true for the macroscopic potential:

$$\Psi(\lambda \underline{\underline{\Sigma}}) = \lambda^p \Psi(\underline{\underline{\Sigma}}) \quad \text{if} \quad \varphi(\lambda \underline{\underline{\sigma}}) = \lambda^p \varphi(\lambda \underline{\underline{\sigma}}) \tag{2.90}$$

Gurson's criterion does not satisfy this condition.

A number of approximate expressions for the potential of a porous medium have been proposed. Michel & Suquet (1992), for example, using a self-consistent model, obtained the following for a rigid, perfectly plastic medium

$$(1 + \frac{2}{3} f) \left(\frac{9}{4} C \Sigma_m^2 + \Sigma_{eq}^2 \right)^2 = \sigma_0^2 (1 - f)^2 \tag{2.91}$$

where C is given implicitly by

$$\sqrt{C} \log f = \tan^{-1}(\sqrt{C}) - \pi / 2 \tag{2.92}$$

For a model of spheres they obtained

$$\frac{9}{4} \left(\frac{1 - f}{\log f} \right)^2 \Sigma_m^2 + (1 + \frac{2}{3} f) \Sigma_{eq}^2 = \sigma_0^2 (1 - f)^2 \tag{2.93}$$

which is the exact solution for a hollow sphere under a hydrostatic stress

$$\Sigma_m = -\frac{2}{3} \sigma_0 \log f \tag{2.94}$$

For a viscoplastic material characterised by $\sigma = \sigma_0 \dot{\varepsilon}^m$ Duva & Hutchinson obtained the following approximation to the potential for small values of the porosity *f*:

$$\Psi = \frac{m\sigma_0}{m+1}\left(\frac{\Sigma_{eq}}{\sigma_0}\right)^{\frac{m+1}{m}}\left\{1+\frac{f}{m}\left[\frac{3}{2}m\frac{\Sigma_m}{\Sigma_{eq}}+(1-m)(1+0.4319m)\right]^{\frac{m+1}{m}}\right\} \tag{2.95}$$

This should be compared with (2.82) for the rate of growth of a cavity, given by Budianski, Hutchinson and Slutsky (1982).

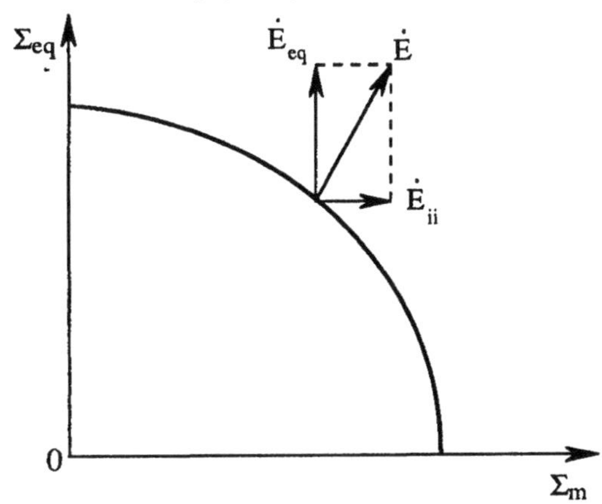

Fig. 2.35 Criterion for yield of a porous material, as a graph of equivalent stress against hydrostatic stress The condition that the direction of increasing deformation is normal to this curve shows that there will be an increase in volume.

The rate of growth of a cavity can be derived from each of these expressions. Fig. 2.35 shows that if the usual relation holds, that the direction of flow is normal to the equipotential surface, there will be in addition to the plastic deformation $\dot{E}_{eq}$ a term $\dot{E}_{ii}$ representing a change of volume; its value is $\dot{E}_{ii} = \partial\Psi / \partial\Sigma_m$. Given that the matrix is incompressible we then have

$$\dot{f}\,/\,(1-f) = \dot{V}\,/\,V = \partial\Psi / \partial\Sigma_m \tag{2.96}$$

Since $\dot{E}_{eq} = \partial\Psi / \partial\Sigma_{eq}$ the ratio $(1/\dot{E}_{eq})dR/R$ follows easily by calculating

$$(1/3f)\frac{\partial\Psi / \partial\Sigma_m}{\partial\Psi / \partial\Sigma_{eq}}$$

The expression for the growth of cavities derived from Gurson's criterion (Gurson (1977)) is an exponential function of the stress triaxiality. It is very similar to that obtained by Rice & Tracey (1969), equation (2.81); the criteria involving Σ_m^2 lead to a linear relation. Duva & Hutchinson's potential (1987) corresponds to a more complicated function of the triaxiality.

Checking of these predictions by experiment requires test specimens subjected to known degrees of stress triaxiality; as we shall see later, in §2.9.3.2, this can be met by using notched specimens. The dimensions of the cavities are measured on micrographs of sections taken after the test, which is stopped before the material ruptures; great care must be taken to avoid enlarging or filling the cavities when polishing the sections. It is not at all certain that results of sufficient precision can be obtained to distinguish between the various models.

Gurson's criterion leads to a value of $(1/\dot{E}_{eq})dR/R$, which is independent of the porosity f, whilst for all the other results for the potential this is a decreasing function of f. Many experiments have shown that the potential *increases* with f_v, the volume fraction of inclusions, which is the same as the initial volume fraction of cavities (see Fig. 2.36); however, experiments with porous nickel obtained by sintering very pure metal have indeed shown this ratio decreasing with increasing f. The observed increase with f can be explained by assuming the existence of a second population of cavities, smaller than and interacting strongly with the first; micrographs of fractured surfaces do indeed show dimples of two different sizes in these cases. The problem is the subject of current research, in which analytic and numerical studies have reproduced such effects.

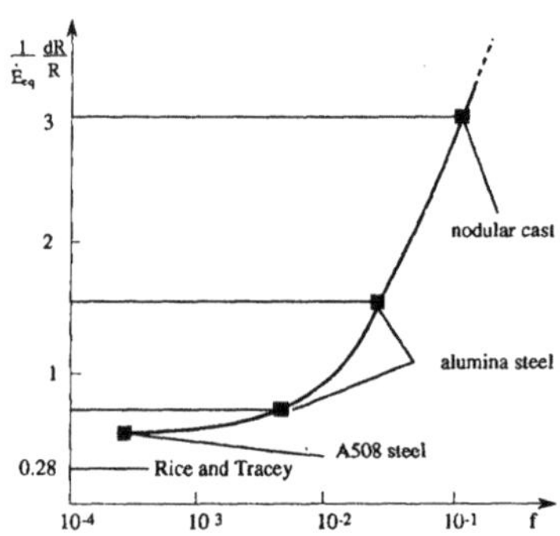

Fig. 2.36 Rate of growth of cavities as a function of the volume fraction of initiating inclusions.

2.8.3 Coalescence of cavities

The final stage in ductile fracture is the coalescence of the cavities. This is a phenomenon of plastic instability that is difficult to express in the form of equations; the

material remaining between the holes forms ligaments which behave somewhat as miniature test specimens, the instability of which is the more delayed the greater the work-hardening exponent. But if the material strain-hardens only slightly the instability is generated more by very localised macroscopic shears; this is the case, for example, for steels of very high yield strength.

The simplest condition for coalescence of the cavities is that the ratio of their size R to their mean distance apart L_0 has reached a critical value $(R/L_0)_c$. This can be written $(R/R_0)_c(R_0/L_0)$, which can be calculated from the various expressions for the rate of growth of the cavities discussed in §2.8.2 above. Thus using the Rice & Tracey model, for example, and assuming that the degree of stress triaxialiy remains constant during the loading, we find

$$\log(R/R_0) = 0.283E_{eq}\exp(3\Sigma_m/2\Sigma_{eq}) \tag{2.97}$$

It follows that coalescence will occur for an equivalent critical stress E_{eqC} such that

$$E_{eqC} = 3.5\log(R/R_0)_C\exp(-3\Sigma_m/2\Sigma_{eq}) \tag{2.98}$$

Similarly, the model of Budianski, Hutchinson & Slutsky (1982) gives

$$E_{eqC} = \left[\frac{3}{2}m\frac{\Sigma_m}{\Sigma_{eq}} + (1-m)(1+0.43m)\right]^{-1/m} \log(R/R_0)_C \tag{2.99}$$

Other expressions correspond to the other criteria that have been proposed. As the deformation is still rather small when the cavities begin to form, the calculations give a good approximation to the elongation at fracture. With the relatively small volume fractions of inclusions that can usually be achieved, the calculated values of $(R/R_0)_C$ remain very modest. In reality, however, the distribution of inclusions is never homogeneous and there will be regions where the density is high; these will start to break, entailing the complete failure of the material (Fig. 2.37). The statistical features of the distributions are absolutely fundamental for the understanding of coalescence; as Fig. 2.37 shows, in certain regions the softening due to damage by growth of cavities can lead to instability even though the overall elongation, calculated on the assumption that all the volume elements work in series, is still relatively small. With the assumptions previously made, the probability P_R of fracture for a volume V of material is

$$P_R = 1 - \exp(-PV/V_0) \tag{2.100}$$

where P is the probability of there being an elementary region of volume V_0 in which the volume fraction exceeds f_0^C.

This approach to the problem enables the importance of various parameters involved in ductile fracture to be brought out:

- the effect of inclusions, and of their distribution. Note in particular that the above equation (2.100) predicts a size effect (see §2.6.1); but in general it is found that such an effect is smaller for ductile than for brittle fracture,

- the effect of the stress-strain behaviour for the material, especially for work-hardening.

It has been used by Mudry to derive a quantitative model for ductile fracture. He assumes a Poisson distribution of inclusions and a work-hardening law of the form $\sigma = \sigma_0\, \varepsilon^n$. The results given in Fig. 2.38 show that the ductility at fracture, for a fracture probability of *50%*, decreases with increasing triaxiality and with increasing values of the ratio *f/n*.

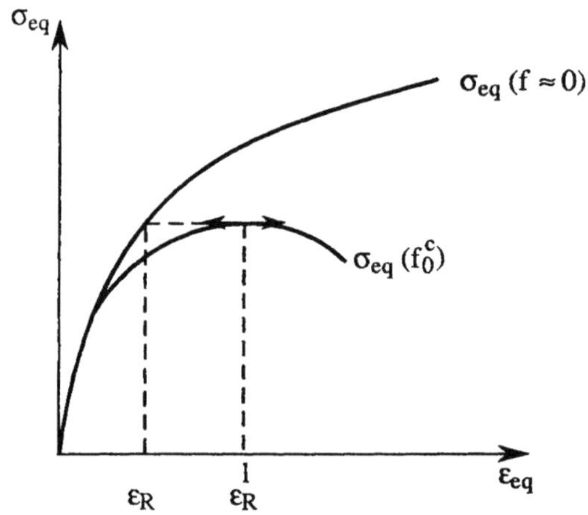

Fig. 2.37 Schematic stress-deformation curves, showing how a local high volume fraction of cavities can lead to rapid coalescence. It is assumed that all elements work in series.

As we have already noted, the presence of cavities reduces the effective modulus of the material - see §2.5.3, equation (2.53) - and results in so weakened a work-hardening that the behaviour can become that of softening; this is a source of instabilities which we shall study in the next section. The variation of the effective modulus with porosity has the form

$$E_{eff} = E(1 - \alpha f) \qquad (2.101)$$

in which αf acts as a damage parameter; the formalism of *damage mechanics* is entirely applicable, with laws of evolution derived from the laws of growth which we have been discussing.

These heterogeneities in volume fractions have little effect on the apparent elastic modulus and thus scarcely change the laws governing the overall behaviour of the damaged material, derived from the growth models.

Ductility, determined by the elongation E_R at fracture, is in general a decreasing function of the stress triaxiality (Fig. 2.38). This effect is easily explained: compressive hydrostatic stress will tend to close any holes that form, whilst a strong hydrostatic tension, as is generated within a confined plastic zone such as the root of a notch or the tip of a crack, will greatly increase the rate of growth of the cavities, with the result that the elongation at fracture is small.

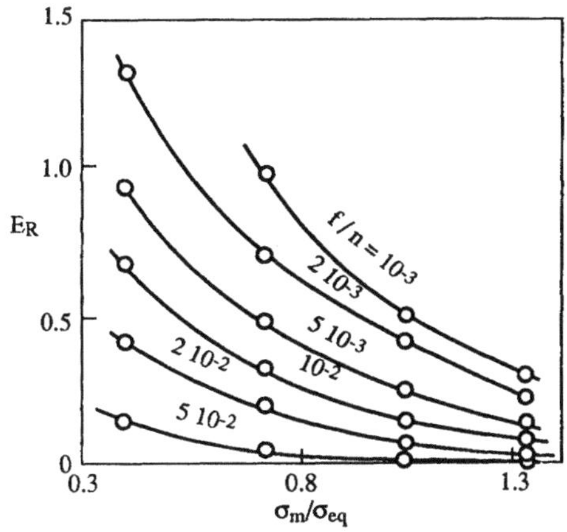

Fig 2.38 Variation of ductility with stress triaxiality, for various volume fractions *f* of cavities. *n* is the strain-hardening exponent.

The ductility of a material improves as the volume fraction of inclusions decreases, and important advances in steel fabrication have been achieved by making use of this effect. The influence of the Norton exponent, or that of the work-hardening exponent, is not evident in the equations we have given; but ductility is in fact an increasing function of the exponent n.

For fracture due to local shear, a calculation due to McClintock and Argon (1966) gives the following condition, in which *a/b* is the eccentricity of the cavities, assumed elliptical, and l_b is their distance apart, measured perpendicular to the direction of the maximum principal stress:

$$\frac{d\Sigma_{eq}}{\Sigma_{eq} dE_{eq}} \leq \sqrt{\frac{3}{8}}\left[1+\left(\frac{a}{b}\right)^2\right]^{1/2}\left(\frac{b}{b_0}\cdot\frac{l_b}{l_{b_0}}\right)^2\left(2b_0 / l_{b_0}\right)^2 \tag{2.102}$$

2.9 THE BRITTLE-DUCTILE TRANSITION

Materials which break by cleavage are very brittle at low temperatures, but above a certain transition temperature their ductility increases greatly. This effect has been studied particularly in the case of steels, for which it has great practical importance.

2.9.1 Smooth specimens

Consider first a tensile test performed on a smooth test piece of a BCC ferritic steel. If the temperature is lowered the yield strength will increase (Fig. 2.39), the cleavage stress will remain approximately constant and the two curves will meet again at a temperature T_D. When loaded at a little above this temperature the sample will begin to deform plastically when the stress reaches the yield strength R_p; the metal will then work-harden and cleavage will occur when the stress reaches σ_f. It is found that if the test is carried out at increasingly higher temperatures the elongation at fracture increases as the divergence between R_p and σ_f, increases. Below T_D fracture does not occur before the yield strength is reached. We saw in §2.2.5 that it is plastic slips that generate stress concentrations sufficient to initiate cleavages; but fracture is then sudden, the elongation is zero and the fracture strength is the same as the yield strength. The results obtained by Low (1954), who compared the tensile fracture strength with the compressive yield strength for a steel at low temperature, are given in Fig. 2.40 and show this effect clearly. T_D is the nil-ductility temperature for the smooth specimen.

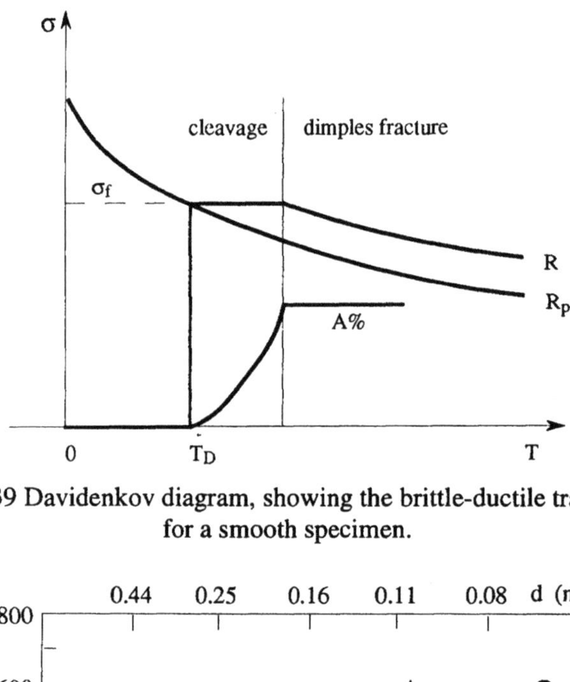

Fig. 2.39 Davidenkov diagram, showing the brittle-ductile transition for a smooth specimen.

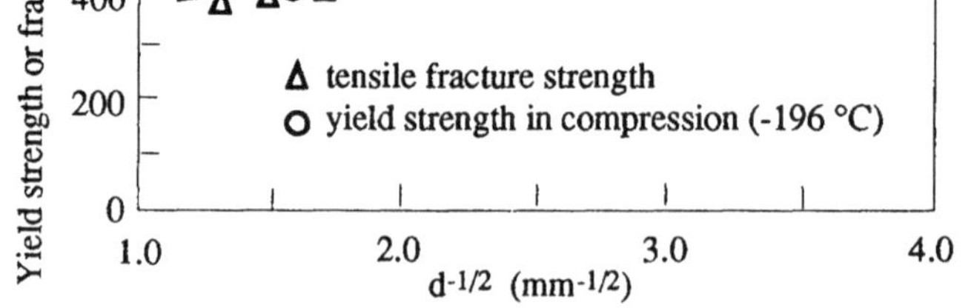

Fig. 2.40 Comparison of tensile stress at fracture with yield strength in compression for mild steel at *-196°C*, for various grain sizes. (Low (1954)).

Returning to tests at higher temperatures, the elongation at fracture increases until finally it reaches that corresponding to ductile fracture by coalescence of cavities. Thus there is an upper level of ductility, at which the evolution of the fracture strength parallels that of the yield strength and a *fracture appearance transition temperature* (FATT) can be defined, above that for nil ductility (see also §2.9.4 below).

2.9.2 Effect of loading rate

Increasing the loading rate increases the yield strength, but the effect on the cleavage stress is less well understood. Equation 2.21 suggests that this stress should fall if the speed v_c of cleavage increases, but it is not clear that this speed is related to that of loading and some observations suggest that σ_f is independent of loading rate. It is certainly true that an increase in the yield strength shifts the transition temperature; if σ_f is constant

$$T_{D_1} / T_{D_2} = \log \dot{\varepsilon}_1 / \log \dot{\varepsilon}_2 \qquad (2.103)$$

2.9.3 Effect of a notch

2.9.3.1 *General*

We consider a notched specimen, either a cylinder loaded in tension or one of Charpy type loaded in bending. So long as the yield strength is not exceeded, the maximum stress will occur at the root of the notch: this is the classic example of stress concentration, and the yield strength is reached first in this region. If we continue to pull or bend the specimen the zone of plastic deformation will spread until finally it will have invaded the entire notch, and the load when this occurs will be the limit load P_{GY} that the piece can bear. Formulae are known from which the value of the corresponding limit load P_L for a perfectly plastic material can be calculated; in general, $P_{GY} > P_L$. This limit load is much greater than what could be borne by a working section of the same area but without a notch, that is, R_pS.

Consider a cylinder of the material in the notched region; if it were isolated it would extend in the direction of its axis and contract laterally so as to keep the volume constant. Within the notch, between the two un-notched parts which deform little and remain below the yield strength, it cannot contract laterally and therefore will be acted on by radial tensions. Since the condition for plasticity - Tresca's, for example -must hold, the axial stress must increase correspondingly; thus the plastic deformation in the confined region has the effect of raising the general stress levels and the stress triaxiality Σ_m/Σ_{eq} ratio can reach very high values.

2.9.3.2 *Mechanics of notched test pieces*

(a) *Axisymmetric notch* Fig. 2.41 is a diagram of a notched cylindrical piece, such that the radius of curvature at the root of the throat is R and the minimum section there is a circle of radius a; the throat can either be machined or be produced by necking. The problem of computing the distributions of stress and deformation is complicated and a complete analytic solution has not yet been found. Simplifying assumptions have to be made, and we shall assume that at the minimum section (where we take $z = 0$) the radial and tangential deformations are uniform and are equal; it follows that the radial and tangential stresses are equal and that the stress deviator is independent of the radial coordinate r there.

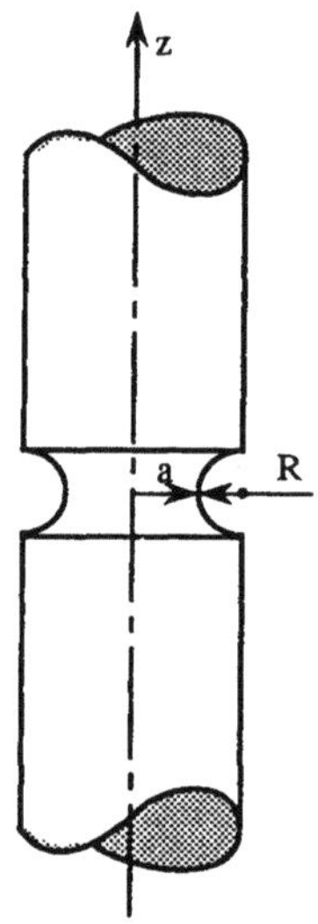

Fig. 2.41 Notched tensile specimen.

Given these assumptions, it follows from the plasticity condition and the equilibrium equations that

$$d\sigma_{rr} / dr = -\bar{\sigma} / \rho \qquad (2.104)$$

where $\bar{\sigma}$ is the equivalent stress and ρ is the radius of curvature of the isostatic lines (lines parallel to the directions of the principal stresses) where they cut the plane $z = 0$ (Exercise 18).

We have to make a further assumption, concerning these contours. Bridgman (1952) has assumed that they are circles cutting the toroidal surface of the notch at right angles. It then follows that

$$\rho = (a^2 + 2aR - r^2)/2r \qquad (2.105)$$

and integration of (2.104) gives

$$\sigma_{rr} = \sigma_{\theta\theta} = \bar{\sigma} \log\left(1 + \frac{a^2 - r^2}{2aR}\right), \quad \sigma_{zz} = \bar{\sigma}\left[1 + \log\left(1 + \frac{a^2 - r^2}{2aR}\right)\right] \qquad (2.106)$$

These results were quoted in the Introduction to Volume I (§1.3.3); numerical values are plotted in Fig. 2.42

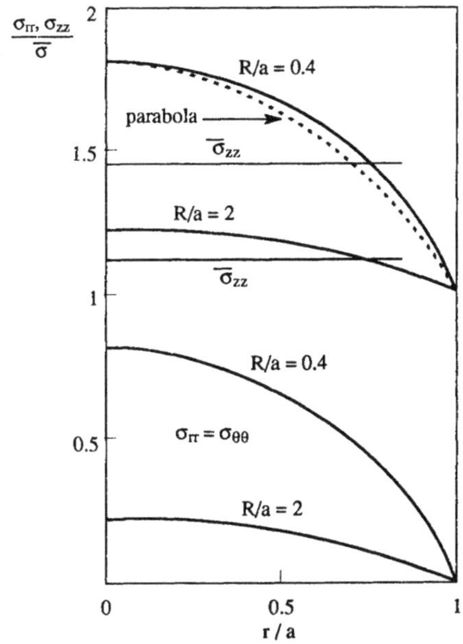

Fig. 2.42 Distribution of axial (σ_{zz}), radial (σ_{rr}) and tangential ($\sigma_{\theta\theta}$) stress in a notched cylindrical test piece, for two values of *R/a*. Full curves are for Bridgman's method, the dotted curve for Davidenkov & Spiridonova's (1945).

The mean axial stress $\bar{\sigma}_{zz}$ follows easily from these, and hence the applied load:

$$\bar{\sigma}_{zz} = \bar{\sigma}(1 + 2R/a)\log(1 + a/2R)$$

The stress triaxiality ratio is maximum on the axis, where its value is

$$\sigma_m / \bar{\sigma} = 1/3 + \log(1 + a/2R) \tag{2.108}$$

For notches with $a = 5\ mm$ this gives *0.55* for $R = 10\ mm$ and *1.14* for $R = 2\ mm$.

It follows from Bridgman's result (2.106) that if $a << R$ the shape of the stress profile is approximately parabolic:

$$\sigma_{zz} / \bar{\sigma} = 1 + 2\alpha[1 - (r/a)^2] \tag{2.109}$$

where $\alpha = \frac{1}{2}\log(1 + a/2R) \approx a/4R$; and

$$\sigma_{rr} = \sigma_{\theta\theta} = 2\alpha\bar{\sigma}[1 - (r/a)^2] \tag{2.110}$$

Integration of the equilibrium equations gives for the axial stress (Exercise 18)

$$\sigma_{zz} / \bar{\sigma} = 1 + 2\alpha - 2\alpha(r / a)^2 - 4\alpha(z / a)^2 \tag{2.111}$$

a distribution obtained by Davidenkov and Spiridonova by assuming $\rho = R(a/r)$.

The strain field can be determined similarly; thus if $\bar{\varepsilon}_0$ is the mean deformation in the minimal section it is found that

$$\varepsilon_{zz} / \bar{\varepsilon}_0 = \exp\left[-12\alpha(z / a)^2\right] \tag{2.112}$$

The graphs of Fig. 2.43, for an aluminum alloy for which the work-hardening exponent is about *0.10*, compares values calculated from these expressions with those found by the method of finite elements; it will be seen that there is little difference between the two.

(b) *Bending test.* Notched test pieces are easy to make and to use, and therefore are much used in practice. The Charpy test piece in particular is the subject of an international standard: it is of square section, *10 x 10 mm*, the notch is V-shaped with angle *45°*, depth *2 mm* and radius of curvature at the root of *0.25 mm.* We now give an approximate treatment of this under static loading, although Charpy specimens are more usually subjected to impact testing.

The limit load for a test piece with a deep notch can be determined by imagining a kinematically admissible stress field consisting of two plastic hinges around two circular arcs of angle 2α (Fig. 2.44). If k is the yield strength in shear, an upper bound for the limit moment is found to be

$$M_L = 0.69kB(w\text{-}a)^2 \tag{2.113}$$

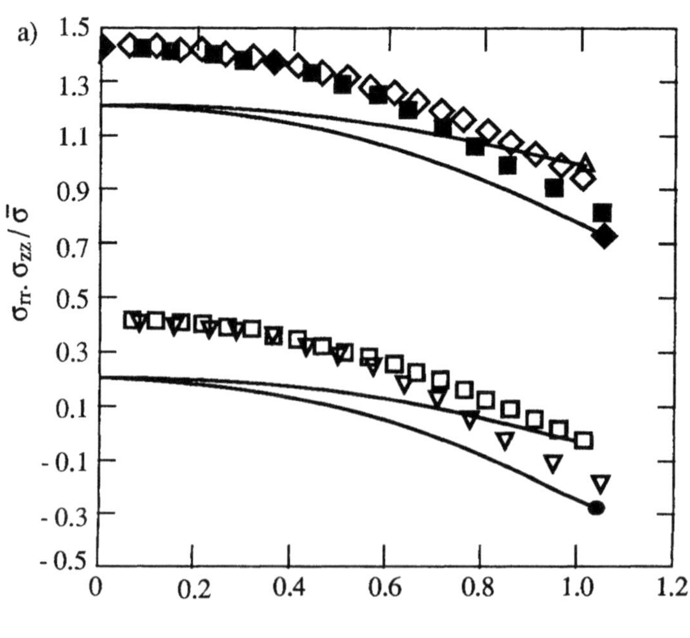

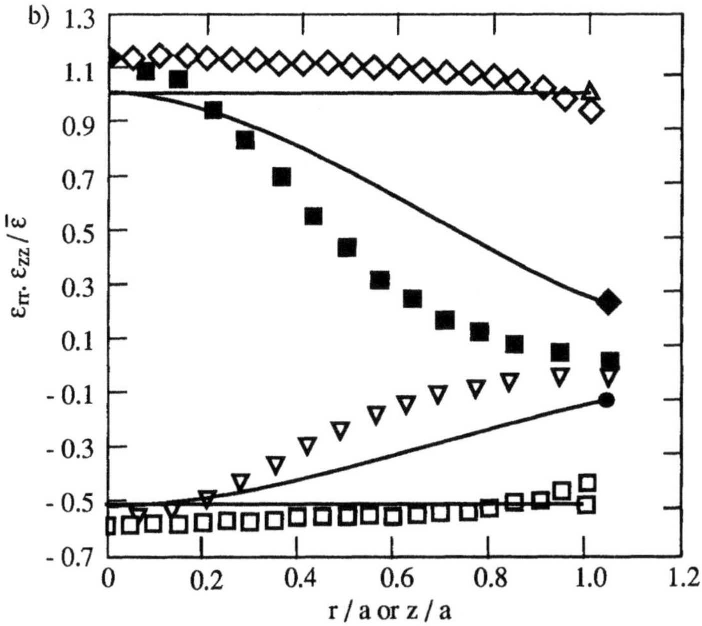

Fig. 2.43 Comparison of distributions calculated from Bridgman's formulae with those from finite-element calculations. Plain curves (Bridgman); Symbols (F.E. calculations). (a) stresses; (b) strains.

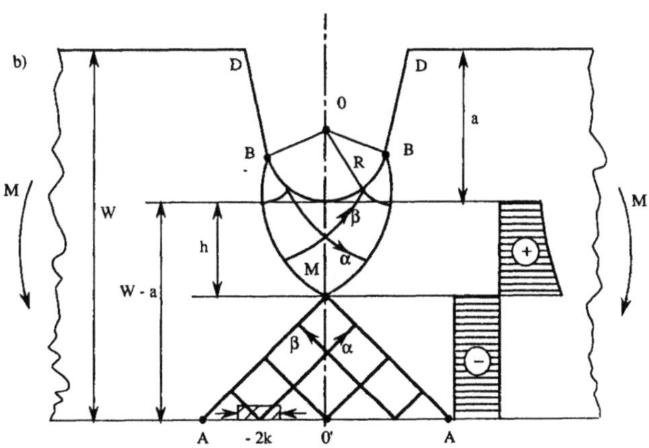

Fig. 2.44 Slip lines in a bend test piece with a deep notch; two suggested solutions, (a) and (b).

For comparison, the limiting moment for a beam of height *w-a* is

$$M_{LO} = \tfrac{1}{2} kB(w-a)^2 \tag{2.114}$$

With this geometry the value of the plastic constraint factor M_L/M_{LO} is *1.38*. A more complex field of slip lines can be constructed, as in Fig. 2.44b, which leads, by more complex computation, to a value *1.28*.

Knowledge of the slip line field in the region of the root of the notch enables the stress distribution to be calculated there, taking for granted of course the usual assumptions of plane deformation and perfect plasticity. This consists of logarithmic spirals, forming an angle of *45°* with the isostatic lines normal to the surface of the notch. The result is that on the axis of symmetry

$$\sigma_{\theta\theta} = 2k[1 + \log(1+x/R)] \tag{2.115}$$

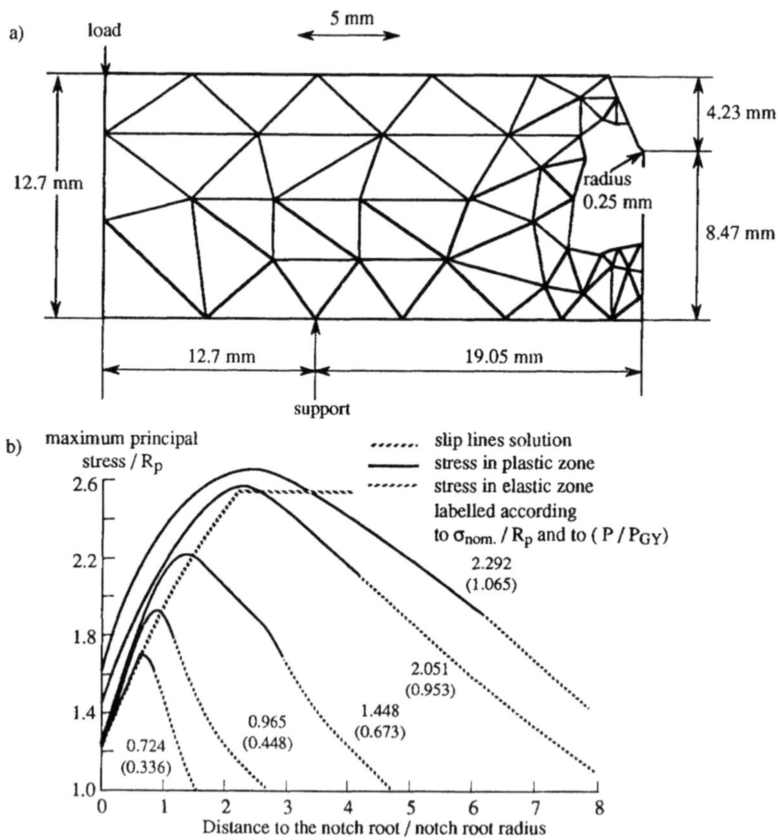

Fig. 2.45 (b) Variation of the maximum principal stress with distance from the base of the notch; full curves are for finite-element calculations, others are based on slip lines. (a) gives the FE mesh used.

This stress reaches a maximum at a distance x_t which depends on θ, the angle of opening of the notch:

$$\sigma_{\theta\theta(max)} = 2k(1 + \pi/2 - \theta/2) \quad (2.116)$$

Figure 2.45, for a mild steel, compares results calculated in this way with those of finite-element calculations (Griffiths and Owen (1971).

2.9.3.3 *Effect on notches on the brittle-ductile transition.*

Below the nil-ductility temperature, at which temperature the yield strength is greater than the cleavage stress, a notched test piece will break when the former is reached (see Fig. 2.46). Fracture starts at the root of the notch, where the stress is maximum before any plastic deformation occurs; since $\sigma_{max} = K_T\sigma_{nom}$ for a notched cylindrical test piece the load at fracture is given by

$$P = (\pi a^2 / K_T)R_p \quad (2.117)$$

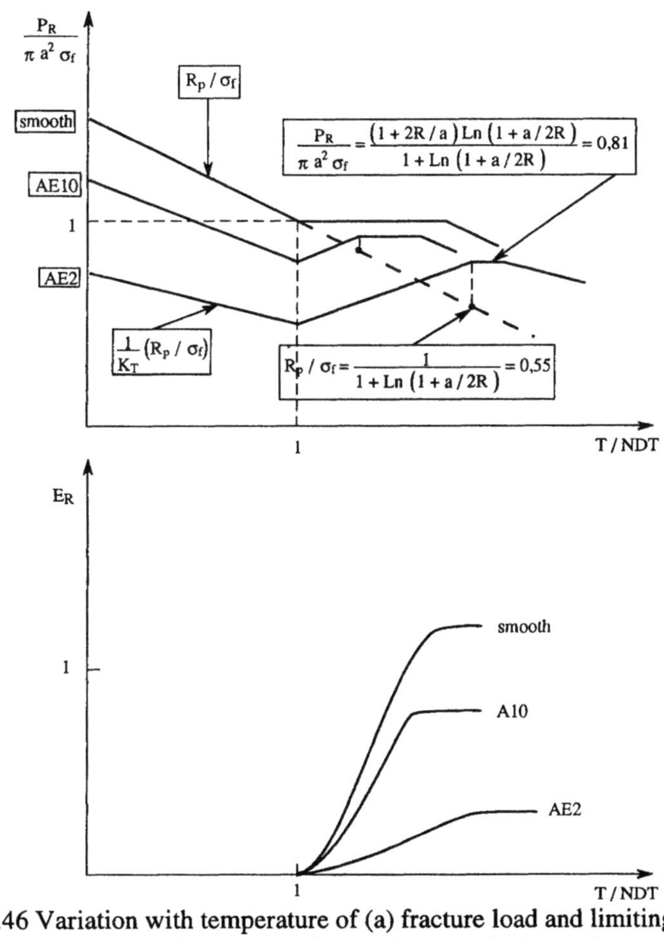

Fig. 2.46 Variation with temperature of (a) fracture load and limiting load (b) extension at fracture for three axisymmetrical test pieces .

Above this temperature, fracture occurs when the maximum stress, at the front of the plastic zone, reaches the cleavage stress; since σ_{max} is then greater than the elastic limit, the fracture load increases in proportion to the reduction in R_p caused by the increase in temperature. This continues until plastification is complete, when the stress is maximum on the axis of the test piece where it has the value (cf. 2.106)

$$\sigma_{zz(max)} = R_p[1+\log(1+a/2R)] = \sigma_f \tag{2.118}$$

and the fracture load is (cf. 2.107)

$$P_R = P_L = \pi a^2 R_p(1+2R/a)\log(1+a/2R) \tag{2.119}$$

Above the temperature corresponding to these conditions, the maximum stress, always on the axis, continues to increase as a result of work-hardening.

The fracture load will remain constant since we have assumed σ_f to be independent of temperature.

However, the deformation of the test piece at the minimum section continues to increase and cleavage fracture is replaced by ductile fracture with dimples when the deformation on the axis, where the stress triaxiality is highest, reaches the corresponding fracture value.

Cavities first start to form and grow in the center of the test piece. After the brittle-ductile transition temperature the variation of the fracture load parallels that of the limit load, the difference between the two depending on the elongation at fracture.

This difference decreases with increasing stress triaxiality, which itself is a function of the geometry (cf. 2.108)

Fig. 2.46 is a schematic representation of the variation of fracture loads with temperature, derived from the above considerations. It shows especially that if the nil-ductility temperature is constant then the transition, especially that of fracture appearance, increases with the sharpness of the notch. It can happen that the elongation at fracture is so great that the load corresponding to plastic instability is reached before fracture.

Fig. 2.47 concerns the Charpy V test piece, to which similar considerations can be applied; the limiting load for this is such that $P_L = 38.8\,k$, or $P_L = 22.4R_P$ (in N and MPa) and the maximum stress is $\sigma_{\theta\theta} = 4.36k$, or $\sigma_{\theta\theta} = 2.52R_P$. Here k is the yield strength in shear.

With such notched test pieces, especially if axisymmetrical, cleavage stress and elongation at fracture can be determined experimentally as functions of stress triaxiality. The Weibull stress can also be determined, and hence the dispersion estimated.

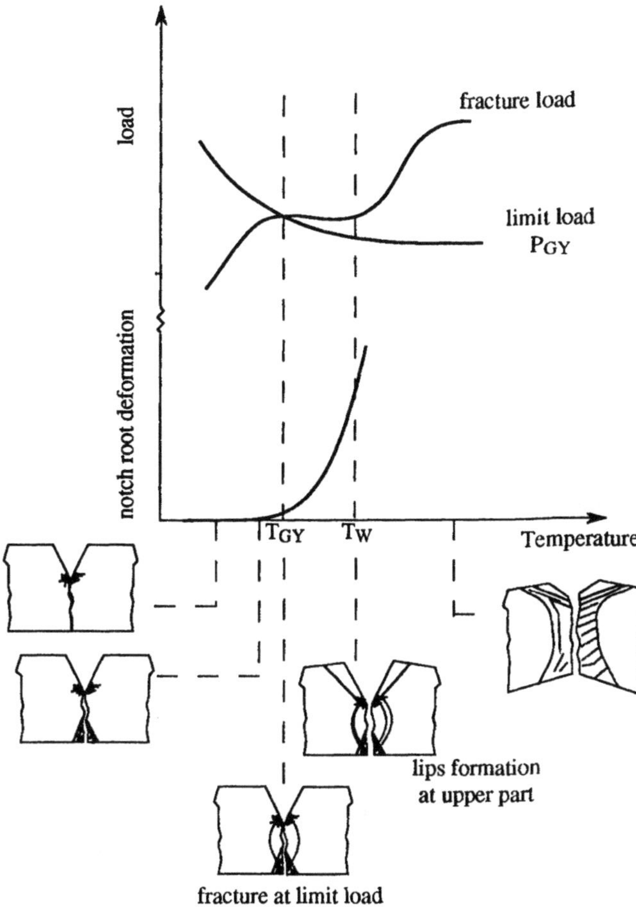

Fig. 2.47 Variation with temperature of limit load, fracture load and elongation at the base of the notch, for a Charpy test piece.

2.9.4 Impact testing of notched bars

The above discussion will have shown why fracture toughness is measured by subjecting a notched piece to impact testing: the high rate of deformation and the effect of the notch combine to raise the brittle-ductile transition temperature. This acts as a factor of safety in the choice of a steel that will have to withstand a certain minimum temperature in working conditions; the choice will be made so that the transition temperature is below the service temperature.

It is clear that this transition temperature can be defined in different ways, according as it is intended to refer to a level of ductility or an aspect of fracture. In practice, what is measured is the energy absorbed by the fracture; the transition temperature can then be taken at a given toughness. Alternatively the proportion of crystalline fracture (corresponding to cleavage) is found by examination and a *fracture appearance transition temperature* (FATT) is defined.

These transition temperatures depend on the sharpness of the notch and the size of the test piece, being higher the more confined the plastic zone. This leads to the need to use test pieces of the same thickness as the product that is to be assessed, which in turn requires test machinery that is more bulky than the simple Charpy pendulum. In the *Pellini test* a vertical pendulum is used to break a plate in which there is a sharp notch within a brittle zone created by a weld bead. According to the temperature it may or may not break. The transition temperature thus determined is called the *Nil Ductility Temperature (NDT).*

2.9.5 Effect of metallurgical factors

(a) The *brittle-ductile transition* is associated with a change in the physical process of fracture; it does not exist in FCC materials, for which there is no cleavage. On the other hand, it does occur in BCC and CPH metals.

(b) *Grain size* is the most important metallurgical parameter. Reducing this raises the yield strength (the Petch relation, see I-§3.101)) and the cleavage stress even more (Fig. 2.17), resulting in a lowering of the transition temperature (Fig. 2.48).

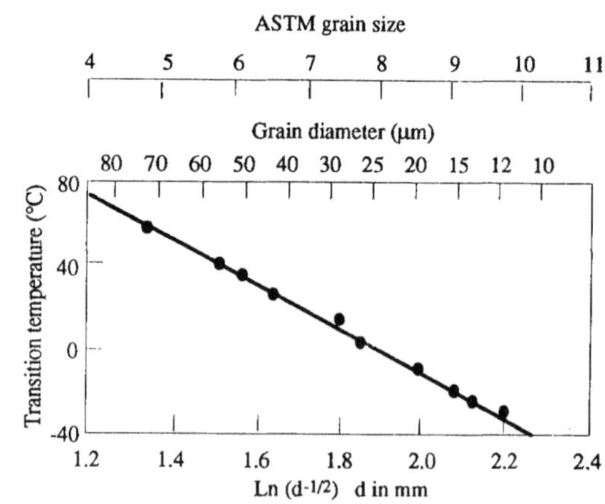

Fig. 2.48 Effect of grain size on the brittle-ductile transition temperature for a *0.11%* carbon steel

(c) *Gamma-forming elements*, by lowering the temperature of the austenite transition, favour the formation of small grains and consequently lower the brittle-ductile transition temperature. An example is *nickel* (Fig. 2.49), which is added to steels intended for use at low temperatures; *manganese* has a similar effect, and is less costly. In contrast, *alpha-forming elements* such as chromium and molybdenum raise this transition temperature.

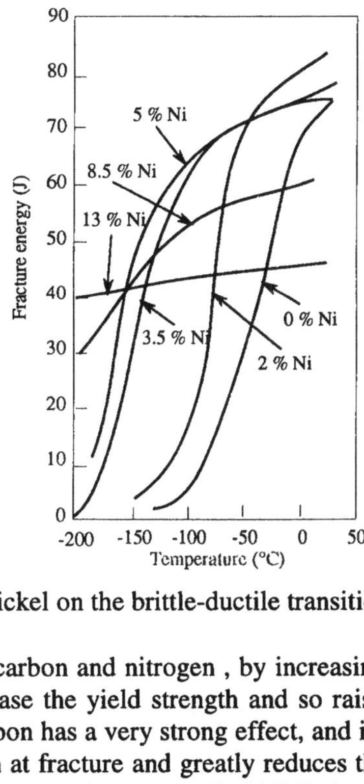

Fig. 2.49 Effect of nickel on the brittle-ductile transition temperature for steels.

(d) *Interstitials* such as carbon and nitrogen , by increasing the value of the σ_i term in the Petch relation, increase the yield strength and so raise the transition temperature. Fig. 2.50 shows that carbon has a very strong effect, and it will be seen that in addition it reduces the elongation at fracture and greatly reduces the level of the upper shelf. If the steel has been heat-treated at around *300°C* the C and N atoms can form *atmospheres* which anchor the dislocations firmly , as we saw in the first volume (I-§3.3.8.4.) This effect, known as *blue embrittlement* because of the color of the oxide FeO which forms at this temperature, makes the steel very brittle (see §2.2.9).

(e) Grain size is dependent on the temperature hot rolling: the lower the temperature at the end of this operation, the smaller will be the grain size and the lower the transition temperature (Fig. 2.51)

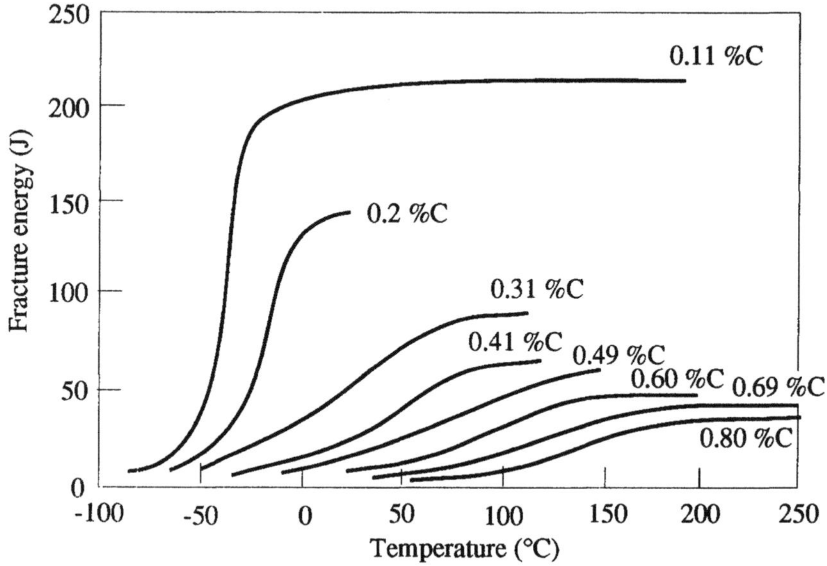

Fig. 2.50 Effect of carbon on the brittle-ductile transition temperature for steels.

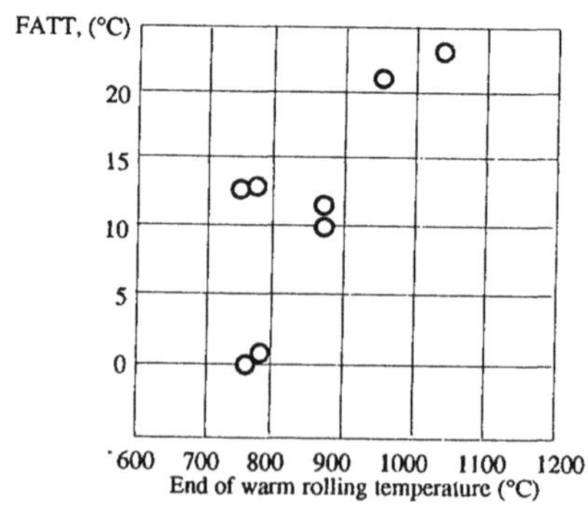

Fig. 2.51 Variation of the Fracture Appearance Transition Temperature (FATT) with final temperature of hot rolling, for a steel.

(f) In general, every factor that tends to distribute slips more uniformly by reducing local deformation heterogeneities will be beneficial. Thus a high stacking-fault energy, which will favour cross-slips, will be valuable; increasing the degree of long-range order will decrease the waviness of slips and make cleavage easier; hardening particles, which cause dislocations to proliferate greatly, will be beneficial.

2.10 FRACTURE OF POLYMERS

Fracture of amorphous polymers very often results from *crazing* (Fig. 2.52). This is the opening of cracks, the edges of which are bridged by fine ligaments, polymeric chains drawn out of the matrix; they are approximately aligned, as in a strongly-stretched polymer. If the edges of a craze separate more widely, the length of the ligaments increases, not by creep but by more material being drawn from the matrix; finally they give way and a true crack is formed. Thus crack propagation is preceded by crazing.

Consider a crack and its associated craze moving at a speed V_c; let τ be the length of time a molecule of a ligament spends in the craze. If S is the length of the craze the longest such time is that for molecules which enter the craze at its front (see Fig. 2.52a), $\tau_0 = S/V_c$, during which time it is subjected to a stress σ. The process is thermally activated and it has been established that it is associated with the β-relaxation of the polymer: the two phenomena have in fact the same activation energy. At the temperature T_β of this β-relaxation peak there is a transition between cracking accompanied by low-temperature multiple crazing, and cracking with single crazing at higher temperature.

There is also a change in the type of fracture, rough at low temperatures and smooth above T_β, and a sudden increase in breaking energy when the temperature falls below this; further, the stiffness of the ligaments is found to change.

The interpretation of these various observations is that the disappearance of the β-relaxation above T_β suppresses the fracture of the ligaments by creep, giving a more disordered fracture aspect. Fracture often takes place at the interface between the matrix and the craze.

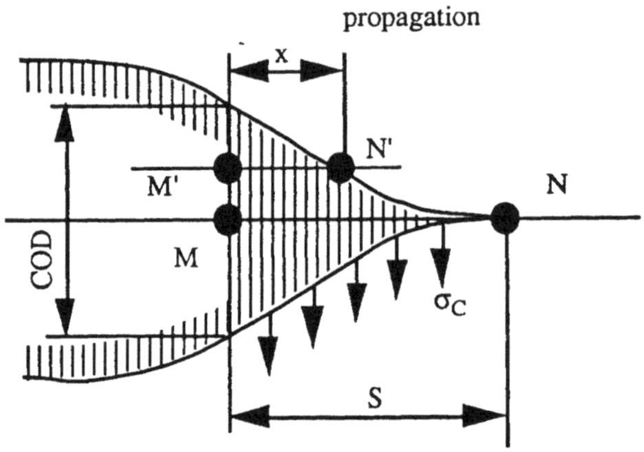

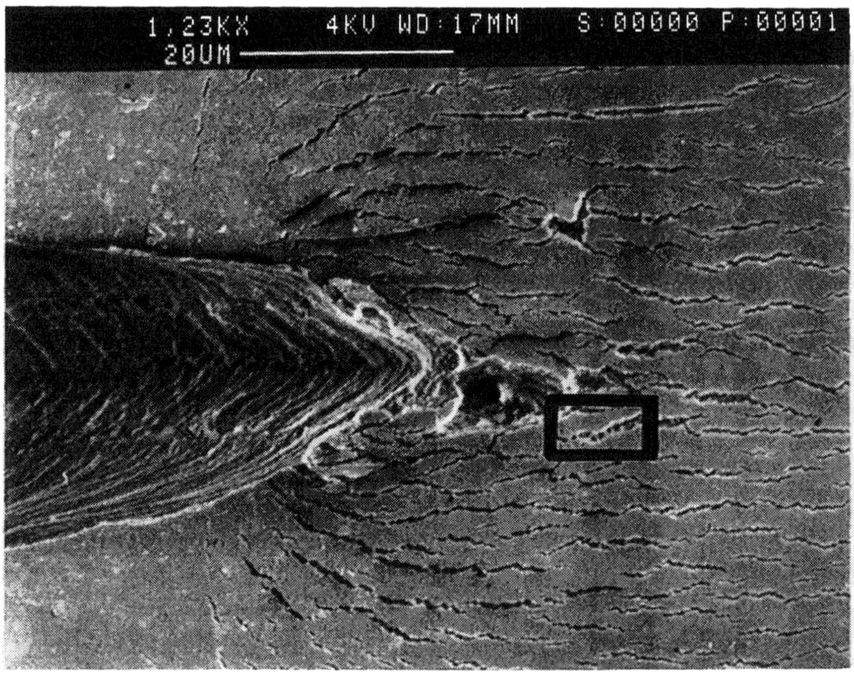

Fig. 2.52 (a) Definition of the "age" τ of a molecule in a ligament of a craze which is being propagated with speed V_c. Molecule M' enters the craze at N' and supports the stress for a time $\tau = X/V_c$. The greatest age is that of molecules arriving by the medium N for which $X = S$. (b) Scanning electron micrograph showing the end of a crack surrounded by crazes.

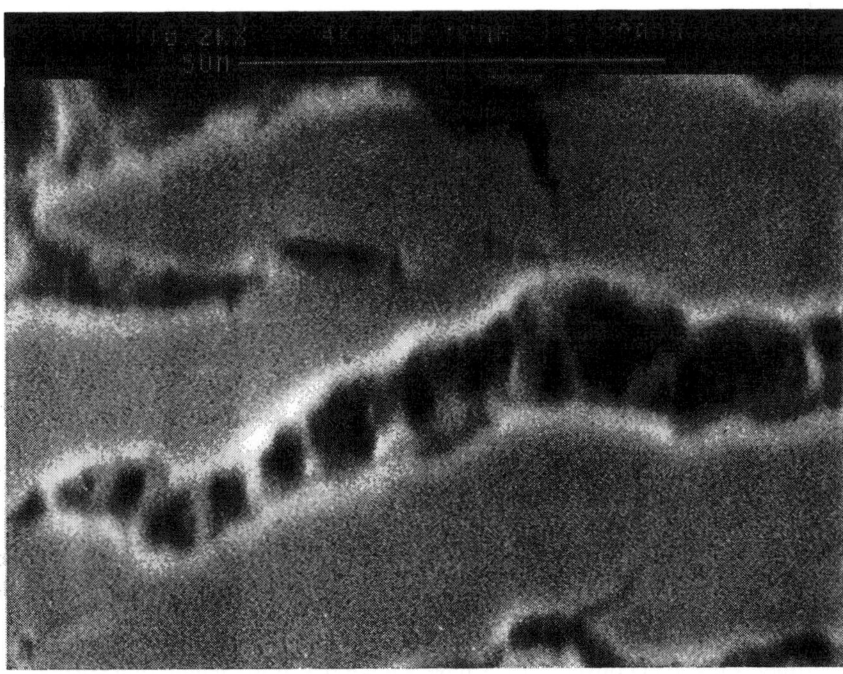

Fig. 2.52 (c) View at higher magnification, showing a craze with bridging ligaments.

The microscopic creep of the ligaments can be described by a classical law of the type

$$\dot{\varepsilon} = \dot{\varepsilon}_0 \exp\left(\frac{\Delta H - \sigma v_a}{kT}\right) \tag{2.120}$$

where ΔH and v_a are the activation enthalpy and volume respectively.

The lifetime τ_0 and propagation speed V_c can be written

$$\tau_0 = \tau_{0i} \exp(-\sigma v_t/kT), \qquad V_c = V_{ci} \exp(\sigma v_c/kT) \tag{2.121-2}$$

where v_t, v_c are the activation volumes corresponding respectively to growth and fracture of ligaments. The length of the craze is thus

$$S = V_c\tau_0 = V_{ci}\tau_{0i} \exp[(\sigma/kT)(v_c - v_t)] \tag{2.123}$$

Eliminating σ we get

$$\log\tau_0 = A \log V_c + B \tag{2.124}$$

where

$$A = -v_t/v_c = constant$$

$$B = v_t/v_c \log V_{ci} + \log\tau_{0i} = constant \text{ if the temperature is constant}$$

For the particular case of equal activation volumes

$$\log \tau_0 = -\log V_c + \log (V_{ci}\tau_{0i}) \tag{2.125}$$

and the length of the craze is constant, whatever the speed of propagation, or whatever the stress.

If the activation volumes are different the length of the craze increases or decreases with increasing propagation speed (or stress) according as the ratio v_t/v_c is less than or greater than *1*.

Experiments made at constant stress have shown that if the length of a craze is constant independent of temperature then the activation enthalpies are the same for growth and fracture of the crazes.

For many polymers it is found that the activation volumes for growth and for fracture of the ligaments are equal; this expresses the fact that the same molecular processes are responsible for both. An outstanding exception is polyvinyl chloride (PVC), for which the growth of crazes often exhibits a periodic behaviour, due to adiabatic heating.

Finally, crazing is strongly dependent on the environment, which acts as a plastifier; this accelerates the growth of the crazes both by drawing more ligaments from the matrix and by increasing the creep rate.

It is known that polymers can be reinforced by incorporating particles of rubber. If these are large they will act as sites for the development of crazes, but small particles, if numerous enough, will greatly affect the progress of cracking and will give rise to branched cracks.

2.11.1 General

Deterioration by fatigue occurs as a result of a material being subjected to cyclic stressing; a structure can fail by fatigue after a certain number of cycles even if the amplitude is well below the yield strength. In Fig. 2.53 the lifetime in number of cycles to failure is plotted as abscissa against the stress amplitude as ordinate, for a particular steel; this is called the Wöhler curve, after the engineer who studied the effect in the 19th century; it is also called the *S-N* (Stress-Number of cycles) curve. The curve will often, but not always, have a horizontal asymptote, called the *endurance limit*, below which the material should never fail in fatigue.

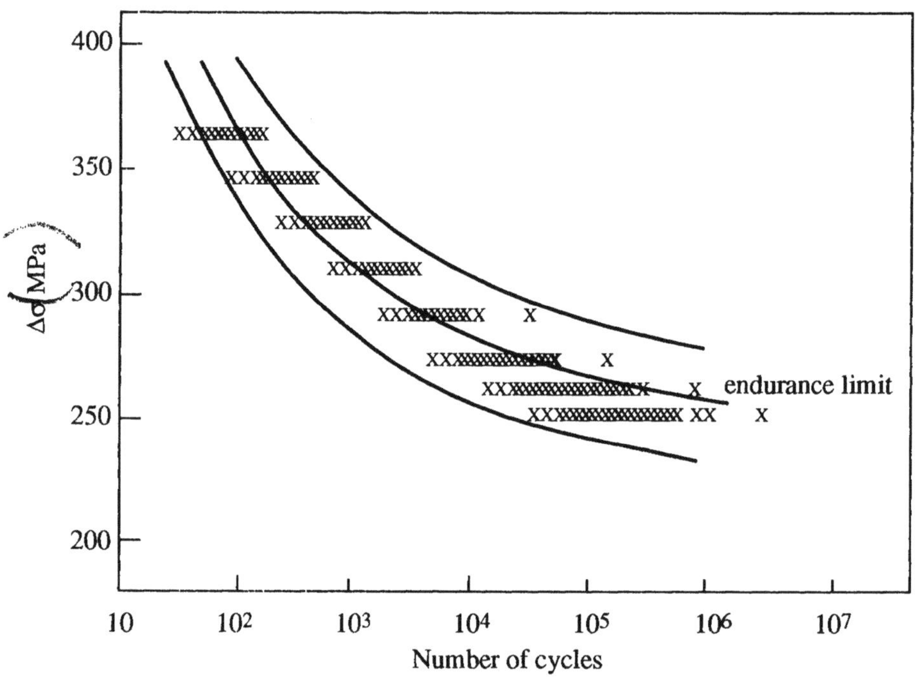

Fig 2.53 Lifetime of an XC10 steel in number of cycles to failure (N) for given amplitude of cyclic stress ($\Delta\sigma$). The three curves correspond to *10%*, *50%* and *90%* probabilities of failure. (Wôhler curve)

The results of fatigue tests will always be distributed statistically, consequently different *S-N* curves can be constructed, corresponding to different probabilities of failure. This variability seems to be an essential feature of all materials, impossible to remove however great the precautions taken in conducting the experiments. The distribution of lifetime for given stress amplitude is not normal, at high stresses being log normal; but for a given number of cycles to failure the amplitude is distributed normally.

The stress cycle is characterised by the amplitude $\Delta\sigma$, the maximum, minimum and mean stresses σ_{max}, σ_{min} and σ_{mean} respectively and the ratio $R = \sigma_{max}/\sigma_{min}$. Clearly these are not all independent and in fact $\sigma_{mean} = \frac{1}{2}(\sigma_{max} + \sigma_{min})$ and

$$\Delta\sigma = \sigma_{max} - \sigma_{min} = \sigma_{max}(1 - R) \tag{2.126}$$

Industrial tests are often performed with a rotating-bending machine, for which $R = -1$ and in which case the stress is said to be *alternating*. In another common test R is set to 0, or close to this, that is, $\sigma_{min} \cong 0$; in this case the stress is repeated. In practice of course the cycling is seldom perfectly regular; most often there will be random variations, which can have an effect on the behaviour of the material.

As a general rule, the *frequency* of the cycling does not affect the lifetime; but this will not hold in the presence of other influences whose damaging effects are a function of time, such as corrosion fatigue or creep fatigue, or when the frequency is so high as to generate over-heating. This last possibility must be borne in mind particularly in the case of polymers.

The effect of the mean stress on the endurance limit is often represented by *Goodman's diagram*, (Fig. 2.54) (Goodman (1914)), in which σ_{max} and σ_{min} are plotted as ordinate against the mean stress σ_m as abscissa; the non-damaging cycles will all lie in the area bounded by the two curves, which meet at the point corresponding to the fracture strength. In Goodman's diagram these curves are straight lines; other representations have been proposed, for example Gerber's parabola.

With multiaxial stresses the von Mises equivalent stress $\Delta\Sigma_{eq}$ is often taken as the parameter. It is immediately clear that the fatigue criterion cannot depend on this stress alone: the Goodman diagram shows that the mean hydrostatic stress Σ_m must also have an influence. Sines has proposed a criterion of the form

$$\Delta\Sigma_{eq} = A + B\Sigma_m \tag{2.127}$$

whilst Crossland introduces the maximum hydrostatic stress

$$\Delta\Sigma_{eq} = A + B\Sigma_{m(max)} \tag{2.128}$$

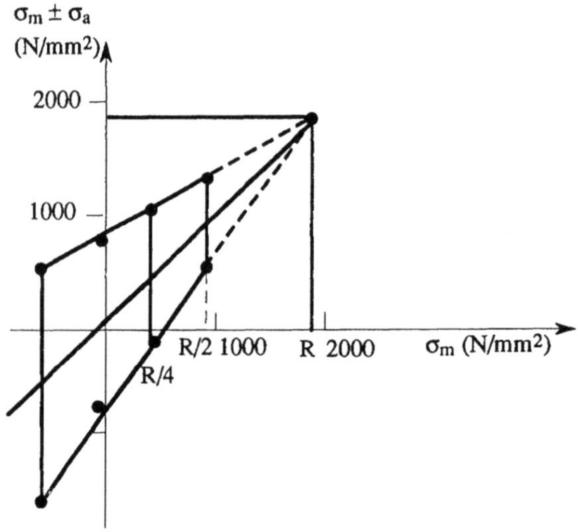

Fig. 2.54 Goodman's diagram for the steel 35 NCD 16, giving the variation of endurance limit with mean stress σ_m; σ_a is the semi-amplitude.

In certain cases, rolling fatigue for example, the variation of stress at a given point in the material can follow a complex path in stress space. According to Dang Van (1973) the endurance limit will not be exceeded provided that this path remains within a conical region defined by two lifetime tests, such as alternating rotating-bending and torsion (Fig. 2.55).

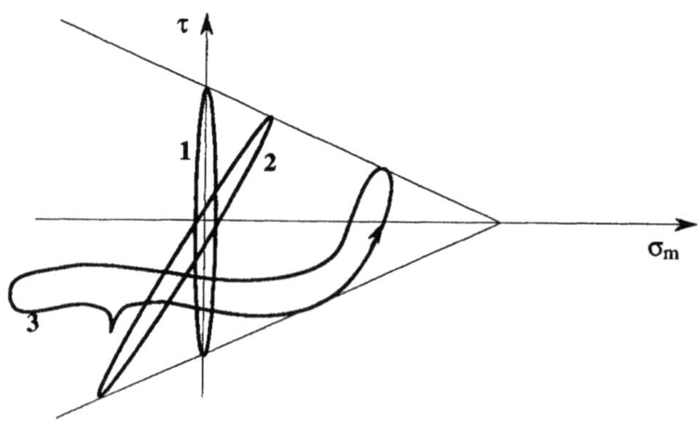

Fig. 2.55 Dang Van's criterion for endurance under complex load trajectories.

Table 2.5 gives the endurance limit for several materials, showing that for steels it is of the order of $R_m/2$.

So far we have been giving a macroscopic treatment of fatigue damage; this is still very much used, although it is empirical and provides little information concerning the influence of any parameters, metallurgical in particular. We now look in more detail into the way damage can arise, distinguishing between the initiation of fatigue cracks and their propagation.

Alloy	*Yield strength* R_p *MPa*	*Fracture strength* R_m *MPa*	*Endurance limit* σ_D *MPa*	σ_D/R_m
mild steel	*260*	*395*	*232*	*0.59*
semi-hard steel	*360*	*590*	*270*	*0.46*
hard steel	*580*	*735*	*410*	*0.56*
GS cast iron		*600*	*230*	*0.38*
10NC6 steel	*900*	*925*	*450*	*0.49*
35NC6 steel	*900*	*1000*	*530*	*0.59*
35CD4 steel	*1430*	*1640*	*660*	*0.40*
brass		*460*	*130*	*0.28*
2024 aluminum	*275*	*395*	*125*	*0.31*

Table 2.5 Endurance limits and other properties for various alloys.

2.11.2 Initiation of fatigue cracks

Observations with test pieces subjected to fatigue show that damage appears essentially on the surface. Even when the amplitude of the stress cycling is below the ordinary yield strength, sources of dislocations can be activated in grains that are most favourably oriented for slip. Those grains in the interior will be rapidly work-hardened and will not be deformed, but slip lines in surface grains will reach the surface, partly or completely eliminating the dislocations and greatly reducing the work-hardening. There will however continue to be some work-hardening, and these grains will be gradually invaded by the increasingly-dense tangles of dislocations.

When a certain level of work-hardening is reached, more-or-less free bands of dislocations begin to appear, forming channels between the cells along which dislocations can slip easily: these are the so-called ***Persistent Slip Bands, PSB***. The dislocations that travel along these will be screw dislocations, whilst on the sides debris will accumulate in the form of dipoles, packed into very dense tangles in a ladder-like structure as shown in Fig. 2.56.

In this process the cyclic deformation is concentrated in the bands, where the amplitude is of the order of 10^{-2} in contrast to the body of the material where it does not exceed 6.10^{-5}; but this gradually decreases as what is called secondary work-hardening develops, and seems to stabilise at around 10^{-3}. The imposed cyclic deformation has to be accommodated and for this, new persistent slip bands begin to form, ultimately invading the whole volume, which is then filled with cells whose size is inversely proportional to the amplitude of the stress. However, if the stacking fault energy is less than about $10\ mJ/m^2$ the dissociation of the dislocations will favor the formation of planar arrangements rather than of cells.

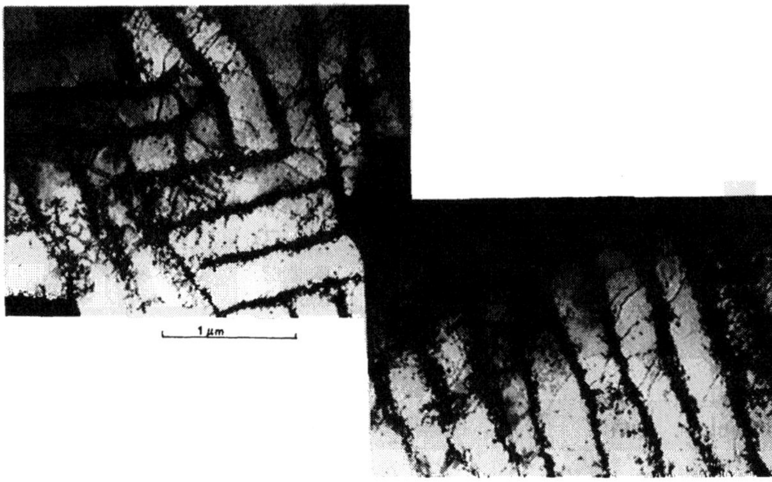

Fig. 2.56 Transmission electron micrograph of an austenitic stainless steel subjected to low-cycle fatigue ($\Delta\varepsilon = \pm\, 0.5\%$) at *20°C*, showing formation of walls and labyrinths.

Since the deformation is concentrated in the bands the surface does not remain plane: irreversible steps form in the bands, especially at their edges. These steps are the sites of stress concentrations, which can only make the situation worse; microscopic flakes, *extrusions,* form, with parallel grooves, *intrusions,* soon becoming micro-cracks (Fig. 2.8).

These are the beginnings of cracks; as the cycling continues they gradually penetrate the grains, following the slip bands, and thus become oriented at *45°* to the direction of maximum tensile stress. This is Stage I of fatigue cracking, and continues until a grain boundary is reached. When this occurs the orientation changes progressively, becoming perpendicular to the maximum principal stress; this is Stage II, which we shall study in §2.11.4.

It is important to be aware that Stage I depends on the stress state at the surface. Under alternating tension the cracks will penetrate the test piece along planes inclined at *45°* to the surface (Fig. 2.57); under alternating torsion they are perpendicular to the surface, remaining longer on the surface, and the transition to Stage II occurs later. This indicates that whilst the initiation is mainly a function of the maximum surface shear strain, it depends also on the stress normal to the plane on which this strain acts.

Thus the initiation of fatigue cracking, a surface phenomenon, is very sensitive to the nature of the surface of the material, to the residual stresses there and to the environment. Rough machining will leave ridges; these will be sites for stress concentration, which will favor the formation of persistent slip bands and thus greatly reduce the duration of the initiation. Resistance to fatigue can be increased by polishing or grinding; such treatment, applied after a certain number of cycles, can prolong the life considerably by suppressing the formation of surface steps. Residual compressive stresses will reduce the mean stress during fatigue and thus will have a beneficial effect on the duration of the initiation phase: this is why *shot-peening* can increase resistance to fatigue. In steels benefits can result similarly from surface quenching, which dilates the surface layers through the martensitic transformation and so generates compressive stresses. The environment, by reducing the reversibility of the slips - not to mention the corrosive activities which accelerate cracking - can affect the initiation time greatly; air or humidity will suffice for this, water or corrosive liquids will have a greater effect. The length of time that fatigue can be supported is greatly increased under vacuum.

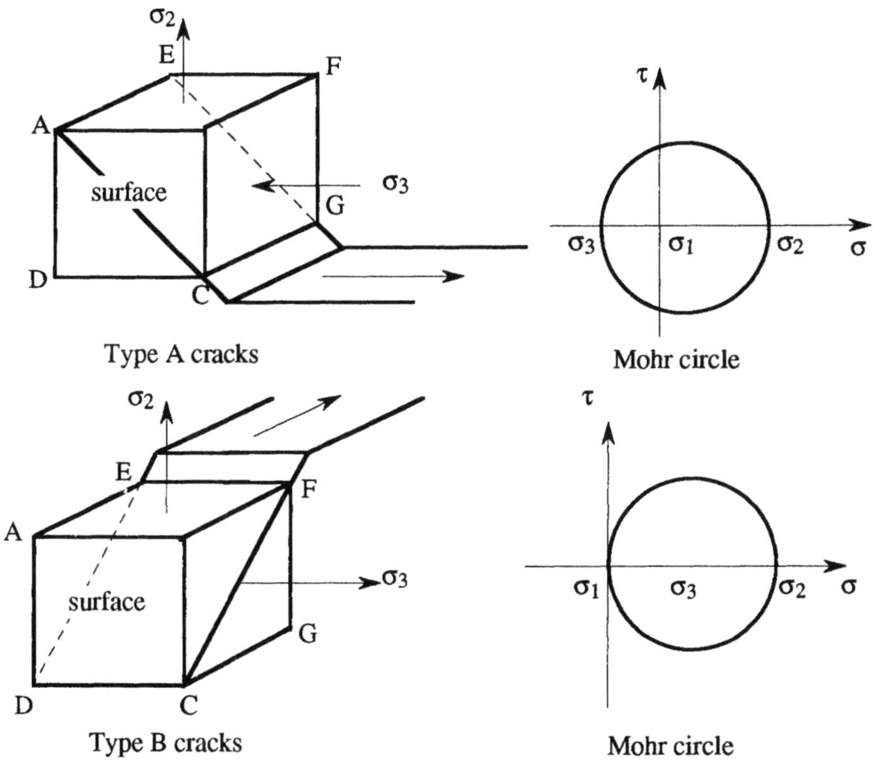

Fig. 2.57 Diagram showing how, according to the stress system, Stage I of fatigue cracking occurs either along facets of type A (case of torsion) or of type B (tension).

2.11.3 Nucleation and continuous growth of cracks; conservation equation for the number of cracks

Treating the development of cracks as nucleation followed by growth is something of a schematic view which does not entirely represent the reality. In any homogeneously-loaded volume element there will be a distribution of cracks whose density is smaller the smaller the load; in general the treatment will have to concern a population of defects - fatigue cracks, cavities due to ductile fracture (§2.8) or to creep (§2.12) - and the various stages will overlap one another. Thus we shall have to study the effect of a continuous nucleation of cracks, and whilst the modeling will certainly be more difficult it will also be richer. We shall show in particular that by following the evolution of a distribution of fatigue cracks we can establish a conservation relation which can prove very useful in determining the speed of propagation of cracks in a sample in which there is a large number of cracks.

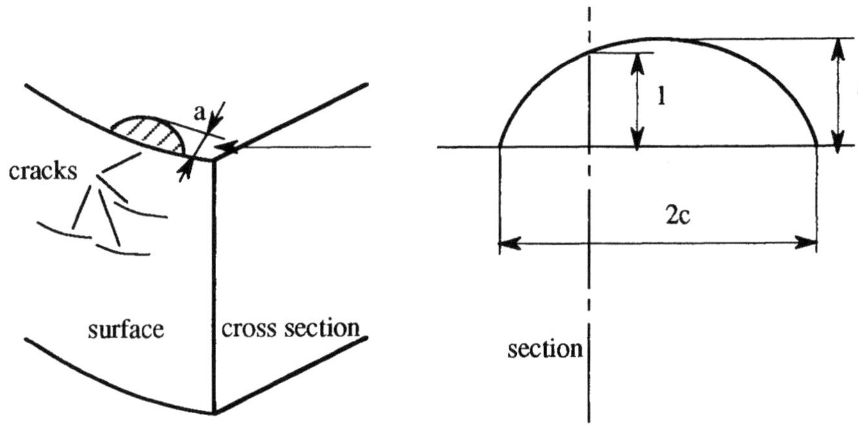

Fig. 2.58 Surface and section views of cracks.

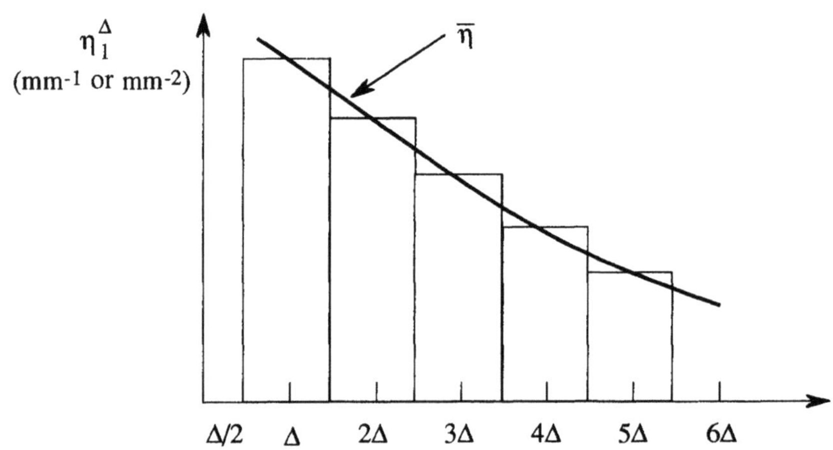

Fig. 2.59 Histogram showing (linear or surface) distribution of crack lengths

Surface measurements will give us the lengths c of the cracks, measurements made on metallographic sections enable us to find their mean depth $\bar{l}$. If we know the shape of a crack we can write $\bar{l} = \lambda c$, where, e.g. $\lambda = \pi / 4$ for a semi-circular crack (Fig. 2.58). We can then construct a histogram giving the distribution $\eta_{l\Delta}$ (either linear or surface) of cracks of depth between $\bar{l} - \Delta / 2$ and $\bar{l} + \Delta / 2$ where Δ is the class interval (Fig. 2.59): in practice this will be the resolution with which the measurements can be made with the method used, but could equally be that for the transition between initiation, that is, the first appearance of a crack, and its propagation.

The density distribution is defined as the ratio

$$\bar{\eta} = \eta_l^{\Delta}(N) / \Delta \tag{2.129}$$

and letting Δ tend to zero we have

$$\eta_l^{\Delta} = \int_{\bar{l}-\Delta/2}^{\bar{l}+\Delta/2} \bar{\eta}\, du \tag{2.130}$$

Fig. 2.59 shows the significance of η.

A starting assumption is that cracks do not coalesce, which implies that changes in the overall density η of the cracks are determined by the initiation stage alone. Support for this is all the firmer if the resolution of the method of measurement allows a "reasonable" value to be taken for $\Delta/2$ - for example, the approximate size of a grain as seen with an optical microscope or with a low-magnification scanning electron microscope.

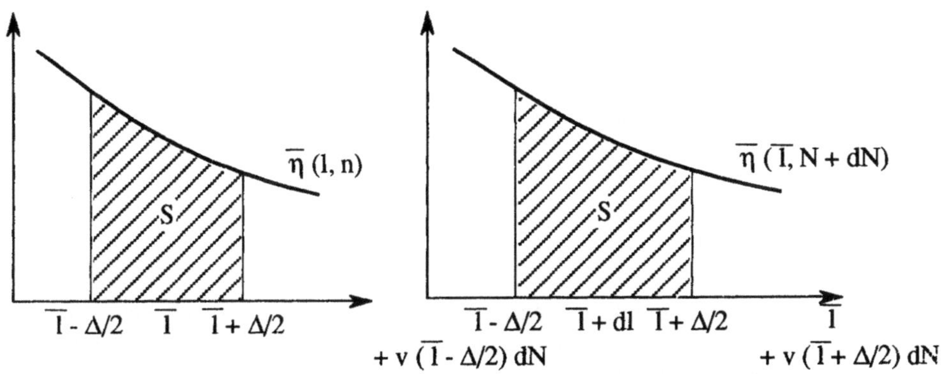

Fig. 2.60 Diagram showing on the density histogram the conservation of the number of cracks.

A second assumption is that the speed of propagation of a crack depends only on its length l and the applied load: that is, that we can neglect possible interactions between cracks themselves and between cracks and microstructural features, for example grain boundaries. This assumption can be removed, but the relations then become more complicated.

Given these assumptions we can derive a relation between the partial derivatives of $\bar{\eta}(\bar{l}, N)$ with respect to $\bar{l}$ and N on the one hand and the propagation speed v on the other;. we do this by drawing up a balance sheet for the number of cracks. (Fig. 2.60)

After N cycles there are $\eta_l^{\Delta}(N) \approx \Delta\bar{\eta}(\bar{l}, N)$ cracks with lengths between $\bar{l} - \Delta/2$ and $\bar{l} + \Delta/2$; after $N+dN$ cycles their lengths will be between $(\bar{l} - \Delta/2) + v_{\bar{l}-\Delta/2}dN$ and $(\bar{l} + \Delta/2) + v_{\bar{l}+\Delta/2}dN$. The conservation of the number of cracks is expressed by the condition that the areas S and S' of Fig. 2.60 are equal, and evaluation of these areas gives the relation

$$\partial\bar{\eta} / \partial N + \partial(\bar{n}v) / \partial\bar{l} = 0 \tag{2.131}$$

This relation becomes useful when the rate of nucleation of cracks $\partial\bar{\eta} / \partial N$ is known, for then the speed of propagation can be determined from the evolution of the distribution of lengths. An important special case is when a steady state has been established, expressed by

$$\partial\bar{\eta} / \partial N = 0 \tag{2.132}$$

from which it follows immediately that

$$\bar{\eta}v = k, \text{ constant independent of } \bar{l} \tag{2.133}$$

This holds reasonably closely for an austenitic stainless steel subjected to low cycle plastic fatigue ($\Delta\varepsilon_p/2 = 0.37\%$ at $600°C$), for which the histogram is given in Fig. 2.61. The value of k is deduced from the flux of cracks entering at $\Delta/2$ in this steady state:

$$k = v(\Delta/2)\bar{\eta}(\Delta/2, N) \tag{2.134}$$

If, as Fig. 2.61 suggests, cracks nucleate continuously during the steady state, the total density η increases linearly at a rate k, since $d\eta/dN = k$. It follows that, provided that the initial assumptions hold, the variation of propagation speed with length between the intervals $\lambda(\bar{l} - \Delta/2)$ and $\lambda(\bar{l} + \Delta/2)$ is

$$v_\lambda(\bar{l} - \Delta/2) - v_\lambda(\bar{l} + \Delta/2) = k\lambda / \bar{\eta} = k\lambda\Delta / \eta_l^{\Delta}.$$

2.11.4 Low cycle plastic fatigue

A test piece cycled beyond the elastic limit has a short lifetime - hence the term "low cycle fatigue». This effect calls for study because it can appear locally, in surface slip bands when cracks are being initiated and at the ends of cracks that are propagating, possibly together with the fatigue resulting from a large number of cycles.

It is usual to employ test pieces cycled in tension/compression ($R = -1$) in this study. Equipment for recording displacements, either longitudinal or diametral, enables the deformation to be monitored and, if required, the test to be controlled so that the amplitude of this is held constant. Results can be recorded in the form of stress/strain cycles, as in Fig. 2.62. If the machine is displacement-controlled so that the value of R is different from -1 it is found that as the test proceeds the mean stress tends to zero. If $R = -1$ the amplitude of the stress cycle does not remain constant

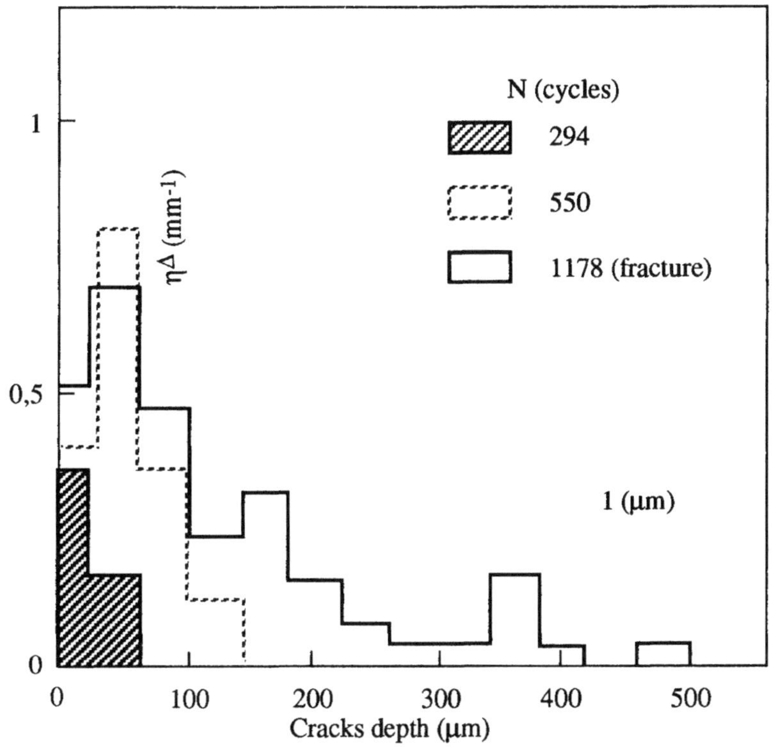

Fig. 2.61 Evolution of the distribution of lengths of secondary cracks as a function of the number of cycles. Notice that a steady state is established at the half-life stage.

but in general increases: this is *cyclic work-hardening*, clearly associated with the changes to structure of the dislocations which we have described previously (§2.11.2). In some cases, however, the amplitude decreases, showing *cyclic softening*; this occurs for example in metals which have previously been work-hardened, when the dislocation structure resulting from monotonic work-hardening is replaced by that characteristic of cyclic loading. Another example of cyclic softening is that of alloys that have been hardened by coherent precipitates, sheared by dislocations. The back- and forth crossing by dislocations can destroy the order or dissolve the precipitates, thus making deformation much easier.

After a certain number of cycles the hardening or softening tends to settle down to a value that depends on the amplitude of the plastic deformation (Fig. 2.63). Using, for example, Hollomon's representation (Fig. 2.64) we can define a *cyclic work-hardening coefficient*:

$$\Delta\sigma = \Delta\sigma_0(\Delta\varepsilon^p)^{n'} \qquad (2.135)$$

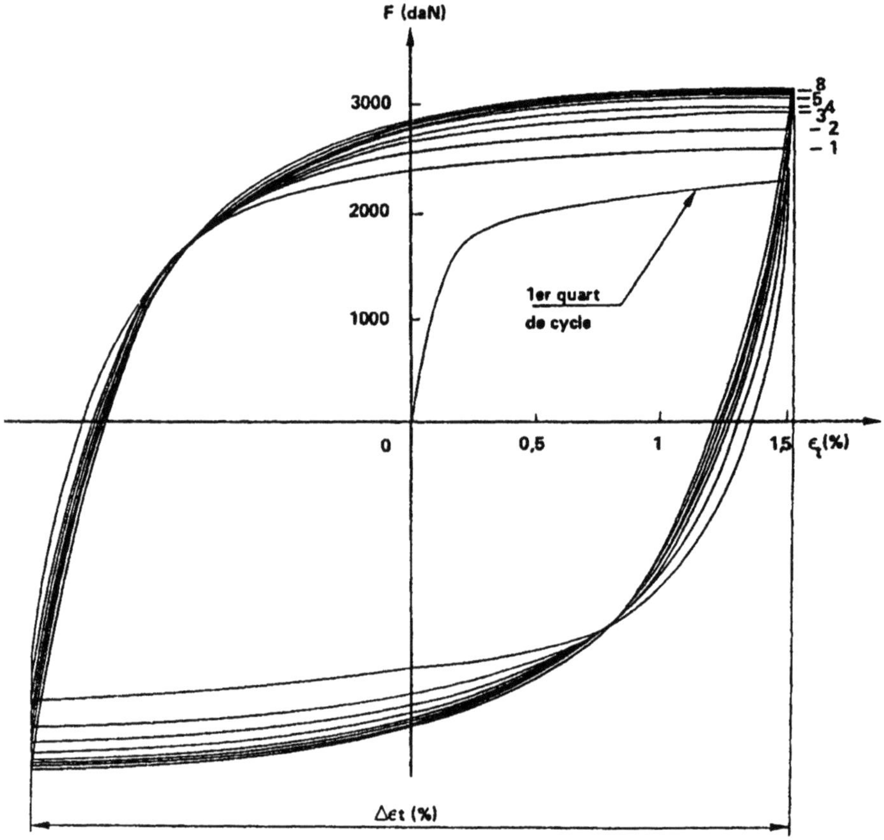

Fig. 2.62 Load/deformation cycles in low cycle fatigue for an austenitic stainless steel at ambient temperature.

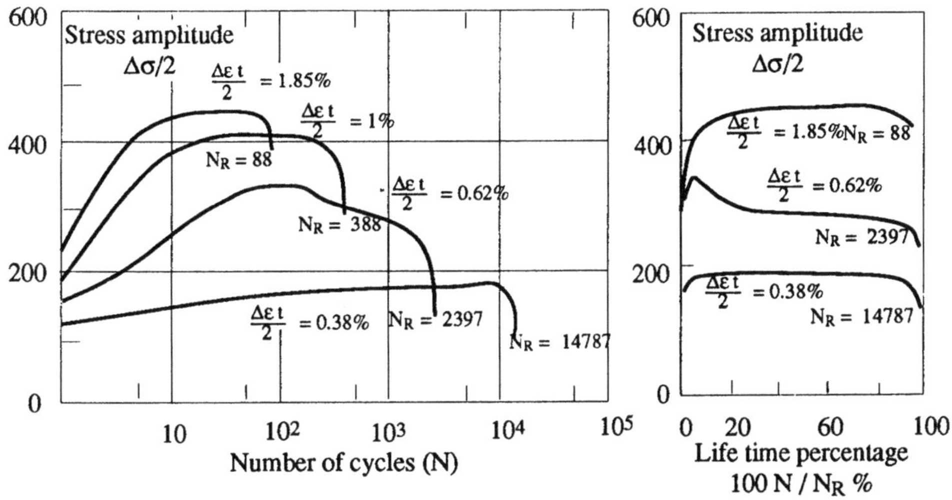

Fig. 2.63 Change in stress amplitude with number of cycles for the steel Z02 CND 17.12. (316). Logarithmic scale at left, linear scale at right.

After some number of cycles the stress amplitude is seen to decrease sharply (Fig. 2.63). This is due to the appearance of surface microcracks; one of these will become dominant, will propagate into the center of the piece and will finally cause fracture.

The lifetime under fatigue is a function of the amplitude of the deformation. This can be separated into an elastic part and a plastic part; if the amplitude is plotted logarithmically against the number of cycles to rupture two straight lines are obtained, as in Fig. 2.65, corresponding to elastic (Basquin's law) and plastic (Manson-Coffin's law) deformation respectively. The latter, of slope approximately *-0.5*, crosses the former, which covers a much greater number of cycles.

We thus have $\Delta\varepsilon = AN^{-a} + BN^{-b}$ (2.136)
in which b, the exponent for Basquin's law, is about *0.12* and a, for Manson-Coffin's law, usually does not differ much from *0.5*.

Consider now the limiting form of this relation when the number of cycles to fracture is very small. Suppose a monotonic tensile test leads to fracture at an extension D; we can regard this as a low cycle fatigue test lasting for *1/4* cycle (!). We can neglect the Basquin term in relation to the Manson-Coffin, so

$$D = A(1/4)^{-a} \tag{2.137}$$

and if a = 0.5 we have A = D/2.

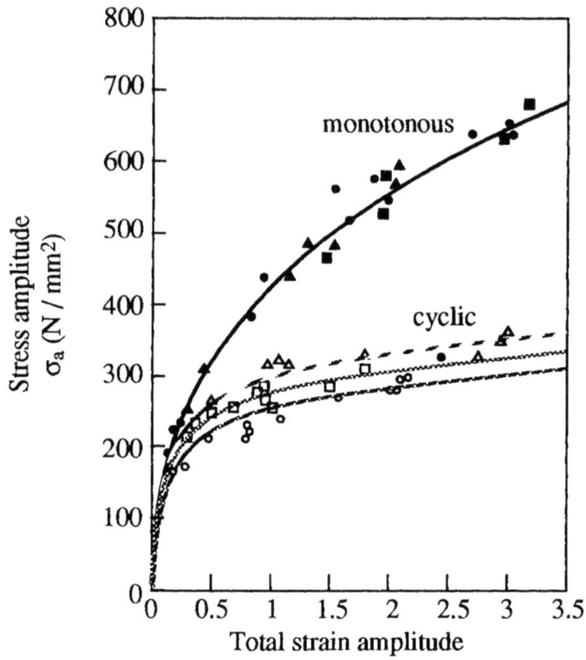

Fig. 2.64 Comparison between cyclic and monotonic work-hardening for the stainless steel Z02 CND 17.12 (316).

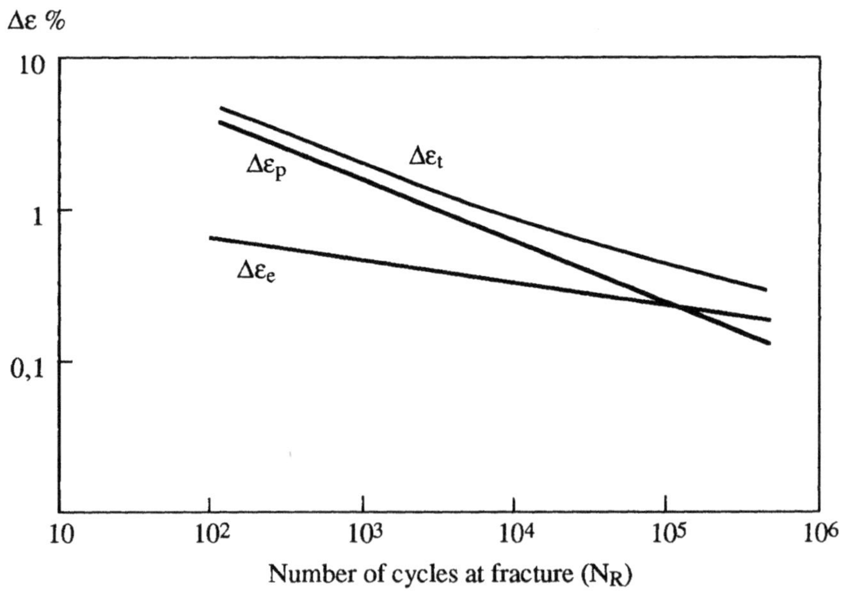

Fig. 2.65 Lifetime as a function of cyclic deformation amplitude (Manson-Coffin diagram) for the stainless steel Z02 CND 17.12 (316).

It is found that taking $D = \log[100/(100\text{-}Z)]$, where Z is the percentage necking; $a = 0.6$, $b = 0.12$; and $A = D^{0.6}/2$, $B = 3.5\ R_m/E$ where R_m is the ultimate tensile strength gives the best agreement with experiment (Fig. 2.66). Expressed as

$$\Delta\varepsilon = (D^{0.6}/2)N^{-0.6} + (R_m/E)N^{-0.12} \qquad (2.138)$$

this is known as the *universal slopes law*; it gives a first approximation to the properties of fatigue as deduced from those of traction, which latter are much easier to determine.

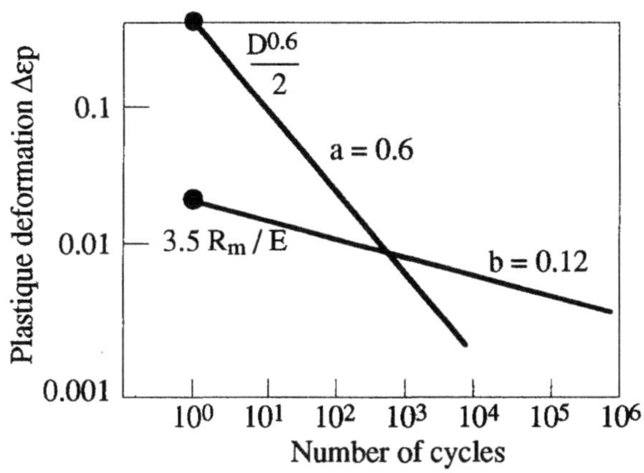

Fig. 2.66 Law of Universal Slopes. Basquin's law is based on the ultimate tensile strength R_m, Manson-Coffin's on necking $D = \log[100/(100\text{-}Z)]$.

The initiation of fatigue and the subsequent damage occur in *stress concentration zones*, such as notches, throats, holes etc. The results derived above enable a method for predicting effects to be developed. We imagine a microscopic-sized element cut from the critical zone and equate its lifetime to the initiation time in this zone, which can be calculated from equations 2.135-6 if the deformation amplitude and the Basquin and Manson-Coffin exponents are known. In the absence of very elaborate finite-element calculations, the amplitude can be found by Neuber's rule, which states that if the elastic limit is exceeded then $K_\sigma . K_\varepsilon = K_T^2$ where K_σ, K_ε and K_T are the concentration factors for the stress, the deformation and the elastic stress respectively. This requires K_T to be known, for which there are handbooks, for example Peterson (1974). Using equation 2.135 for cyclic work-hardening and assuming that the sample as a whole is subjected to a cyclic stress whose amplitude $\Delta\sigma_{nom}$ does not exceed the yield strength, we then have

$$\Delta\varepsilon\Delta\sigma/\Delta\varepsilon_{nom}\Delta\sigma_{nom} = \Delta\varepsilon\Delta\sigma_0(\Delta\varepsilon)^{n'}E/\Delta\sigma_{nom}^{\ 2} = K_T^2 \qquad (2.139)$$

from which, provided that we have determined the cyclic work-hardening law 2.135, we can calculate $\Delta\varepsilon$.

2.11.5 Propagation of fatigue cracks

At the end of Stage I one of the cracks that have been initiated will have become dominant and will propagate in a direction perpendicular to that of the maximum principal stress. During that part of the cycle in which the stress is increasing this crack will open and the stress concentration at its head will be so high that a plastic zone is created there. The plastic deformation will blunt the crack and cause it to open by an amount δ, and will be accompanied by a small advance of the tip, of the order of $\delta/2$ (Fig. 2.67).

In the following part in which the stress is decreasing, the crack will close again and the surrounding material, which will have remained elastic, will exert a pinching action on the plastified zone.

This zone, which has suffered an irreversible deformation, will now be compressed to the extent that it replastifies in compression. Thus at the head of the crack there will be a *cyclic plastic zone*, which will behave like a test piece under low cycle plastic fatigue. The local compressive stresses, which will remain after the load has been removed, will keep the crack closed and in the next cycle a positive stress will be needed to reopen it. However, since plastic deformation is not perfectly reversible, at each cycle the crack will propagate a little further; this will be shown by striations on the fracture surface, each corresponding to one cycle (Fig. 2.68). The overall appearance is of a silky texture

If the stress amplitude is large the opening of the crack can be accompanied by ductile tearings which increase the speed of propagation over that due to the simple mechanism of striation; striations and dimples, both characteristic of ductile fracture, are then seen on the broken surfaces.

At small amplitudes several cycles may be needed to form a single striation and this kind of structure may not be seen at all. As the amplitude decreases it may become too small to reopen a crack that has been closed by the force exerted by the plastic zone, and cracking will stop: thus there is a *propagation threshold.*

Fig. 2.69 shows how two additions to the closure effect of the forces due to the local plastification at the tip of the crack make reopening impossible below a certain applied load.

The first is related to the tortuous profile of the crack, which impedes complete reclosure in the absence of load because the two faces are not brought accurately into contact with each other. This is especially acute at low amplitudes of load, when cracking tends to follow the crystallographic slip planes; thus in metals with large grains, which lead to very irregular crack profiles, there will be high barriers to propagation.

The second effect is due to the formation of oxides on the surfaces of the cracks, which also impedes complete reclosure.

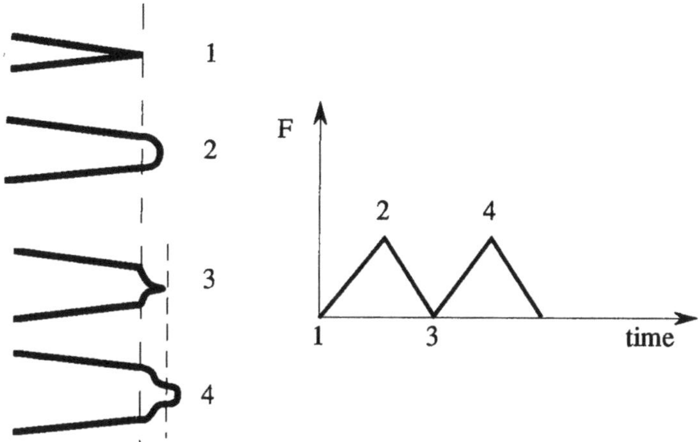

Fig. 2.67 Tip of a fatigue crack propagating under cyclic loading.

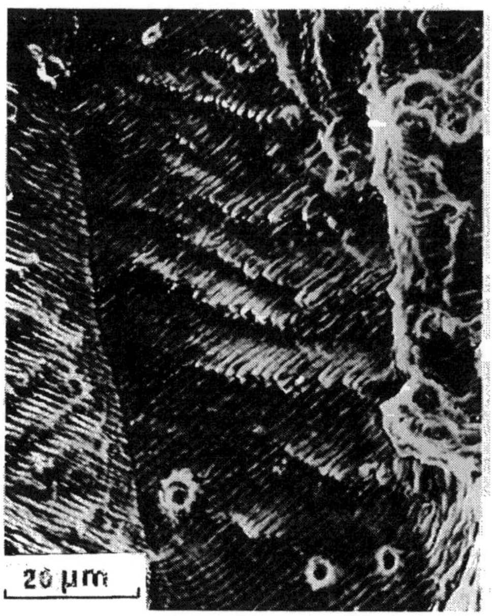

Fig. 2.68 Scanning electron micrograph showing fatigue striations in an austenitic stainless steel; note the change in orientation across the grain boundary.

The mechanics of fracture which we shall develop in the next chapter enables us to treat the propagation of fatigue cracks more quantitatively.

The theory shows that if the plastic zone at the crack tip is of very small extent then the stresses and deformations there are a function of a single parameter, the stress intensity factor K_I. This in turn is a function of the applied stress σ_{nom}, the square root of the length of the crack and the geometries of the item and of the crack.

In fatigue K_I will have an amplitude, ΔK_I; it was first shown by Paris (see Paris, Gomez and Anderson (1961)) (Fig. 2.70) that a log-log plot of propagation speed *da/dN* against ΔK_i will give a straight line, and therefore that

$$da/dN = C(\Delta K_I)^m \tag{2.140}$$

This is found not to hold for small values of ΔK_I, and there is a propagation threshold ΔK_{th}.

For large values of K_I the propagation accelerates greatly as a result of ductile tearing, ending with fracture at a value $K_{max} = K_{IC}$

The opening δ (§2.11.5) is a function of K_{max}. The theory developed in Chapter 3 gives the relation

$$\delta = K_{max}^{\ 2} / 2ER_p \tag{2.141}$$

which together with (2.126) gives,

$$da/dN \approx \delta/2 = \Delta K_I^2 / 4ER_p(1-R)^2 \tag{2.142}$$

Thus the exponent *m* in Paris's relation should be *2*; but observation gives a range of values, often *3*, *4* or greater.

The discrepancy arises because (2.142) gives the microscopic rate of propagation, and this is not the same as the macroscopic rate which is observed. The head of the crack does not in fact advance as a whole, especially when the value of ΔK_I is small. On the other hand, the rate is increased by ductile tearing when ΔK_I is large.

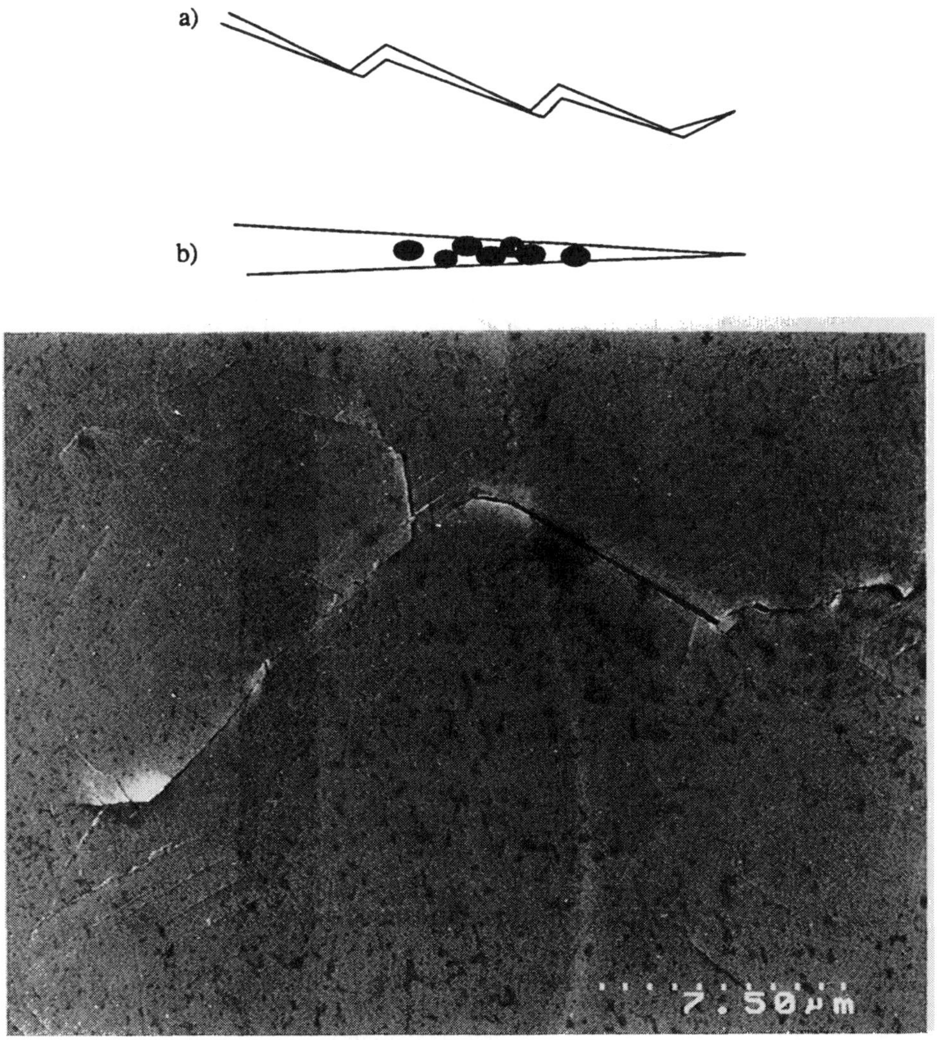

Fig. 2.69 Closure of a crack: by (a) the tortuous form of its profile (b) oxides. (c) Tip of a fatigue crack in a nickel-based alloy, showing propagation along crystallographic planes with failure to reclose.

a)

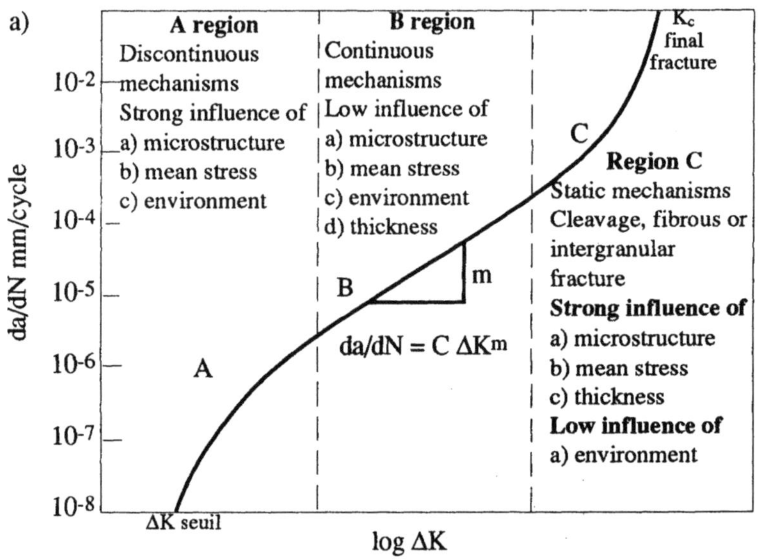

b)

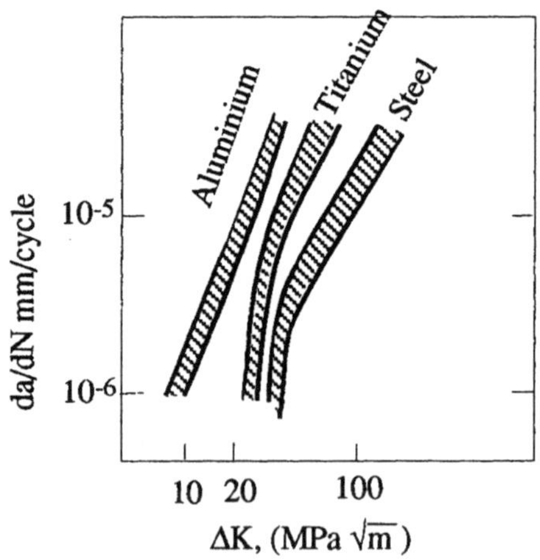

Fig. 2.70 Propagation rate of fatigue cracks as a function of amplitude of stress intensity factor. Paris relation (2.140) gives a linear log-log plot. Three domains are distinguished. In each domain, the relevance of microstructural details and environment is indicated. (a) Diagram showing threshold for cracking and acceleration as final rupture is approached. (b) Observed results for various metals.

Observations have shown that the speed does indeed vary like *1/E*, but no correlation with the elastic limit has been observed; this latter is due in part to the effect of work-hardening in changing the plastic limit.

Consideration of the effects of the forces which act to close the crack suggest that in (2.142) ΔK_I should be replaced by

$$\Delta K_{eff} = K_{max} - K_{op} \tag{2.143}$$

where K_{op} is the stress intensity factor needed to start the crack to open. This concept is essential for the understanding of the effect of *overload*. Experiments with fatigue cracking have shown that a suddenly-applied overload stops propagation and that this does not start again until after a large number of load cycles. This is due to the formation of a large plastic zone, with strong closure forces; the effect is to increase K_{op} to such an extent that it exceeds K_{max}.

Many experiments have shown that short cracks behave differently from long: the latter do not propagate at all below the threshold, but this is not the case for the former. The reason is that the plastic zone in the neighbourhood of a short crack has a different shape from that at the tip of a long crack and that it extends to the surface at which the (short) crack originates, surrounding this completely. Thus there are no closure forces acting on short cracks and K_{op} is small, zero even.

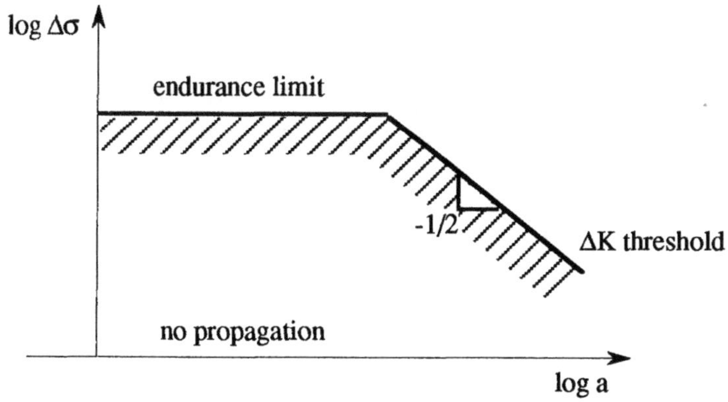

Fig. 2.71 Kitagawa diagram: log-log plot of amplitude of stress variation against crack length, defining the region of non-propagation (hachured).

Another manifestation of this difference in behaviour is represented by the *Kitagawa diagram* (Fig. 2.71) (Kitagawa and Takahashi (1976)), in which the two regimes are clearly separated. The log-log plot of stress amplitude against crack length gives two straight lines, one horizontal and the other with slope *-1/2*. The first corresponds to the threshold for endurance, the second to that for propagation of long cracks; thus we can speak of *mechanically short* cracks, meaning of size smaller than that corresponding to the intersection of the two lines, and *microstructurally short* cracks, meaning cracks which are short in relation to structural obstacles to their propagation such as grain boundaries or pearlitic colonies. The rate of propagation of the latter will be profoundly affected by these obstacles.

2.11.6 Influence of microstructure on fatigue

The effects of the microstructure are often complex and can be mutually inconsistent with respect to different criteria: thus a structure that seems desirable when initiation of cracks is being considered can seem undesirable so far as propagation is concerned. Consequently if we are concerned with the resistance of a material to fatigue we must specify very clearly what aspect of the lifetime is important to us. As Fig. 2.72 shows, the part of the lifetime spent in the initiation stage (Stage 1) is of greater importance the greater the number of cycles to fracture. Fig. 2.73 gives more precise data for austenitic stainless steels at ambient temperature and between *550* and *750°C*. Further, if there are stress concentration zones, especially as in welded joints, initiation is very fast and it is propagation which controls the lifetime.

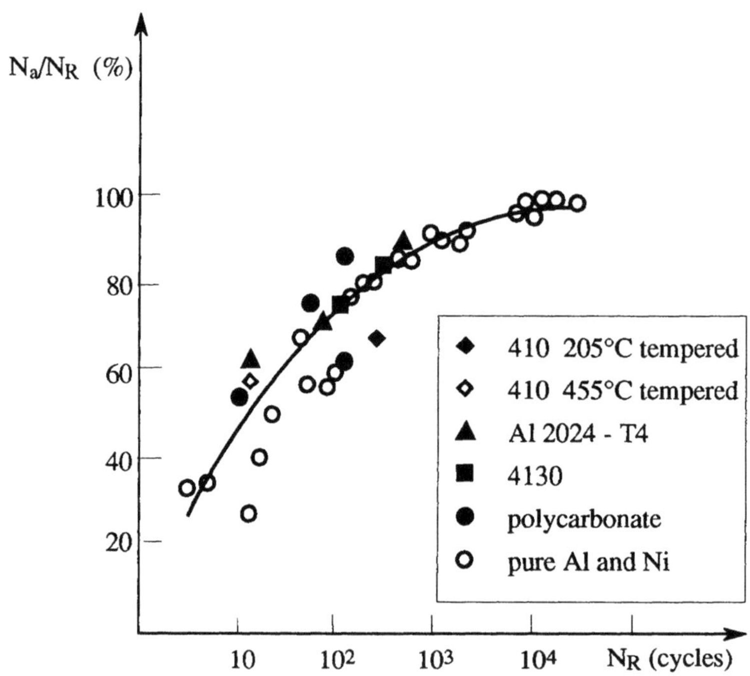

Fig. 2.72 Ratio of number N_a of cycles needed to create a crack of length *50μm* to the number N_R to fracture, as a function of N_R, for various materials (Manson (1972)).

These two figures show that in considering endurance all the metallurgical factors that could favor resistance to the formation of cracks must be taken into account; in this connexion, any increase in resistance to deformation will in general improve the endurance limit. Decreasing the grain size will increase the resistance to fatigue; as we have seen, initiation of cracks is associated with persistent slip bands and is slower the shorter these bands. Thus the endurance limit follows a law of the Hall-Petch type, varying inversely with the square root of the grain size. On the other hand, any modification that tends to concentrate the deformation in narrow bands has a bad effect on the endurance limit; thus in general this limit is greater for over-aged hardened alloys, in which the dislocations bypass the precipitates, than for under-aged alloys in which they shear these.

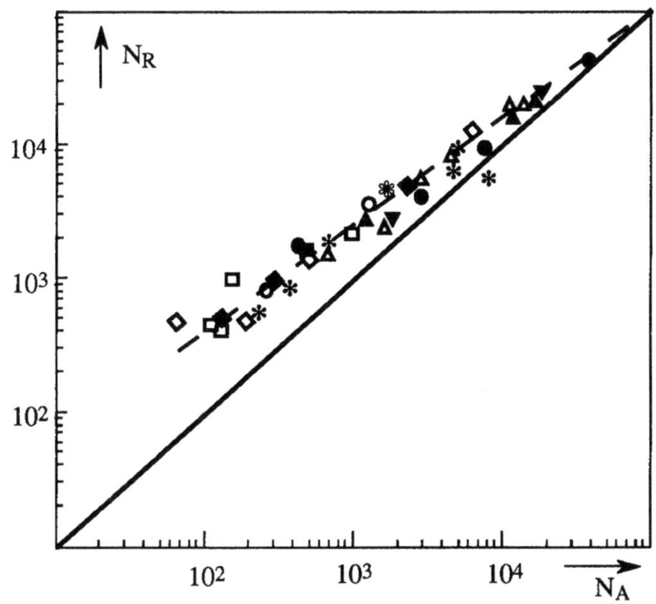

Materials	T° C		Materials - tests	T° C	
316L (virgo)	600	○	316L :		
(Z03CND 17-13)			cyclic tensile test	20	▲
			with torque	20	△
316 (ICL)	550	◇			
(Z03 CND 17-13)		◆	800 alloy	20	▼
304	593	●	Inconel 718 :		
(Z05 CN 17-12)			cyclic tensile test	20	∗
			with torque	20	❋
316	600	□			
(Z03 CND 17-13)	700	■			

Fig. 2.73 Relation between number N_a of cycles needed to create a crack of length $20 \mu m$ and number N_R to fracture, for austenitic stainless steels at various temperatures.

In many industrial alloys *inclusions* play a vital role in the formation of cracks since they constitute stress-concentration sites: there is increasing reason to believe that the endurance limit for high-resistance alloys is related to a *non-propagation limit* for very small cracks that initiate on inclusions. The metallurgical and mechanical factors which control the propagation of the defects we have mentioned in this and the preceding paragraph will play a dominant role. The ones that should be favored are those which lead to a rapid development of the crack-closure effects and a reduction in the effective part of the stress intensity factor.

When looking for good features at the propagation stage it is useful to distinguish the region of the threshold and the low speeds from that of the intermediate stage of striations. At the threshold stage, all the metallurgical factors which lead to a high value for the opening stress intensity factor K_{op} are favorable. Thus very tortuous profiles, deeply marked by crystallographic features, are to be encouraged. Increasing the grain size, reducing the stacking fault energy and shearing the particles in alloys that have been hardened by precipitation will all, in general, lead to an increase in the value of stress intensity factor at the threshold, K_{th} In some cases one might consider making use of the phase changes brought about by the cyclic plasticity at the crack tip, in order to increase the resistance to propagation. This could be so for certain metastable stainless steels which, by deformation, give rise to martensitic transformations of the type γ(FCC) $\rightarrow$ ε(CPH) $\rightarrow$ α'(BCC); since the specific volume of the last phase is greater than that for the first (the mother phase), it gives rise to compressive stresses at the tip, which tend to close the crack.

As we saw in §2.11.5, at greater propagation rates, around *0.1 - 1* μm per cycle, the elementary mechanism for crack propagation results in the formation of striations. Equation 2.142 shows that an essential factor here is the elastic modulus; if in this region we divide the stress by the modulus we obtain rates which are basically the same for all materials. However, the equation shows that the resistance of the alloy is also an important factor, and so leads to the thought that the crack growth rate might be greatly reduced by increasing the resistance of the material. In making these decisions it must not be forgotten that the yield strength that appears in the equation is that for cyclic stressing, which for most high-resistance materials is lower than that for monotonic loading because of the softening effect of the cycling.

The general results we have given so far apply only when the cyclic loading is strictly periodic. As we have seen, occasional overloading can produce retarding effects; under such conditions therefore the best materials are those that can retain a memory of this over-hardening effect: in the law describing their behaviour the isotropic term must be important. In practice, it often happens that the best materials are those that maintain as great as possible a plastic difference, measured by the ratio R_m/R_p, between ultimate strength and yield strength. Thus in metallurgical terms the favorable factors are those which contribute to increasing the work-hardening: reduction of the stacking-fault energy in FCC alloys, hardening by precipitates which deflect dislocations rather than are sheared by these.

2.11.7 Cumulative damage by fatigue

So far in studying the effects of load cycling we have assumed that the amplitude remains constant, whereas in real life it is often random; we must therefore consider the possible effects of variations in this.

In order to investigate the effect of N_1 cycles at amplitude $\Delta\sigma_1$ followed by N_2 at amplitude $\Delta\sigma_2$ etc (a succession N_i at $\Delta\sigma_i$) we need to define a quantitative measure of damage; the simplest is the ratio N_i / N_{R_i} where N_{R_i} is the number of cycles to fracture at amplitude $\Delta\sigma_i$. It is important to bear in mind that this definition is not tied to the physical process of fatigue: the rate at which the area of a fatigue crack increases is a very non-linear function of the stress amplitude.

Further, the ratio of this area to that of the section of the component being stressed, which could provide a better physical measure of damage, varies with their shapes and sizes. So there remains a problem.

These reservations notwithstanding, the simplest method for predicting the effect of variable-amplitude stressing is Miner's linear sum: at fracture

$$\sum_i N_i / N_{R_i} = 1 \tag{2.144}$$

The rule often fails; for example, if there are two amplitudes with $\Delta\sigma_1 < \Delta\sigma_2$ it can happen that

$$N_1 / N_{R_1} + N_2 / N_{R_2} > 1 \tag{2.145}$$

which expresses an adaptation effect. Conversely, starting with the greater amplitude is more damaging. Nevertheless, the rule is often used, because of its simplicity.

2.12 VISCOPLASTIC DAMAGE, INTERGRANULAR FRACTURE AT HIGH TEMPERATURE

2.12.1 Introduction

Measuring temperatures in degrees absolute, we shall take "high temperature" to mean above half that of fusion for the material considered. At such temperatures the dominant mechanisms are those of diffusion, and dislocations, for which we had envisaged only slip at low temperatures, can climb, as we showed in Volume I; in fact, diffusion of vacancies towards the dislocations makes non-conservative displacements possible.

In keeping with this, when metals and metallic alloys are cycled at high temperature their ductility is usually greatly reduced, the amount of reduction increasing with increasing temperature and with decreasing rate of deformation. The fall in ductility observed in creep is most often associated with intergranular fracture, which is itself linked to the formation of intergranular cavities.

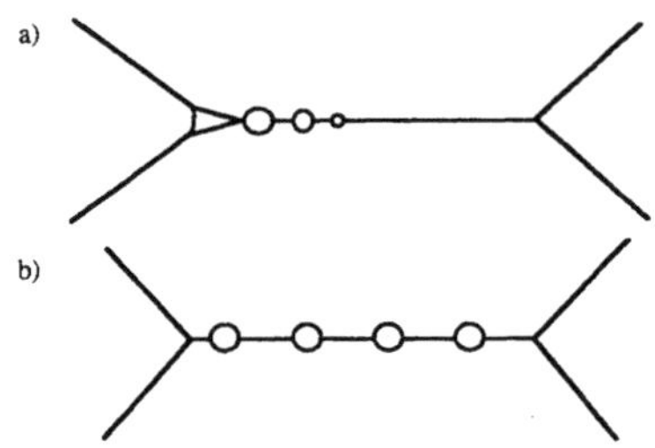

Fig. 2.74 Diagram showing the two types of intergranular cavity due to creep: (a) wedge (b) round.

It is usual to distinguish between two forms of intergranular damage: "wedge cracks" and "round cracks" (Figs. 2.9 and 2.74). Recent work has shown that there is no essential difference between the two, wedge cavities being formed by accelerated coalescence of round cavities under the effect of stress concentrations generated at triple intersections of grain boundaries. Whilst intergranular cavitation is associated specifically with creep damage, such cavities can in certain cases be generated transgranularly by inclusions and precipitates, as at fracture in the plastic regime; this however is rather rare and is observed only under conditions of high stress. As in the case for ductile fracture, three stages can be distinguished in the formation of cavities: nucleation, growth and coalescence; there is in fact a strong similarity between the subject and that of ductile fracture.

In this section of the chapter we shall first describe the mechanisms that give rise to intergranular cavitation, giving for each the laws that have been established; this is a field in which great progress has been made recently and the reader could with profit consult some of the latest publications, such as Argon et al (1980), Beere (1981) and Evans and Wilshire (1985). Only the first two stages, nucleation and growth respectively, have been studied in depth. After this we shall discuss the results of measurements of damage that have been made with various alloys and attempt to locate these in a framework of phenomenological approaches.

2.12.2 Mechanisms of intergranular cavitation

2.12.2.1 *Nucleation of cavities*

(a) *Nucleation not activated thermally.* As in plasticity, separation or fracture of inclusions or precipitates can take place under the effect of local stress concentrations; the corresponding mechanisms were described in §2.7. This occurs less frequently in the case of creep, because of the low levels of applied stress then obtained.

(b) *Thermally activated nucleation.* Creation of a cavity corresponds to the appearance of a surface energy proportional to the square of the radius and the vanishing of what we shall show to be a negligible amount of deformation energy. The applied stress provides a quantity of work proportional to the cube of the cavity radius and therefore the energy balance defines a critical radius for which the energy is maximum. With the classical treatment of nucleation, the thermal agitation enables cavities to form at a rate which increases with both temperature and applied stress.

(b1) In a homogeneous medium under hydrostatic tension σ_m the change in free enthalpy when a spherical cavity of radius R and surface energy γ_S appears is

$$\Delta G = 4\pi R^2 \gamma_S + \Delta E_e - \sigma_m (4/3)\pi R^3 \qquad (2.146)$$

where ΔE_e is the change in elastic energy, equal to $\alpha[3(1-2\nu)/E]\sigma_m^{\ 2}\,(4/3)\pi R^3$. α is a coefficient representing the perturbation of the stress field by the cavity; its value is only slightly greater than 1 since this effect extends over only a short distance. Since the creep stresses are not high this term, proportional to $\sigma_m^{\ 2}$, is negligible.

The energy barrier to be overcome in order to create a cavity is the ΔG^* corresponding to the radius R^* for which $\partial \Delta G/\partial R = 0$. Neglecting ΔE_e in (2.146) we find

$$R^* = 2\gamma_S / \sigma_m, \quad \Delta G^* = \frac{16\pi}{3}\frac{\gamma_S^3}{\sigma_m^2} \qquad (2.147\text{-}8)$$

(it is easily checked that this energy is much greater than ΔE_e). The radius R^* thus found is in fact the radius of curvature of a surface in equilibrium under pressure σ_m and surface tension γ_S.

The rate of nucleation of cavities will be proportional to $\exp(-\Delta G^*/kT)$, which must be multiplied by the frequency with which vacancies cross the cavity surface, of area $4\pi R^{*2}$. For a single vacancy the frequency is $\exp[-(\Delta E_L - \sigma_m \Omega)/kT]$, where ΔE_L is the energy of formation of the vacancy and Ω is its volume; if D_v is the coefficient of self-diffusion this is $(D_v/\Omega^{2/3})\exp(\sigma_m \Omega/kT)$. Since there are $4\pi R^{*2}/\Omega^{2/3}$ atomic sites on the surface of the cavity we have finally for the rate

$$\dot{N} = \frac{4\pi R^{*2}}{\Omega^{2/3}} \cdot \frac{D_v}{\Omega^{2/3}} .\exp\frac{\sigma_m \Omega}{kT} .\exp\left(-\frac{\Delta G^*}{kT}\right) \qquad (2.149)$$

(b2) Cavities nucleate most easily on the grain boundaries because these have a surface energy γ_j which vanishes when a cavity is formed. Thus for a spherical cavity under hydrostatic stress

$$\Delta G = 4\pi R^2(\gamma_S - \gamma_j/4) - \sigma_m (4/3)\pi R^3 \qquad (2.150)$$

However, at equilibrium the cavity is not spherical, since at the point where it meets the grain boundary the surface tensions are in equilibrium (Fig. 2.75); therefore

$$\gamma_j = 2\gamma_S \cos\Psi \tag{2.151}$$

where Ψ is the angle at which the two halves of the cavity meet; for the usual surface energies of grain boundaries this is about *1* radian.

It is convenient to put S = area of cavity surface = $F_S R^2$,
S_j = area of boundary intersected by cavity = $F_j R^2$,
V = volume of cavity = $F_V R^3$,

where R is the radius of curvature of the cavity. Then

$$\Delta G = 2\gamma_S(F_S/2 - F_j \cos\psi)R^2 - \sigma_n F_V R^3 \tag{2.152}$$

where σ_n is the normal stress acting on the boundary. The critical radius R^* is the value for which $\partial\Delta G/\partial R = 0$; hence

$$R^* = \frac{4}{3}\frac{\gamma_S}{\sigma_n}\frac{F_S/2 - F_j\cos\Psi}{F_V}, \quad \Delta G^* = \frac{32}{27}\frac{\gamma_S^3}{\sigma_n^2}\frac{(F_S/2 - F_j\cos\Psi)^3}{F_V^2} \tag{2.153-4}$$

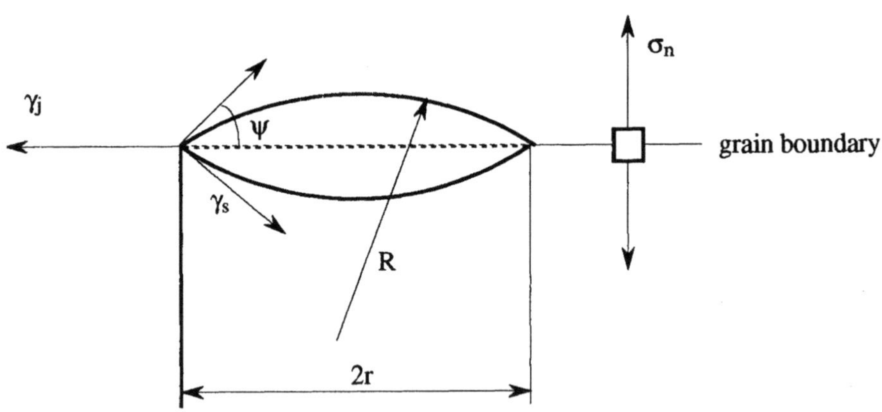

Fig. 2.75 Lenticular cavity of surface energy γ_S on a grain boundary of surface energy γ_j.

For a cavity on a plane joint, as in Fig. 2.75,

$$F_S = 4\pi(1-\cos\Psi),\ F_j = \pi\sin^2\Psi,\ F_V = (2\pi/3)(1-\cos\Psi)^2(2+\cos\Psi) \tag{2.155}$$

$$R^* = \frac{2\gamma_S}{\sigma_n},\quad \Delta G^* = \frac{8\pi}{3}\frac{\gamma_S^3}{\sigma_n^2}(1-\cos\Psi)^2(2+\cos\Psi) \tag{2.156}$$

It follows that R^* and ΔG^* decrease with decreasing ψ, that is with increasing values of the ratio γ_j/γ_S.

This explains the role played by *segregation of impurities* on the grain boundaries: by changing the surface energy these can greatly reduce the activation energy for nucleation of intergranular cavities, as is the case for example for antimony in copper and sulfur and phosphorous in steels.

Another important effect is that of a gas under pressure in the cavity: this pressure will add to the stress σ_n, resulting again in a large decrease in the activation energy. An example is the helium generated by nuclear reaction with boron atoms during experiments on irradiation-induced creep.

It follows also that the critical size and the activation energy are less if the cavity forms at the intersection of two boundaries which meet at an acute angle; this explains why cavities form more easily at triple nodes and on precipitates or particles lying along the boundaries.

Once the activation energy ΔG^* is known the nucleation rate can be calculated; the method is analogous to that for the homogeneous case, with the difference that the transport of vacancies is now taken to be by diffusion along the grain boundary , with coefficient D_j and over an effective thickness δ_j. The result is

$$\dot{N} = \frac{2\pi R^*\sin\Psi}{\Omega^{2/3}}\frac{\delta_j D_j}{\Omega^{2/3}}\exp\left(\frac{\sigma_n\Omega}{kT}\right)\exp\left(-\frac{\Delta G^*}{kT}\right) \tag{2.157}$$

Fig. 2.76 gives the results of the calculation for γ-iron, taking the values:

$\Omega = 1.21.10^{-29}\ m^3$, $\delta_j D_j = 7.5.10^{-14}\exp(-159.10^3/RT)\ m^3 sec^{-1}$, $\Psi = 1$ *radian*, $\mu = 1.40.10^2$ *GPa*. R, is the gas constant, in *kJ/mole*,

These show that $r^* = R^*\sin\Psi$ decreases with increasing σ_n/E.

Fig. 2.76 shows also that the nucleation rate reaches a constant value, varying only slightly with temperature, for values of σ_n/E of the order of 5.10^{-3}, that is for σ_n about *700 Mpa*.

Creep stresses are about ten times smaller than this, from which we might conclude that for reasonable rates of nucleation of cavities on the boundaries there must be stress concentration factors of the order of *10*. However, we must recall that, as has just been shown, segregation of impurities on the boundaries or presence of gas in the cavities can greatly reduce the nucleation threshold.

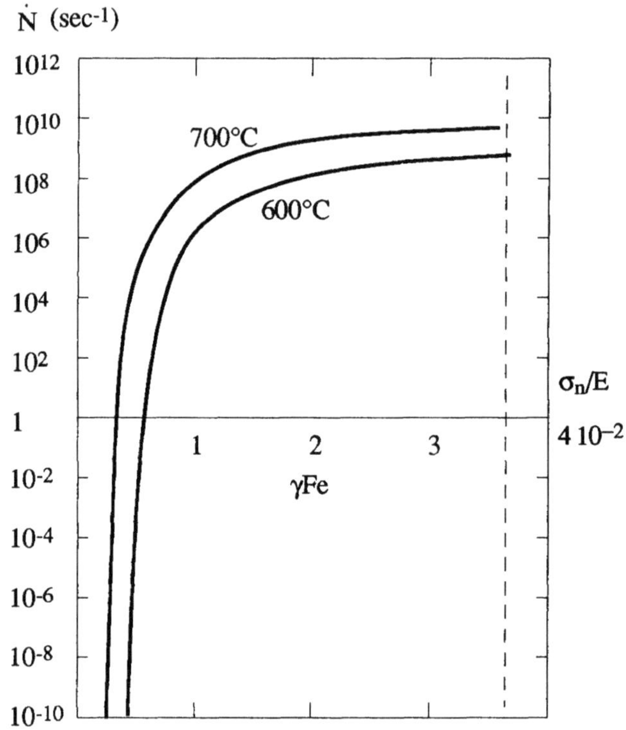

Fig. 2.76 Results of the theory of nucleation of creep cavities, applied to iron. Nucleation rate (sec^{-1}) is plotted against reduced normal stress σ_n/E on the boundary, where E is Young's modulus.

(b3) Stress concentration.

Several mechanisms have been postulated to account for concentrations high enough to nucleate cavities at certain sites along the grain boundaries (Fig. 2.9): pile-up of dislocations at the head of slip lines, meeting of two intergranular slips at a triple node, interference between intergranular slips and kink bands, twins or precipitates. However, it is important to take into account the ease with which the stresses can relax in the presence of creep.

This is a very complex problem, involving transport of material along the boundaries (Coble creep, see I-§1.3.2.2b) and creep of the grains themselves, to which only approximate solutions have been found. What is first found is that only *intergranular particles* provide sites at which the relaxation time is not much too short. In the following, simple calculations to evaluate this relaxation time are presented.

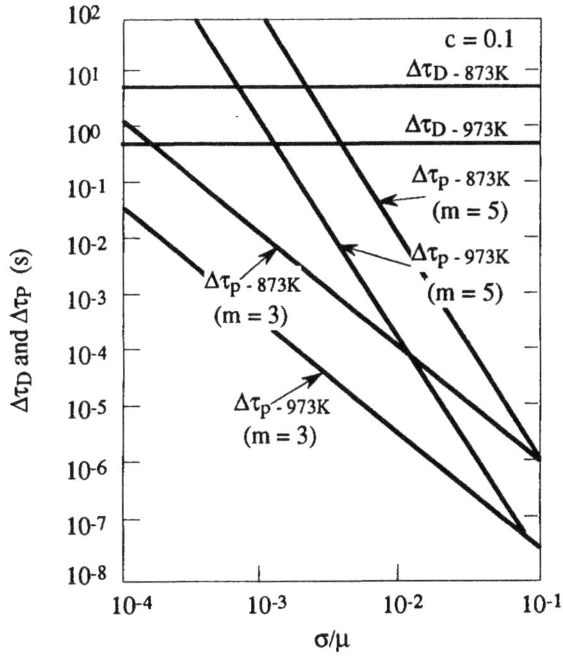

Fig. 2.77 Relaxation time for *0.4μm* particles, *4μm* apart, in γ-iron at *600°C* and *700°C*. The relaxation time $\Delta\tau_D$ for small particles (Eqn. 2.160) is independent of the shear stress, $\Delta\tau_p$ for large particles (Eqn. 2.161) decreases at a rate that increases with increasing values of the index m in the creep law.

There are three stages, each more or less long lasting, to the development of the stress along the boundary. In the first it becomes uniform between the particles; in the second a stationary stress field forms around the particle; the third sees the complete relaxation of the stress concentrations acting on the particles, the homogenisation of the stress along the whole length of the boundary and the transfer of deformation incompatibilities to the triple nodes. Thus for particles of size p, at distances L_p along the boundary, there will be a characteristic time $\Delta\tau_p$ for the establishment of a shear stress τ, uniform between the particles. The calculation of this characteristic time starts from the statement that the elastic slip along the boundary is canceled by the creep slip, and the result (Exercise 13) is

$$\Delta\tau_p = \frac{\eta}{\delta_j}\frac{L_p}{k} = \frac{1}{\pi\mu}\frac{kT}{\delta_j D_j}\frac{L_p}{\delta_j}\beta(p/L_p) \tag{2.158}$$

where η is the coefficient of viscosity for the boundary, μ is the shear modulus, $k = \pi\mu / L_p\beta$ is the rigidity of the boundary and β (a function of the ratio p/L_p) is a parameter for the interaction between segments of the boundary. The time is very short, a fraction of a millisecond for stainless steels in the region of *600-700°C*.

Thus there is a local concentration of stress on the particles of the order of $\tau L_p/p$; but this is relaxed either by diffusion along the interface or by creep of the grains around the particles, the latter following a power law with respect to the stress. The first of these mechanisms will predominate if the particle size is less than some characteristic distance Λ, the second if it is greater than this. The value of Λ is found by equating the creep rate to the diffusion flow along the interface to the particle:

$$\Lambda = (\delta_j D_j \Omega\sigma/kT\varepsilon)^{1/3} \tag{2.159}$$

where σ is the local (concentrated) stress and y is the volume of the vacancies. It is a few nanometers.

The characteristic time for the relaxation is found by stating that the flux of vacancies along the particle destroys the void which the slip would tend to create there; for very small particles ($p << \Lambda$) this is

$$\Delta\tau_D = p^3 kT\beta/8\pi\Omega\mu\delta_j D_j \tag{2.160}$$

For large particles, $p >> \Lambda$, if c is the surface concentration of particles the result is

$$\Delta\tau_p = (1/\pi\,\dot{\varepsilon}_0)[c/(\sigma/\mu)]^{m-1}\beta \tag{2.161}$$

For these particles, relaxation is brought about by creep, with the power law

$$\dot{\varepsilon} = \dot{\varepsilon}_0(T)\,(\sigma/\mu)^m = A(\mu b/kT)D_v(\sigma/\mu)^m \tag{2.162}$$

where D_v is the coefficient of self-diffusion in the volume.

It is important to note that in this latter case the relaxation affects only a region of size of the order of Λ, which is much smaller than p; thus there is a high concentration of stress acting on the particles, so that cavities can form there.

Fig. 2.77 gives these relaxation times for γ-iron at *600* and *700°C*, for particles of size *0.4 μm* a distance *4μm* apart.

After these relaxation times have expired a regime is gradually established in which the stress is constant along the whole length of the boundary; this occurs over a time which depends on the mechanism in play, and if the length is d an approximation is given by multiplying the appropriate relaxation time by d/L_p The entire shear stress is then transferred to the triple node, with nucleation of r-type (round) cavities replaced by that of w-type (wedge).

(c) *Conclusion.* The models we have described in this discussion of nucleation emphasize the roles played by normal stress and intergranular slip. If the second factor dominates it seems reasonable to expect that in a uniaxial test the maximum damage will occur at boundaries oriented at *45°* to the axis of the traction; but in practice most of the cavitation is found to occur on boundaries at right angles to this axis. Two explanations for this have been suggested. The first is related to the fact that both the normal stress and the intergranular slip are important; the second, equally plausible, that the methods of observation that are available do not enable cavities to be examined at the nucleation stage, when the critical radius R^* is a few nanometers, but only at the later stage of growth.

2.12.2.2 *Growth of cavities*

Many more laws describing the growth of cavities have been established for the viscoplastic than for the plastic regime; this is because of the diversity of mechanisms than can be in action:

(i) viscoplastic deformation,
(ii) diffusion,
(iii) coupling between (i) and (ii),
(iv) constrained growth by creep in the matrix.

We shall discuss each of these in turn, giving their physical bases and the relevant laws.

(a) *Growth controlled by viscoplastic deformation.* If the rate of deformation is great enough there will not be time for the diffusion effects to appear; consequently the laws established for the plastic regime are easily transposed to apply here, with strain replaced by rates. Thus the Rice-Tracey law (§2.8.3) becomes, approximately,

$$\frac{1}{R}\frac{dR}{dt} = \alpha\dot{\varepsilon}_{eq} \exp\left(\frac{3}{2}\frac{\sigma_m}{\sigma_{eq}}\right) \tag{2.163}$$

where R is the radius of the (spherical) cavities.

(b) *Growth controlled by diffusion.* This is specific to creep. It is more likely to occur at low rates of creep than at high, implying that it is more likely to be seen in real components tested for long periods at low stress amplitudes than in laboratory tests, which are more often performed under conditions of accelerated creep.

The mode was first studied by Hull and Rimmer (1959), for a spherical cavity; their treatment, though approximate, is interesting. We start with an account of this and follow it with indications of some improvements, always for cavities in equilibrium; afterwards we shall consider the case in which the cavities have the shape of a crack.

(b1) Cavities in equilibrium: (i) Hull Rimmer model

This is based on the following assumptions

- On the surface of a cavity of radius R the chemical potential of a vacancy is $2\gamma_S\Omega/R$; on a grain boundary on which the normal stress is σ_n it is $\sigma_n\Omega$. If $\sigma_n\Omega > 2\gamma_S\Omega/R$ there will be a flux $\vec{J}$ of vacancies along the boundary, increasing the size of the cavity.

- Diffusion of vacancies at the surface of the cavity is sufficient to keep the radius always at the equilibrium value.

- The boundaries are inexhaustible sources of vacancies.

- The grains are rigid and nothing prevents their relative displacement.

If g is the potential of the vacancies, L the distance between the centers of the cavities and ρ a radial co-ordinate for the latter, then

$$\vec{J} = \frac{D_j}{\Omega kT}\overrightarrow{grad}\, g, \quad \left|\overrightarrow{grad}\, g\right| = dg/d\rho \approx [\sigma_n\Omega - 2\gamma_S/R]/(L/2) \qquad (2.164\text{-}5)$$

The vacancies diffuse in the cavity across a surface $2\pi R\delta_j$, where δ_j is the width of the boundary; the change in volume of the cavity is thus

$$dV/dt = 2\pi R\delta_j\left|\vec{J}\right|\Omega = 4\pi RD_j\delta_j(\sigma_n - 2\gamma_S/R)\Omega/kTL \qquad (2.166)$$

If the stress is broadly sufficient to cause growth the term $2\gamma_S/R$ can be neglected and the final result is

$$\frac{dR}{dt} = \frac{D_j\delta_j}{kT}\frac{\sigma_n\Omega}{LR} \qquad (2.167)$$

As we have seen, it follows from the equilibrium conditions for the surface tensions in cavities in equilibrium that the latter are not spherical but lenticular; it is also possible to calculate a better approximation to the potential gradient for the vacancies.

(b1) Cavities in equilibrium: lenticular cavities (Fig. 2.78)
In the stationary state each element of the surface of the boundary emits the same number of vacancies per unit time . If this rate is λ the conservation of mass gives

$$\text{div}\,\vec{J} + \lambda = 0 \qquad (2.168)$$

and the flux $\vec{J}$, proportional to the potential gradient, is given by (2.164). This is a differential equation for the potential g with boundary conditions

$$dg/d\rho = 0 \text{ for } \rho = L/2,\ g = -2\gamma_S\Omega/R \text{ for } \rho = r \qquad (2.169)$$

where $r = R\sin\Psi$ is the radius of the cavity in the plane of the boundary.

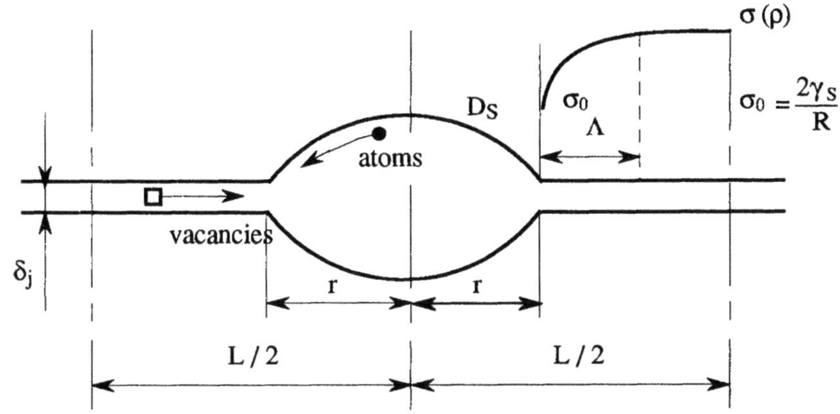

Fig. 2.78 Model for the growth of intergranular cavities by intergranular diffusion.

Integrating the equation we find (Exercise 14)

$$g = \frac{\lambda kT\Omega}{4D_j\delta_j}\left[\rho^2 - r^2 + (L^2/2)\log(r/\rho)\right] - \frac{2\gamma_S\Omega}{R} \tag{2.170}$$

with a corresponding normal stress on the plane of the boundary given by

$$g = -\sigma_n\Omega \tag{2.171}$$

These stresses, which vary along the length of the boundary, are in equilibrium with the applied stress σ_a; the equilibrium equation is

$$\pi L^2\sigma_a/4 = \int_r^{L/2} 2\pi\rho\sigma_n d\sigma \tag{2.172}$$

from which we can calculate the rate λ of emission of vacancies.

The rate of change of volume of the cavity is given by

$$dV/dt = \pi\lambda\Omega(L^2/4 - r^2) \tag{2.173}$$

which if R_c $(= 2\gamma_S/\sigma_a)$ is the critical radius for growth becomes

$$\frac{dV}{dt} = 2\pi\Omega\frac{D_j\delta_j}{kT}\frac{[1-(2r/L)^2]\sigma_a\{1-(R_c/R)[1-(2r/L)^2]\}}{\{-(3/4)+\log(L/2r)+(2r/L)^2[1-(r/L)^2]\}} \tag{2.174}$$

It is interesting to calculate the rate of change of the surface fraction (ratio of cavity surface to total surface) A_j for the cavities; we find

$$\frac{dA_j}{dt}=\frac{32}{3\sqrt{\pi}}\Omega\frac{D_j\delta_j}{kT}\frac{\sigma_a}{L^3}\left(\frac{F_j^{3/2}}{F_V}\right)f(A_j) \tag{2.175}$$

where

$$f(A_j)=\frac{1-A_j}{\sqrt{A_j}}\frac{1-2(F_j/\pi)^{1/2}(R_c/L)(1-A_j)/\sqrt{A_j}}{-(3/4)-\log\sqrt{A_a}+A_j(1-A_j/4)} \tag{2.176}$$

From the rate of growth of cavities given by (2.175) we can find the time needed for these to cause fracture at the boundary, which will occur when A_j, with initial value A_{j0}, reaches a critical value A_{jc}; this is

$$t_R=\frac{3\sqrt{\pi}}{32}\frac{kT}{\Omega D_j\delta_j}\frac{1}{\sigma_a}\frac{F_V}{F_j^{3/2}}\int_{A_{j0}}^{A_{jc}}dA_j/f(A_j) \tag{2.177}$$

and is thus inversely proportional to the applied stress σ_a.

(b2) Flat cavities

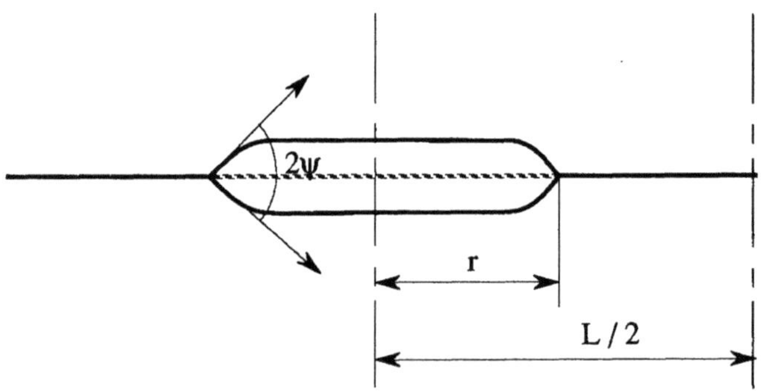

Fig. 2.79 Diagrammatic representation of a flat intergranular cavity.

In deriving the above results we have assumed that the rate of diffusion of the atoms at the cavity surface is great enough to compensate for the flux of vacancies from the boundary, and therefore to keep the radius of curvature always at the equilibrium value. If this rate is too low the cavities will flatten and become cracks; if δ_S is the atomic diameter $\Omega^{1/3}$ and D_S is the surface diffusion coefficient this will be the case when the ratio $\Delta = D_j\delta_j/D_S\delta_S >> 1$.

Table 2.6 gives the values of Δ for various pure metals; it is important always to remember that the diffusion coefficient is very sensitive to the presence of impurities.

Metal	*Ag*	*Cu*	*Fe* α	*Fe* γ	*Ni*	*Zn*
Δ	*570*	*0.08*	*3*	*2*	*28*	*13000*

Table 2.6 Values of the ratio $\Delta = D_j\ \delta_j / D_s\ \delta_s$ for various metals.

Chuang and his collaborators (Chuang 1979) have calculated the rate of growth of such flat cavities, controlled by surface diffusion. The equations of conservation of matter and continuity of flux everywhere on the surface determine the form of the cavity for a given flux at its extremity; the result is as follows:

$$\frac{dr}{dt} = \frac{3}{(2\sin\Psi/2)^3}\frac{D_S\delta_S\Omega\gamma_S}{(L/2)^3 kT}\left(\frac{\sigma_a}{2\gamma_S/L}\right)^3\left[1-(2r/L)^2\right]^{-3} \qquad (2.178)$$

(recall that the angle Ψ is determined by the equilibrium of the surface tensions, cf. Fig. 2.79). This shows that the time to fracture varies like σ_a^{-3}.

(b3) Coupling between diffusion and viscoplasticity

In the models which we have developed no part is played by creep of the grains, since the atoms emanating from the cavities cover the entire grain boundaries. We can envisage a treatment in which the vacancies necessary to ensure the growth of the cavities are assumed to come only from the immediate neighbourhood of the latter; the diffusion distance will then be small and the effect will be to increase the rate of growth. This has been the approach of several authors, notably Speight and Beere (1975); but this requires that the deformation far from the cavities is due to creep of the grains themselves.

Fig. 2.80 is a diagrammatic presentation of the model that results from coupling between diffusion and viscoplasticity, with the cavities assumed spherical for simplicity. In region I, close to the cavities, vacancies appear at a uniform rate and diffuse towards these, helping to relax the stresses; here the displacements are due mainly to the formation of vacancies. In contrast, in region II, distant from the cavities, the rate at which vacancies are created is zero and the displacements are due to transgranular creep. The size of the first region, *b-r*, is comparable to the characteristic distance introduced above (Eq. 2.159), defined by equating the creep rate ε to the diffusional flux and calculated to be

$$\Lambda = \left[\frac{D_j\delta_j\Omega\sigma_a}{kT\dot{\varepsilon}}\right]^{1/3} \qquad (2.179)$$

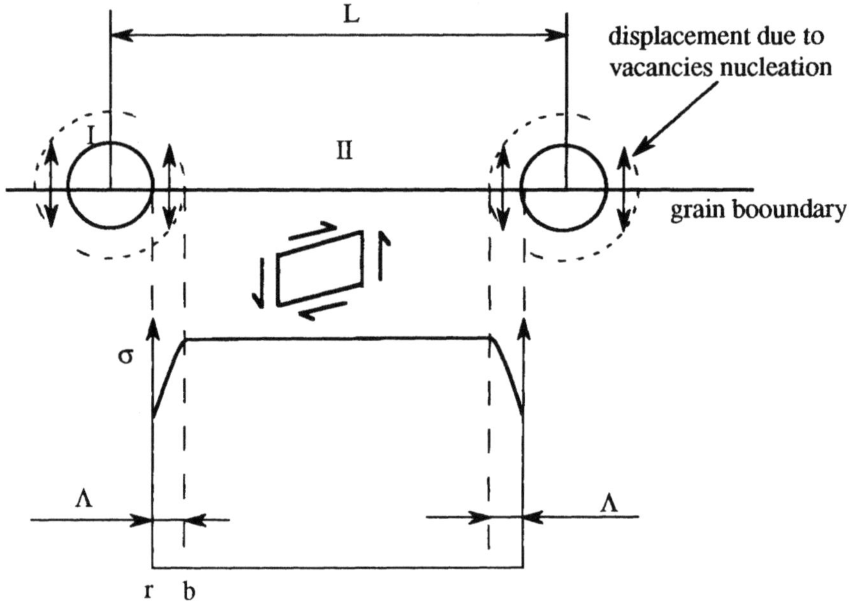

Fig. 2.80 Model of growth of intergranular cavities with coupling between inter-granular diffusion and transgranular viscoplasticity.

A simple modeling is to equate the displacement rate $\lambda\Omega$ in region I to that due to creep, $2b\,\dot{\varepsilon}$, in region II; since we have

$$dV / dt = \pi b^2 \lambda\Omega \qquad (2.180)$$

we find, with $b \approx \Lambda$,

$$dV / dt = 2\pi\Lambda^3 \dot{\varepsilon} = 2\pi D_j \delta_j \Omega \sigma_a / kT \qquad (2.181)$$

This gives a time to fracture inversely proportional to the stress.

Chen and Argon (1981) have given a more exact treatment of the problem in which the reduction in diffusion distance is taken into account; they arrive at the following:

(i) for cavities in equilibrium ($\Delta < 1$)

$$\frac{1}{r}\frac{dr}{dt} = \frac{\dot{\varepsilon}}{2h(\Psi)}\left(\frac{\Lambda}{r}\right)^3 \left\{ \log\left(\frac{r+\Lambda}{r}\right) + \left(\frac{r}{r+\Lambda}\right)^2 \left[1 - \frac{1}{4}\left(\frac{r}{r+\Lambda}\right)^2\right] - \frac{3}{4} \right\}^{-1}$$

(2.182)

where $h(\Psi)=\left(\frac{1}{1+\cos\Psi}-\frac{\cos\Psi}{2}\right)\frac{1}{\sin\Psi}$

(ii) for flat cavities ($\Delta >> 1$)

$$\frac{1}{r}\frac{dr}{dt}=\left(\frac{1}{4\sin\Psi/2}\right)^{3/2}\left(\frac{D_j\delta_j}{D_S\delta_s}\right)^{1/2}\frac{D_j\delta_j\Omega\gamma_S}{\pi^4 kT}\left(\frac{\sigma}{\gamma_S/r}\right)^{3/2}\{\ \}^{3/2} \tag{2.183}$$

where the expression in braces is the same as in (2.182)

Fig. 2.81 gives, on a log-log scale, the rate of growth of cavities $(1/V)(dV/dt)$ as a function of the applied stress for the purely diffusional and the purely viscoplastic regimes; this shows that, as would be expected, the first predominates at low stresses.

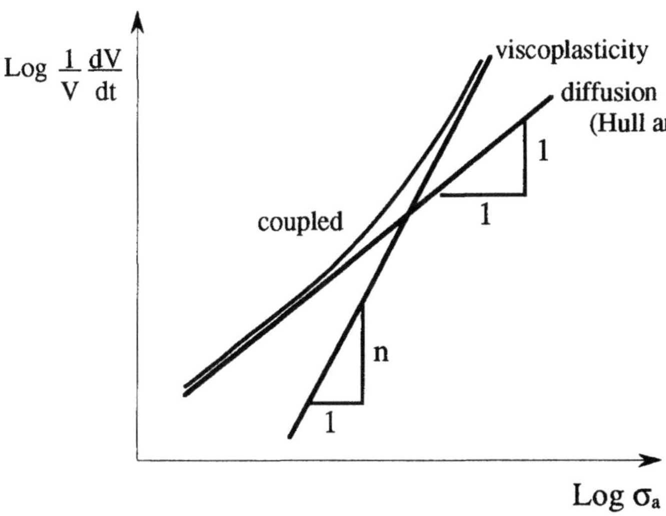

Fig. 2.81 Variation of rate of cavity growth with applied stress for (i) diffusional regime (Hull-Rimmer), (ii) plastic regime (iii) coupling between these.

Fig. 2.82, also on log-log scales, is a plot of a normalised growth rate against a normalised radius.

In this figure, the dotted lines correspond to the simplified model in which the grains are assumed to be rigid, for different values of the ratio r/L; this approximation clearly becomes closer as r/Λ decreases and for small values of this is quite good - again as would be expected. The horizontal asymptote for large values of the ratio corresponds to the purely viscoplastic regime. This figure defines various regimes of cavity growth.

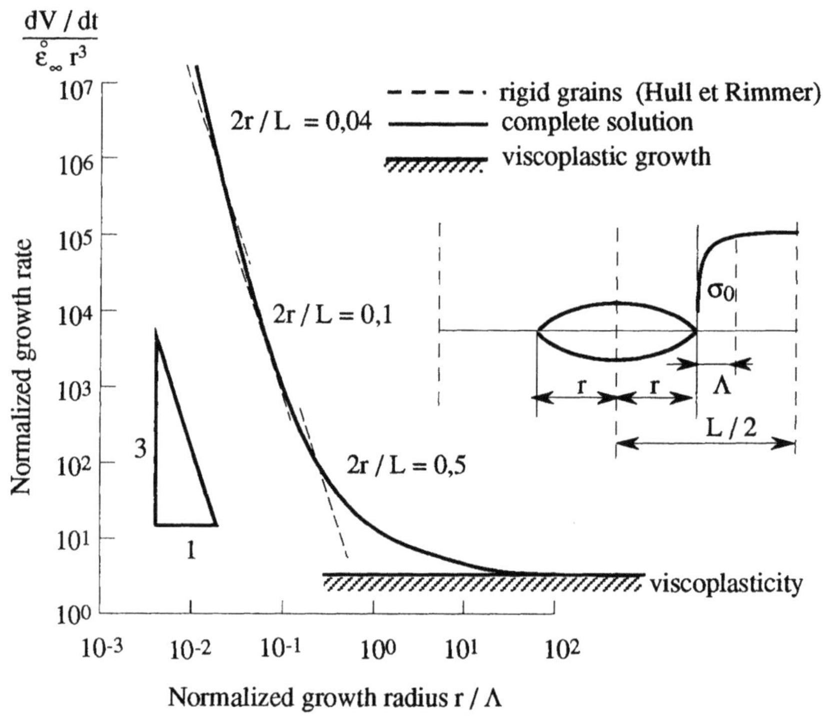

Fig. 2.82 Cavity growth rate normalised by $r^3\ \dot{\varepsilon}_\infty$ as a function of normalised radius r/Λ, where Λ is the diffusion length $(\Omega\delta_j\sigma_d/kT\varepsilon)^{1/3}$.

(c) *Growth retarded by creep in the matrix.* So far we have assumed that intergranular cavitation is the same for all grain boundaries; but practical experience has shown that this is not so and that the situation represented by Fig. 2.83 is closer to what is often observed, especially in alloys with precipitates in the grain boundaries. If the cavitation occurs on an isolated grain, oriented at right angles to the applied stress, it will have the width of the boundary locally; otherwise expressed, the load will be partly removed from the boundary and consequently there will be a transfer of load to the undamaged neighbouring grains. Thus the cavitation will be partly controlled by the manner in which traction forces on the boundary can be maintained by the surrounding grains: this is a problem that has been studied by several authors. In the limit, for high creep rates - that is, for high stresses - cavity growth is not retarded and will proceed at the rates previously established in the purely diffusional regime; whilst for low creep rates it can be partly controlled by viscoplastic relaxation of the stresses.

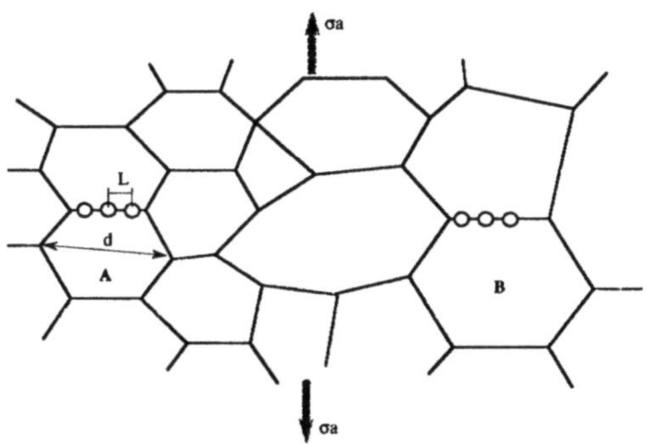

Fig. 2.83 Cavity growth in boundaries damaged at A and B is retarded by creep in the body of the matrix.

Rice has attacked the problem by first writing the equation that expresses the condition that the rate of increase of the width of a length d of a boundary is equal to that of the opening of a circular crack of diameter d under the effect of an unknown tensile stress. This is an equation for the stress, solving which gives a stress which can then be used to calculate the rate of growth of the cavities. The result is

$$\frac{dr}{dt} = \frac{1}{r^2 h(\Omega)} \frac{D_j \delta_j \Omega \sigma_a / kT}{(D_j \delta_j \Omega \sigma_a / kT)(16 / \alpha \dot{\varepsilon} L^2 d) + Q(2r/L)} \tag{2.184}$$

where α is a constant slightly less than 1 (about 0.9 for a creep exponent of 5) and the function Q is defined by

$$Q(2r/L) = \log(L/2r) + (2r/L)^2\left[1 - (r/L)^2\right] - 3/4 \tag{2.185}$$

When the creep rate ε is low the stress relaxation is so marked that the first term in the denominator dominates, giving the asymptotic solution

$$\frac{dr}{dt} = \frac{\alpha d}{16 h(\Psi)} \left(\frac{L}{r}\right)^2 \dot{\varepsilon} \tag{2.186}$$

(e) *Summary of results, comparison with observations.* Table 2.7 summarises the results obtained.

growth controlled by	*growth law*
viscoplastic deformation	$dA_j / dt = 2\alpha\dot{\varepsilon}A_j \exp(3\sigma_m / 2\sigma_{eq})$
diffusion:	
cavities in equilibrium	$\dfrac{dA_j}{dt} = \dfrac{32}{3\sqrt{\pi}} \dfrac{F_j^{3/2}}{F_v} \dfrac{D_j\delta_j\Omega}{kT} \dfrac{\sigma_a}{L^3} f(A_j)$
$(\Delta = D_j\delta_j/D_S\delta_S < 1)$	where if A_j is small $f(A_j) \approx \dfrac{32}{3\sqrt{\pi}} \dfrac{1 - 2(\sqrt{F_j/\pi})R_c / L\sqrt{A_j}}{-\log\sqrt{A_j}}$
flat cavities $(\Delta >> 1)$	$dA_j / dt = \dfrac{12}{(\sin\Omega/2)^3} \dfrac{D_S\delta_S\Omega}{kT} \dfrac{\gamma_S}{LR_c^3} \dfrac{\sqrt{A_j}}{(1-A_j)^3}$
diffusion-viscoplasticity coupling	$dA_j / dt \approx \dfrac{32}{3\sqrt{\pi}} \dfrac{F_j^{3/2}}{F_V} \dfrac{D_j\delta_j\Omega}{kT} \dfrac{\sigma_a}{L^3} \dfrac{1}{\sqrt{A_j}}$
creep	$dA_j / dt = \dfrac{\alpha\dot{\varepsilon}d}{L\sqrt{A_j}h(\Psi)}$, $\dot{\varepsilon}$ small
	$= \dfrac{16}{h(\Psi)} \dfrac{D_j\delta_j\Omega}{kT} \dfrac{\sigma_a}{L^3} \dfrac{1}{\sqrt{A_j}.Q(\sqrt{A_j})}$, $\dot{\varepsilon}$ large
	where $Q(A_j) = \log\sqrt{A_j} + A_j(1 - A_j/4) - 3/4$
	and $h(\Psi) = \left(\dfrac{1}{1+\cos\Psi} - \dfrac{\cos\Psi}{2}\right)\dfrac{1}{\sin\Psi}$

Table 2.7 Approximations for the rate of growth of the surface fraction A_j of the cavities. L = distance between cavities, $R_c = 2\gamma_S/\sigma_a$, $\cos\Psi = \gamma_j/2\gamma_S$, σ is the normal stress on a grain boundary.

In 3 dimensions, with σ_{eq} and σ_m retaining their usual meanings, $\dot{\varepsilon}$ must be replaced by the equivalent deformation rate in the sense of von Mises; if the law of steady state creep under uniaxial stress has the form

$$\dot{\varepsilon} = \dot{\varepsilon}_0 (\sigma / \mu)^m \tag{2.187}$$

we have

$$\dot{\varepsilon}_{ij} = \frac{3}{2}(\dot{\varepsilon}_0 / \mu^m)\sigma_{eq}^{m-1} s_{ij}, \quad \dot{\varepsilon}_{eq} = \sqrt{\frac{2}{3}\dot{\varepsilon}_{ij}\dot{\varepsilon}_{ij}} \tag{2.188}$$

For a polycrystal it is usual to take for σ_a the maximum principal stress Σ_I.

These laws can be used to find the conditions of stress and temperature under which a maximum growth rate dominates. In view of the large number of items of physical data that are required, this can be done for only a limited number of pure metals. Ashby (Frost and Ashby (1982)) has studied copper and silver; Fig. 2.84 gives his results for copper, where the growth rate is taken as the rate of increase of the surface fraction of the cavities, that is,

$$A_j = 4r^2/L^2 \tag{2.189}$$

The relation between dA_j/dt and dV/dt is easily found to be

$$\sqrt{A_j}\, dA_j / dt = \frac{16}{3\pi^{3/2}} \frac{F_j^{3/2}}{F_V} \frac{1}{L^3} dV / dt \tag{2.190}$$

The two charts of Fig. 2.84 correspond to two values of the parameters L and A_j; they delimit the regions in the (Σ, T) plane in which one or other of the previously-described mechanisms is dominant.

Ashby's diagrams make precise the general tendencies we have already indicated. Thus at intermediate temperatures - T/T_f around 0.5, meaning T around $600°C$ for copper - and stresses - $3.10^{-5} \leq \Sigma_I/E \leq 10^{-3}$ - two main modes of growth are in operation: by surface diffusion and by viscoplasticity. It is only at very low stresses that intergranular diffusion becomes important. An interesting comment on the comparison between the various laws is the following. As we shall see later, for the majority of metals there is an approximate relation between the creep rate $\dot{\varepsilon}$ and the time to fracture t_R (the Monkman-Grant law) which states that $\dot{\varepsilon}^q t_R$ is constant, where q is a constant, often approximately 1. The growth mechanisms we have studied enable us to calculate the time to fracture, on the basis of this occurring when the surface fraction of the cavities on the boundaries reaches a certain critical value. Table 2.7 shows that this does indeed follow from the laws derived for growth controlled by viscoplastic deformation or by creep.

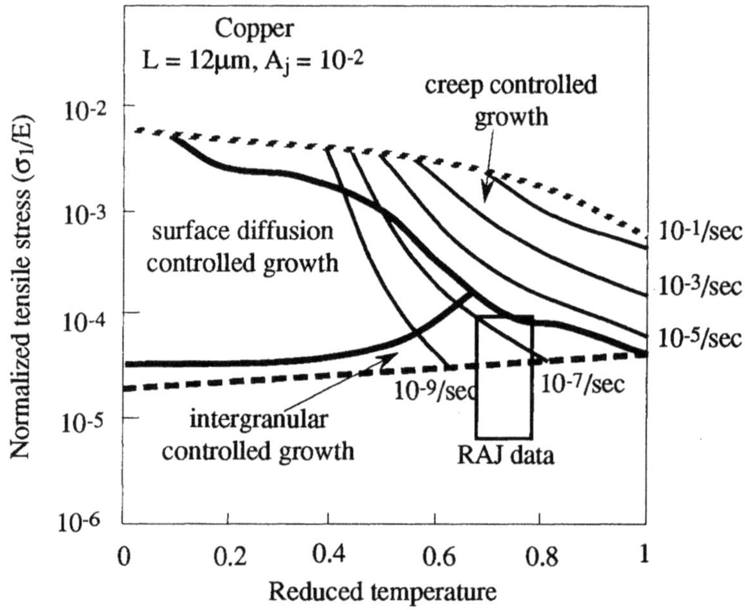

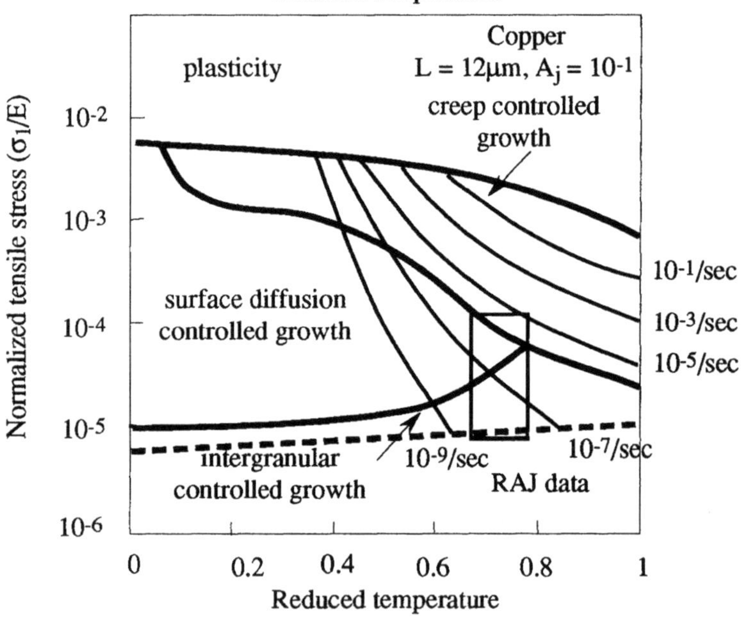

Fig. 2.84 Modes of viscoplastic fracture for two initial values of the surface fraction A_j of intergranular cavities for copper with grain size $L = 12 \mu m$. Temperature is given as the reduced temperature T/T_f, where T_f is the fusion temperature. The charts give lines of equal creep rate. (Frost and Ashby (1982))

However, in the course of creep there is a continuous nucleation of cavities which should be taken into account; if this is assumed to be at a rate proportional to the creep rate it follows that

$$\dot{N} = \dot{\varepsilon} / a \tag{2.191}$$

where a is a constant having the dimension of area. In the equation for creep-controlled growth, (2.184), we can express the changing surface fraction A_j of the cavities and introduce the continuous reduction of the inter-cavity distance L given by

$$L/2 = (\pi N)^{-1/2} = (\pi \dot{\varepsilon}\, t / a)^{-1/2} \tag{2.192}$$

assuming that the creep rate $\dot{\varepsilon}$ is constant.

This leads to two limiting regimes. The first corresponds to low creep rates, which allow the stress to relax completely. Integration of the differential equation for A_j gives for the time t_R to fracture

$$t_R = \left[\frac{2h(\Psi)}{\alpha d \sqrt{\pi / a}} \right]^{3/2} \frac{A_{jc}}{\dot{\varepsilon}} \tag{2.193}$$

where A_{jc} is the critical fraction of cavities which leads to fracture; this is approximately *0.4*.

The second corresponds to high creep rates, when there is very little relaxation of stress. (2.184) now approximates to

$$\frac{dA_j}{dt} = \frac{16}{L^3 h(\Psi)} \frac{D_j \delta_j \sigma_a \Omega}{kT} \frac{1}{\sqrt{A_j}} \frac{1}{Q(\sqrt{A_j})} \tag{2.194}$$

giving on integration

$$t_R = \left[\frac{5h(\Psi) A_{jc} kT}{4KD_j \delta_j \Omega (\pi / a)^{3/2}} \right]^{2/5} \left(\frac{\mu}{\dot{\varepsilon}_0} \right)^{3/5} \frac{1}{\sigma_{eq}^{3m/5} \Sigma_I^{2/5}} \tag{2.195}$$

where $K = (L/2r)/Q(L/2r)$, approximately *10*, and a creep law $\dot{\varepsilon}_{eq} = \dot{\varepsilon}_0 (\sigma_{eq} / \mu)^m$ is assumed. This can be written as a function of creep rate:

$$t_R = \left[\frac{5h(\Psi) A_{jc} kT}{4KD_j \delta_j \Omega (\pi / a)^{3/2}} \right]^{2/5} \frac{\dot{\varepsilon}_0^{1/m}}{\mu} \frac{2}{\dot{\varepsilon}_{eq}^{(2+3m)/5m}} \tag{2.196}$$

The first regime (low creep rate, complete relaxation) gives the Monkman-Grant law; for the second (high creep rate, little relaxation) the value of the exponent $(2+3m)/5m$ is between 0.60 and 0.68, which is much less than that of about 0.85 usually observed.

Comparison of theoretical predictions with experimental observations is made even more difficult in the regime of intergranular cavitation than in that of plastic cavitation by the serious lack of experimental data. Direct measurement is very difficult, consequently most comparisons are indirect, made through the intermediary of the time to fracture; this requires strict control of the nucleation rate, for which various methods are available. One, for the case of silver or copper, involves implanting bubbles of water vapour before the creep test; another, creating cavities by means of a preliminary creep at high stress, followed by creeping under low stress. The charts of Fig. 2.84 give some results obtained with the help of these methods.

2.12.3 Damage by creep; phenomenological approaches

2.12.3.1 *Introduction*

As we have emphasized previously, damage measured in terms of, for example, the average size of the cavities at a given instant or the total volume occupied by them at that instant, consists in general of two components, nucleation and growth. Formally we can write

$$V(t) = \int_0^t \dot{N}(t')dt' \int_{t'}^{t} \frac{d}{dt} V(t',t'')dt'' \qquad (2.197)$$

Here dV/dt is the rate of growth of the cavities which were created at the instant t' and $\dot{N}$ is the rate at which they were created. In view of the variety of regimes that we have described it will be appreciated that there would be great difficulty in deriving from the basic laws of nucleation and growth a law governing the kinetics of damage that could be considered as being at the microscopic level.

Faced with this, the approaches that have been made to the problems of creep damage have been at the macroscopic and phenomenological levels, with damage measured most often in terms of time to fracture. In this section we first describe the main approaches and then review the measurements of intergranular creep with the aim of locating them with respect to the macroscopic approaches.

2.12.3.2 *The Monkman-Grant law*

In §2.12.2.1(d) we said that for many materials the data for creep at fracture under uniaxial tension follow a simple law, established by Monkman and Grant (1956), which can be written

$$\dot{\varepsilon}_S^q t_R = C_{MG} \qquad (2.198)$$

where $\dot{\varepsilon}_s$ is the steady-state creep rate, t_R the time to fracture and C_{MG} a constant for each material, varying little with temperature and applied stress. The exponent q does not differ greatly from *1*, lying between *0.7* and *2*; a value *1* indicates that to a first approximation the ductility of the material in creep is constant, at least when the deformation in the tertiary stage is small.

2.12.3.3 *Time-to-fracture contours in creep under multiaxial loading*

Over the past twenty years or so experimenters have been concerned with extending creep data obtained under conditions of uniaxial loading to the multiaxial case. The existing data refer most often to plane stress on tubes loaded in tension-torsion. If σ_0 is the uniaxial stress under which the time to fracture has a given value, a curve can be constructed in the (σ_1, σ_2) plane on all points of which the time to fracture has this value. The equation for this curve can be written

$$\Delta(\sigma_{ij} / \sigma_0) = 1 \quad \text{where} \quad \Delta = \tilde{\alpha}\frac{\sigma_1}{\sigma_0} + \tilde{\beta}\frac{\sigma_{eq}}{\sigma_0} + \tilde{\gamma}\frac{\sigma_1 + \sigma_2 + \sigma_3}{\sigma_0} \tag{2.199}$$

If $\tilde{\gamma}$ is small the curves will lie between two surfaces which in plasticity correspond respectively to the criteria of equivalent stress ($\tilde{\alpha} = 0$) and of maximum principal stress ($\tilde{\beta} = 0$). These boundaries are shown in Fig. 2.85; as noted there, the two extremes of behaviour are seen in aluminum and copper respectively.

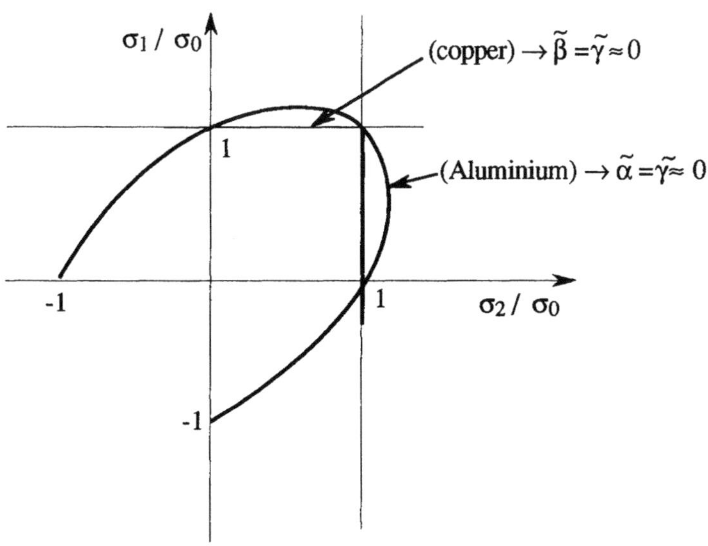

Fig. 2.85 Curves of equal time to fracture in creep, for the extremes of equivalent stress criterion (aluminum) and maximum principal stress criterion (copper).

2.12.3.4 *Bases of the mechanics of damage*

The mechanics of viscoplastic damage [Kachanov (1958), Rabotnov (1969), Leckie and Hayhurst (1977), Lemaitre and Chaboche (1990).] is a continuous approach in which damage is expressed in terms of scalar or tensor variables.

The most recent work is based on the thermodynamics of irreversible processes and the formalism of internal variables; we shall give only an elementary account of the theory and an introduction to the scalar quantities used to provide a unique measure of damage (see also §2.5.3)

Damage, which we have denoted by D, does not necessarily have a precise physical meaning, at least in a first approach; it can include various forms of physical damage such as the nucleation and growth of intergranular cavities previously discussed. We shall assume that damage D and the viscous deformation depend not only on the stress and temperature but also on the value of D itself; thus in simple traction

$$\dot{\varepsilon} = d\varepsilon / dt = f(\sigma, T, D), \quad \dot{D} = dD / dt = g(\sigma, T, D) \tag{2.200}$$

The lifetime under creep, the deformation at fracture and the creep curves are all obtained by integrating these equations between appropriate limits.

Various forms have been proposed for the functions f and g; those due to Kachanov (11958) are usually written as follows:

uniaxial stress:

$$\dot{\varepsilon} = \dot{\varepsilon}_0 (\sigma / \mu)^m (1 - D)^{-n} \quad \dot{D} = \dot{D}_0 (\sigma / \mu)^k (1 - D)^{-q}$$

multiaxial stress:

$$\dot{\varepsilon}_{ij} = \frac{3}{2} \frac{\dot{\varepsilon}_0}{\mu^m} \sigma_{eq}^{m-1} s_{ij} (1 - D)^{-n} \quad \dot{D} = \dot{D}_0 \Delta^k \frac{\sigma_{ij}}{\sigma_0} \left(\frac{\sigma_0}{\mu} \right)^k (1 - D)^{-q} \tag{2.201}$$

The time t_R to fracture is found by integrating the appropriate equation for D with the boundary conditions $D = 0$ (no damage) at $t = 0$, $D = 1$ (fracture) at $t = t_R$; the result for the uniaxial case is

$$t_R = \frac{1}{\dot{D}_0 (q + 1)(\sigma / \mu)^k} \tag{2.202}$$

from which the corresponding deformation at fracture is found to be

$$\varepsilon_R = \frac{1}{q - n + 1} \frac{\dot{\varepsilon}_0}{\dot{D}_0} \left(\frac{\sigma}{\mu} \right)^{m-k} \tag{2.203}$$

Very often $m < k$, so that the ductility in creep is a decreasing function of the stress; and this is what is observed. For this ductility to remain constant, and consequently for the Monkman-Grant law to hold in its simplest form (2.198), we must have $m = k$.

2.12.3.5 *Mechanics of damage; physical measures of intergranular damage*

As we have said, there is very little recorded information on intergranular damage that has a direct bearing on the rate of nucleation of cavities and on the kinetics of their growth. However, there are reasonably complete sets of data for copper and for the nickel-based alloy Nimonic 80; and data have been published more recently for an austenitic stainless steel. In reviewing these we shall show in each case how the results of the purely macroscopic treatment of damage that we have given can be reconciled with the metallographic observations.

(a) *Results for copper.* A detailed study of this metal was made by Hanna and Greenwood (1982), using creep tests under uniaxial tension at *500°C*. They measured the time to fracture and also the number, dimensions and total volume of the cavities. If for a test at stress σ_0 the time to fracture is t_0, the deformation there is ε_0, the number of cavities is N_0 and their average volume is V_0. Hanna & Greenwood's results can be expressed as follows:

$$d(\varepsilon/\varepsilon_0)/dt = (\sigma/\sigma_0)^5/t_0$$

$$d(N/N_0)/dt = (\sigma/\sigma_0)^2 d(\varepsilon/\varepsilon_0) \qquad (2.204)$$

$$d(V/V_0)/dt = (\sigma/\sigma_0)/t_0$$

Comparing these equations with those derived from the theoretical models (cf. 2.174) we see that the kinetics of growth follows essentially the law corresponding to intergranular diffusion. Further, a comparison with the laws postulated in the mechanics of damage (cf. 2.201) shows that in this particular case damage has a negligible effect on creep rate.

We could take as a definition of damage the surface fraction of the cavities:

$$D = NV^{2/3}/N_0V_0^{2/3} \qquad (2.205)$$

If the tests are performed under constant uniaxial tension it follows from (2.204) that

$$D = (N/N_0)(V/V_0)^{2/3} = (t/t_0)^{5/3}(\sigma/\sigma_0)^{23/3} \qquad (2.206)$$

and with the condition for fracture $D = 1$

$$(t_R/t_0)(\sigma/\sigma_R)^{23/5} = 1,\ (\varepsilon_r/\varepsilon_0) = (\sigma/\sigma_0)^{2/5} \qquad (2.207)$$

Equation (2.207) expresses the fact that creep ductility is a decreasing function of stress and therefore does not follow the simple Monkman-Grant law.

Equation (2.206), giving D as a function of time and of load, can be written as a differential equation:

$$dD/dt = (5/3t_0)(\sigma/\sigma_0)^{-23/5}D^{2.5} \tag{2.208}$$

which is a particular form for the kinetics of damage proposed by Kachanov and Rabotnov.

For the case of multiaxial loads Hayhurst, Trampczynski and Leckie (1983) have given a generalisation of these equations, based on simple concepts suggested by the theories and models of the nucleation and growth of intergranular cavities. The equations from which this starts are:

$$d(\varepsilon_{eq}/\varepsilon_0)dt = (\sigma_{eq}/\sigma_0)^5/t_0$$

$$d(N/N_0)/dt = (\Sigma_I/\sigma_0)^2 d(\varepsilon_{eq}/\varepsilon_0)/dt$$

$$d(V/V_0)dt = (\Sigma_I/\sigma_0)/t_0 \quad \text{with } D = (N/N_0)(V/V_0)^{2/3} \tag{2.209}$$

Proceeding as before, we find for the uniaxial case:

- the time-to-fracture contour has the form

$$\Delta = (\Sigma_I/\sigma_0)^{8/23}(\sigma_{eq}/\sigma_0)^{15/23} = 1 \tag{2.210}$$

- the kinetics of the variation of D can be written

$$dD/dt = (2/3t_0)\Delta^{23/5}(s_{ij}/\sigma_0)(1 - D)^{-1/2} \tag{2.211}$$

Fig. 2.86 gives the corresponding time-to-fracture contour; it will be seen that this lies between the two curves corresponding to the criteria of equivalent stress and maximum principal stress respectively.

(b) *Results for the alloy Nimonic 80A*

Dyson and McLean (1977) have made a detailed study of this alloy; making creep tests not only under uniaxial stress but also under tension-torsion; all at *700°C*. They measured the time to fracture, the number of cavities per unit area of grain boundary and the total volume of the cavities. They give the following equations for the multiaxial case:

$$d(\varepsilon_{eq}/\varepsilon_0)dt = 3(\sigma_{eq}/\sigma_0)^4(V/V_0)^{4/9}/t_0$$

$$d(N/N_0)dt = ½(\Sigma_I/\sigma_{eq})^2(\varepsilon_{eq}/\varepsilon_0)^{-1/2}d(\varepsilon_{eq}/\varepsilon_0)/dt \tag{2.212}$$

$$d(V/V_0)dt = (\Sigma_I/\sigma_{eq})^{0.7}d(\varepsilon_{eq}/\varepsilon_0)/dt$$

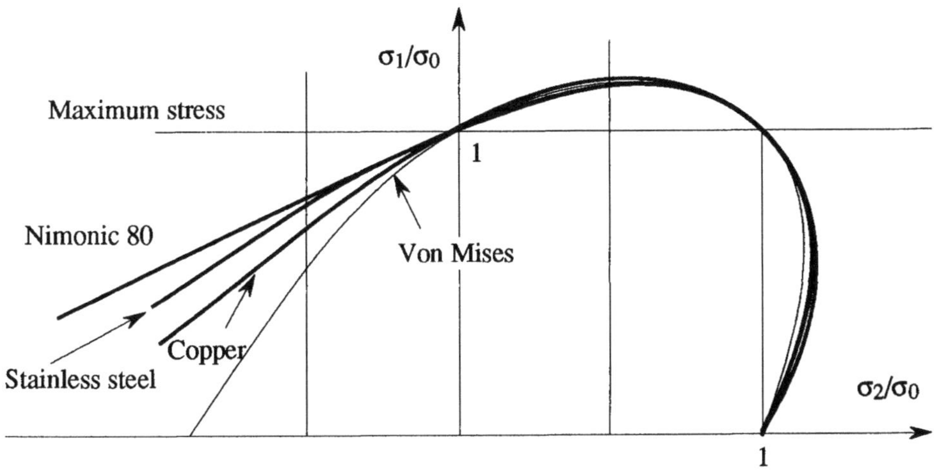

Fig. 2.86 Time-to-fracture contours for copper, austenitic stainless steel and the nickel-based alloy Nimonic 80A, together with the maximum-stress and von Mises criteria.

The first equation expresses the coupling between damage and creep rate and is therefore of the same type as the first equation proposed in the macroscopic treatment of damage mechanics (cf. 2.5.3) The third equation does not seem to match any of those already established for the kinetics of cavity growth; however, in the uniaxial case it can be reduced to the form corresponding to growth retarded by creep in the matrix (cf. 2.186), since then dV/dt is directly proportional to $d\varepsilon/dt$: this assumes that the effect of the damage in accelerating this latter rate of change can be neglected.

The criterion for fracture adopted by Dyson and McLean is

$$NV/N_0V_0 = \sigma_0/\Sigma_1 \tag{2.213}$$

The same procedure as before gives the form of the time-to-fracture contour as

$$(\Sigma_1/\sigma_0)^{1/9}\,\Delta(\sigma_{ij}/\sigma_0) = 1 \quad \text{where } \Delta = [\Sigma_1^{\,1.8}\,\sigma_{eq}^{\,2.2}/\sigma_0]^{1/4} \tag{2.214}$$

This also is shown in Fig. 2.86; it is seen to lie between the two extreme contours. It differs very little from that corresponding to $\Delta = 1$, which means that a result very close to that obtained here is given by taking the still simpler definition of damage

$$D = NV/N_0V_0 \tag{2.215}$$

which in this case represents a linear fraction.

For a proportional loading the ratio Σ_I/σ_{eq} is constant; the Dyson-McLean equations (2.212) then depend only on the single variable D and can be integrated. This gives

$$d(\varepsilon_{eq}/\varepsilon_0)/dt = (3/t_0)(\sigma_{eq}/\sigma_0)^4\, D^{4/9},\; dD/dt = (9/2t_0)\Delta^4 D^{7/9} \qquad (2.216)$$

These are again particular forms of the relations proposed by Kachanov and Rabotnov.

A criticism that could be leveled against this treatment is that according to the first of these equations the creep rate is zero at $t = 0$ (when $D = 0$), and this is contrary to what is found in practice. In fact, the expression concerning creep rate in the Dyson-McLean treatment takes into account only the part corresponding to the tertiary creep; that corresponding to the steady-state creep should be added when calculating the total deformation.

In order to bring these equations to a form closer to that usually adopted in damage mechanics, Leckie and Hayhurst (1977) showed that they are essentially equivalent to the following:

$$\frac{d(\varepsilon_{eq}/\varepsilon_0)}{dt} = \left(\frac{\sigma_{eq}}{\sigma_0}\frac{1}{1-D}\right)^4 \frac{1}{t_0},\quad \frac{dD}{dt} = \left(\frac{\Delta}{1-D}\right)^4 \frac{1}{t_0} \qquad (2.217)$$

These two examples are good illustrations of the fact that links can be established between apparently different approaches to the problem of creep damage.

(c) *Results for an austenitic stainless steel*

The steel concerned is grade AFNOR Z02 CND 17.12; having a controlled nitrogen content of *0.07%* by weight; tests were made at *600°C* with both smooth and notched test pieces, calculations for the latter being made by the finite element method.

In this material, as in the majority of alloys that have carbides in their grain boundaries, it is very difficult to determine the size of the cavities on the boundaries unless these are already at an advanced stage of growth.

The measure of physical damage was taken to be the ratio D of the length of the broken boundaries within a given area to that of the boundaries lying in that area, all this measured on a micrograph of a section at an enlargement of *200x*. Fig. 2.87 shows the very irregular distribution of damage in this steel after creep. The situation is very likely to be one in which the microscopic growth follows a law corresponding to constrained cavitation by creep in the matrix (equations 2.184 and 2.186).

Fig. 2.87 Optical micrograph of a longitudinal section of a sample of an austenitic stainless steel creep-tested at *600°C* (t_R = *370 hr.*), showing intergranular cracks.

The tests on notched pieces made it possible to quantify the influence of the maximum principal stress and of the deformation due to creep on the kinetics of inter-granular damage dD/dt. These effects can be linked with a continuous nucleation of cavities and with the growth law (see §2.12.2.2d). The empirical relation found is

$$dD = A\varepsilon_{eq}^{-0.5}\Sigma_I^{4}\, d\varepsilon \tag{2.218}$$

where A is a constant for a given temperature and Σ_I is the maximum principal stress.

If we assume that in a uniaxial test fracture occurs when the damage measure reaches a critical value D_0, integration of (2.218) leads to a simple relation between the strain to failure ε_R and the stress σ, of the form

$$\varepsilon_R\sigma^8 = const. \tag{2.219}$$

In reality, in steels of this type, which have properties associated with ageing and with precipitation of carbides in the grain boundaries, the development of creep ductility follows a more complex path.

If we can assume that the creep law is not affected by the damage, as is the case for copper, and limit our consideration to steady-state creep, we have also the relation

$$\dot{\varepsilon} = \dot{\varepsilon}_0(\sigma / \mu)^m \quad \text{with } m = 8 \tag{2.220}$$

In the multiaxial case it can be shown that the kinetics of damage, in Kachanov & Rabotnov's form, can be expressed as

$$\frac{dD}{dt} = \frac{1}{2}\left(\frac{\sigma_{eq}}{\sigma_0}\right)^8\left(\frac{\Sigma_I}{\sigma_0}\right)^8 \cdot \frac{1}{D} \cdot \frac{1}{t_0} = \frac{1}{2}\Delta \cdot \frac{1}{D} \tag{2.221}$$

The fracture condition $D = 1$ then leads to a time-to-fracture contour essentially of the form $\Delta = 1$, which is approximately

$$\frac{\sigma_{eq}}{\sigma_0} \cdot \frac{\Sigma_I}{\sigma_0} = 1 \tag{2.223}$$

In this application (to Nimonic 80A), this last equation differs little from that given by Dyson and McLean's treatment (cf. 2.214, 2.215)

In conclusion we have shown, with the aid of three specific examples, that there are strong similarities between two apparently very different approaches to the problem of damage: one, microscopic, based on measuring metallographic damage, the other, macroscopic, postulating the existence of laws having a stated form. In practice it often proves difficult to discriminate between the various laws proposed, especially when the tests performed under biaxial loading, such as tension-torsion, are limited to values in the first quadrant of the stress plane.

A more rigorously-based approach would certainly take account of the directional aspect, but this is an aspect of the problem which, important though it is, we have not considered here and which has still been little studied.

2.13 INTERGRANULAR BRITTLENESS

If, for any of a variety of reasons, the grain boundaries are more brittle than the grain interiors then cracks will develop along the latter. There are many causes of embrittlement; we shall consider that due to tempering and to superheating in steels and to liquid metals.

2.13.1 Reversible-temper embrittlement in steels

This appears in steels that have been quenched and tempered and have spent some time at a temperature in the range *350-550°C*; its occurrence is shown by an increase in the transition temperature of fracture appearance, which can be as much as *100°C*. It is fastest at *550°C* and develops more quickly during cooling than at constant temperature; it is reversible and can be eliminated completely by annealing at above *550°C*; followed by quenching; it is characterised by intergranular fracture which usually, but not always, follows the former austenitic grains.

Analysis of the fractured surfaces shows metalloid impurities, the most harmful of which are, in descending order, antimony, tin, phosphorous and arsenic; but it is found that their embrittling effect is always associated with the presence of alloying elements such as nickel and manganese.

Table 2.VIII (below) gives the embrittling elements, distinguishing those which favour segregation at the boundaries and those which oppose this.

2.13.2 Segregation of impurities at the boundaries

The solubility of elements is not the same in the interior of the grains as in the boundaries: the distortion energies generated by the solute atoms are different in the two locations. A solute that reduces the surface energy of the boundary will diffuse towards this so as to reach an equilibrium composition defined by

$$\Gamma_j/(1 - \Gamma_j) = [c/(1 - c)] \exp(-\Delta G_j/kT) \qquad (2.223)$$

where Γ_j is the fraction of atomic sites in the boundary that are occupied by solute atoms, c is the concentration of the solute in the alloy and ΔG_j is the Gibbs free energy of segregation in the boundary.

If there are several different elements in solution this equation must be modified to take into account their interaction energies. If this is high, as in the case of titanium and phosphorous, precipitates will be formed in the grains, eliminating the impurities and thus inhibiting the segregation; if it is moderate, the two elements will diffuse together in the boundary, reducing the energy more than would have been the case had only one done so (Fig. 2.88). It can even happen that a pair of elements, each of which would not itself have any tendency to concentrate in the boundary, will together make this segregation possible.

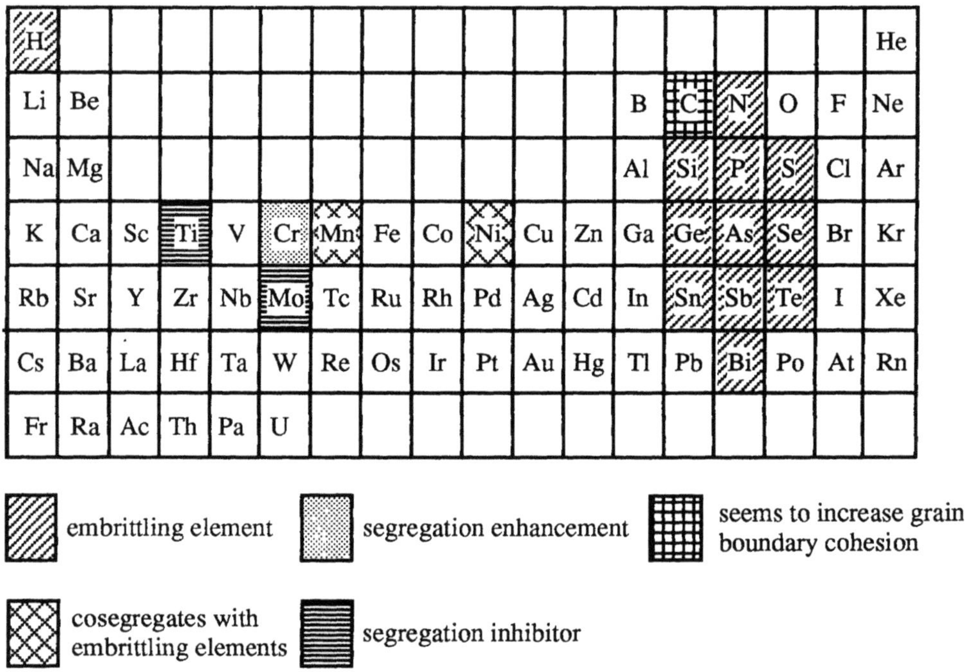

Table 2.8 Intergranular embrittlement - Effect of composition.

There is now the question of why this segregation of impurities results in embrittlement of the boundaries, a question already touched on in §2.2.4. Recall that fracture at a boundary requires the expending of an amount of energy per unit area of

$$\gamma = 2\gamma '_S - \gamma_j \qquad (2.224)$$

where $\gamma '_S$ is the surface energy immediately after fracture and γ_j that of the boundary; γ'_S can differ from γ_S, the equilibrium surface energy. The presence of impurities in the boundary can modify the two terms in this relation and reduce γ sufficiently to favour fracture along the boundary.

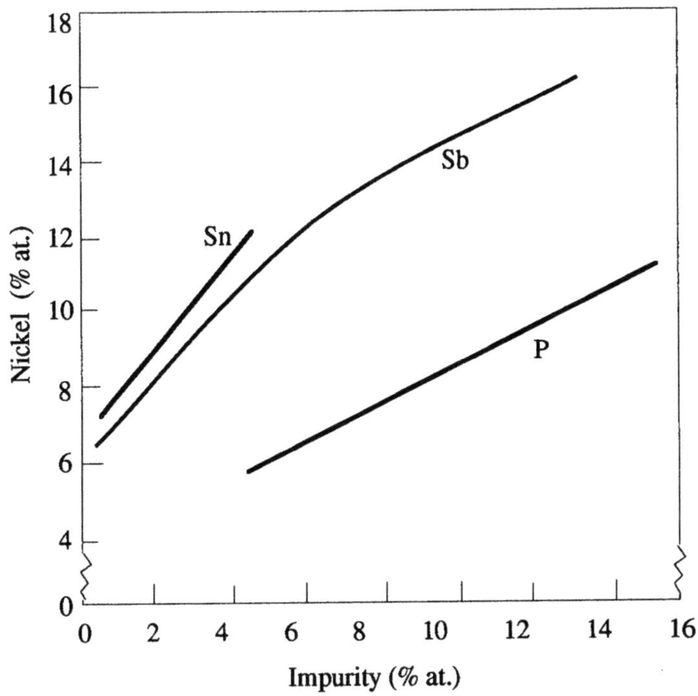

Fig. 2.88 Nickel content as a function of content of metalloid impurities in the grain boundaries, showing the joint enrichment

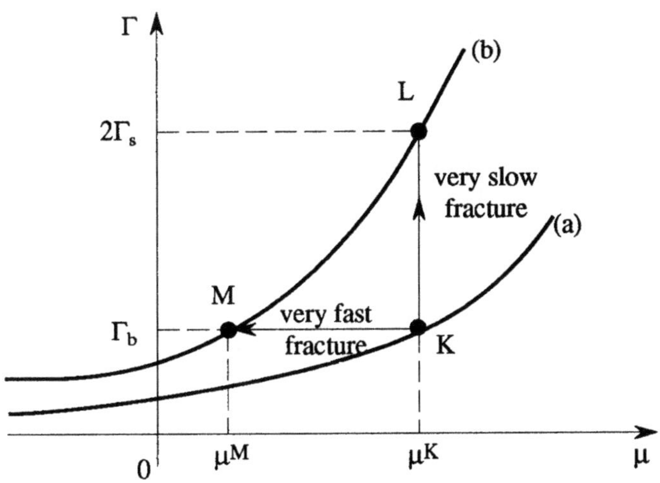

Fig. 2.89 Concentration of foreign atoms (a) in the grain boundaries (b) at the surface, as a function of the chemical potential

The concentration of foreign atoms Γ is not the same on the boundary as at the surface. Both are functions of the chemical potential μ, as Fig. 2.89 shows. Very slow fractures take place at constant values of this, with Γ changing from Γ_j to $2\Gamma_S$; but during the rapid fractures in which we are interested there is not time for the concentration to change, and the potential decreases.

If there are no foreign atoms the fracture energies at constant potential γ_μ and constant concentration γ_Γ are both equal to γ_0:

$$\gamma_\mu = \gamma_\Gamma = \gamma_0 = 2\gamma_S^0 - \gamma_j^0 \qquad (.2.225)$$

where γ_S^0 and γ_j^0 are the surface and grain boundary energies respectively for the pure metal.

If there are impurities we can calculate γ_Γ by going round a closed loop in the Γ-μ diagram: we create the crack by expending an energy γ_Γ (the path KM), bring μ to $-\infty$ by desorption of the surface, re-create the boundary, gaining γ_0, and bring about the segregation along the boundary:

$$\gamma_\Gamma = \gamma_0 - \int_0^{\Gamma_j} [\mu_j(\Gamma) - \mu_S(\Gamma/2)]d\Gamma \qquad (2.226)$$

For low concentrations $\mu = \mu_0 + kT \log c$, $\Gamma_j = c \exp(-\Delta G_j/kT)$, $\Gamma_S = c \exp(-\Delta G_S/kT)$

where ΔG_S is the Gibbs free energy of segregation at the surface.

Substitution of these in (2.226) gives

$$\gamma_\Gamma = 2\gamma_S^0 - \gamma_j^0 + \Gamma_j(\Delta G_S - \Delta G_j - kT \log 2) \qquad (2.227)$$

Very few values are known for ΔG_S and ΔG_j; however, an approximate solution can be developed by using a model based on atomic pairs, as indicated in Fig. 2.90. With this,

$$\gamma_0 = -Z_j\, u_{AA} \qquad (2.228)$$

where Z_j is the coordination number for an atom on one side of the boundary with respect to one on the other side and u_{AA} is the binding energy of an AA pair; on average,

$$Z_j = (1 - \gamma_j^0/2\gamma_S^0)Z \qquad (2.229)$$

where Z is the coordination number of the crystal system and γ_j^0/γ_S^0 has typically the value $1/3$.

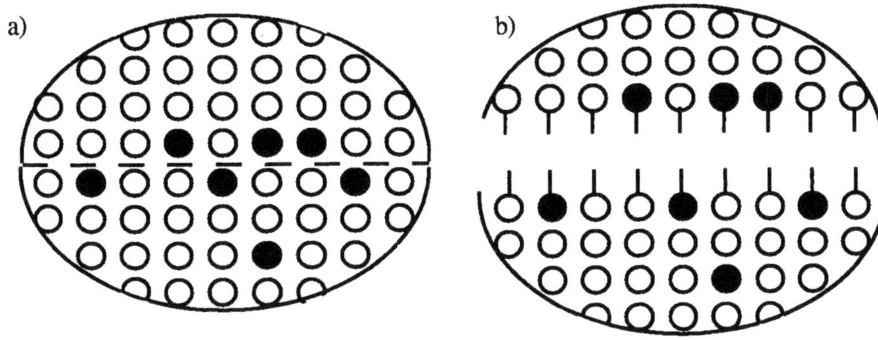

Fig. 2.90 The "pairs" model for evaluating embrittlement of boundaries.

For a boundary containing atoms B

$$\gamma_\Gamma - \gamma_0 = Z_j \Gamma_j(-u_{BB}+u_{AA} - \omega) \tag{2.230}$$

where $\omega = u_{AB} - \frac{1}{2}(u_{AA} + u_{BB})$ (2.231)

The binding energies u_{AA}, u_{BB} of AA and BB pairs can be calculated from the sublimation enthalpies:

$$u_{AA} = -H_A^{sub}/Z,\ u_{BB} = -H_B^{sub}/Z \tag{2.232}$$

By neglecting the regular term ω of regular solutions, Seah (1980) was able to construct the diagram of Fig. 2.91, from which $\gamma_\Gamma - \gamma_0$ can be determined. If the matrix consists of iron all impurities with lower sublimation enthalpies will be embrittling; the diagram shows also the elements that should be added to counter this effect, such as molybdenum. A good illustration of the importance of this diagram is provided by Fig. 2.92, which shows the effect of this particular alloying element on the brittle-ductile transition temperature in the case of embrittlement by phosphorous.

It is equally interesting to return to the Rice-Thomson (1974) analysis described in §2.2.6, in particular the graph of Fig. 2.19 in the plane of the parameters S and R_0; here the curve corresponding to $U = 0$ delimits the region $U > 0$ in which intergranular fracture would be favoured, provided that the boundary energy γ_Γ is included in the expression for U, that is

$$U \approx U_{d\max}(\gamma_\Gamma / \gamma_m \mu b^3) \tag{2.233}$$

In the pure metal $\gamma_\Gamma = 2\gamma_S^0 - \gamma_j^0$ and $\gamma_m = 2\gamma_S^0$; and since $\gamma_0/2\gamma_S^0 \approx 5/6$, $R_0 \approx 5r_0/6b$ and $S \approx 2\gamma_S^0/\mu_b$. We are thus in a region in which a very small change in the extent of the core of the dislocations r_0 or in γ_S^0 can change the behaviour drastically.

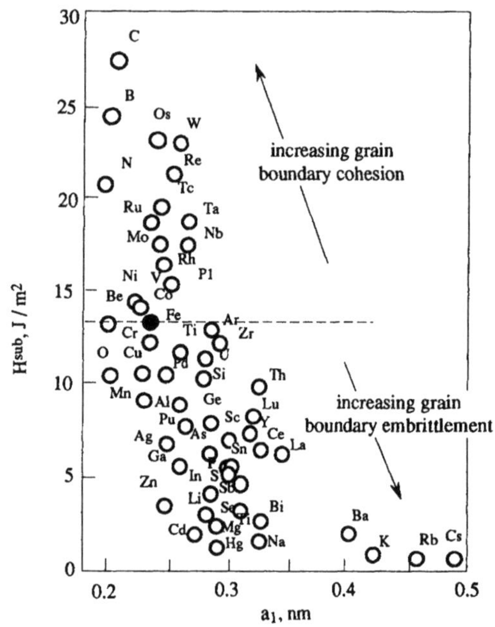

Fig. 2.91 Seah's diagram, showing the tendencies of various atoms to embrittle iron.

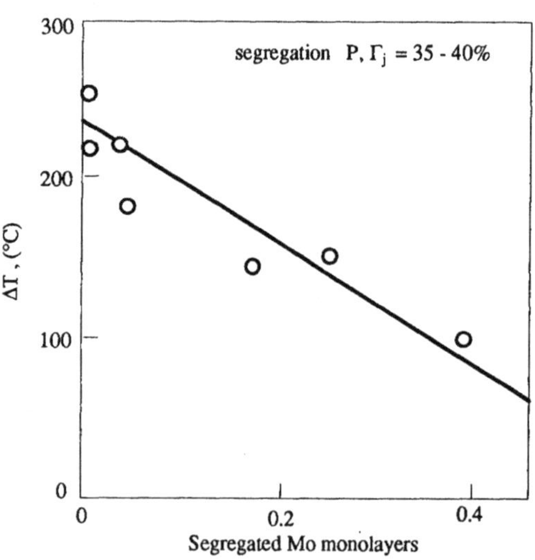

Fig. 2.92 Effect of molybdenum in reducing the brittle-ductile transition temperature, thus countering the embrittlement of steel by phosphorous. The line is the solution given by the idealized model.

In Fig. 2.19 we gave some results concerning fracture of copper and of bismuth-doped copper, the latter giving intergranular fracture. In order to evaluate the effect of this impurity we need to know its influence on both γ_r and γ_m; we may assume

$$\gamma_r(\mathrm{CuBi})/\,\gamma_r(\mathrm{Cu}) \approx H_{Bi}^{sub}/H_{Cu}^{sub} = 0.3$$

whilst

$$\gamma_m(\mathrm{CuBi})/\gamma_m(\mathrm{Cu}) \approx (H_{Bi}^{sub} + H_{Cu}^{sub})/2H_{Cu}^{sub} = 0.65$$

the latter evaluated by counting the bonds broken on steps inclined at $45°$.

2.13.3 Embrittlement by liquid metals

Certain liquid metals can have a dramatic embrittling effect on other metals, for example, indium on zinc, gallium on aluminum, lead on steels; the effect occurs in the region of the melting point of the liquid metal. The fracture is very often intergranular but not always so, and even single crystals can be fractured in this way.

The simplest explanation of this effect is based on the reduction of the cohesion. Equation (2.225) shows that the fracture energy is a function of the surface energy γ_S^0; this is reduced by the presence of a liquid and in terms of the pair model (2.228) becomes

$$\gamma_{0l} = Z_j.2\omega_{AC} \tag{2.234}$$

where $\omega_{AC} = u_{AC} - \frac{1}{2}(u_{AA} + u_{CC})$, C representing the atoms of the liquid.

If ω_{AC} is negative, intermetallic compounds will form on the surface; if positive, metals will dissolve in the liquid. Embrittlement will follow if ω_{AC} is positive, but not too great.

It has been suggested that the presence of liquid on the surface facilitates slip. To assess this it would be necessary to investigate the effect of the core radius r_0 in particular on the parameter R_0 in the Rice-Thomson theory; it could provide the best explanation of embrittlement of amorphous metals,

Finally, it has been suggested that in certain cases the atoms of the liquid metal penetrate the boundaries and embrittle them.

2.13.4 Superheating of steels

This type of embrittlement occurs when a steel has been heated to a high temperature, so as to dissolve all the impurities, and then cooled slowly to a temperature region in which the impurity atoms begin to be concentrated in the grain boundaries; at a lower temperature they precipitate out there again. This particular morphology has no effect on the cleavage phenomena and therefore on the brittle-ductile transition temperature, but it does affect the ductile fracture. Dimples form along the former austenitic grain boundaries and the energy of the upper shelf is lowered (Figs. 2.93, 2.94). This is what happens, for example, in a steel containing nitrogen cooled slowly from *1300°C*, when precipitates of aluminum nitride form in the austenitic grain boundaries. The same effect is seen in maraging steels cooled slowly after heating to *1200°C*, due to precipitates of titanium carbo-nitrides.

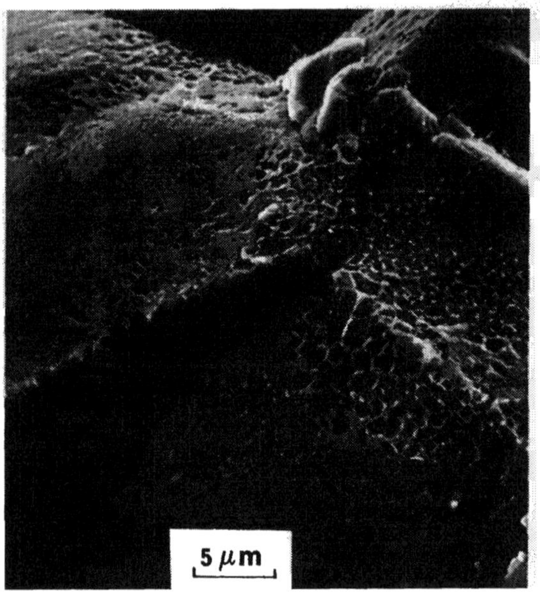

Fig. 2.93 Ductile intergranular fracture of a steel of type 15 MND 6 subjected to superheating. Some of the dimples on the surfaces of the grains contain fine particles of manganese sulphides.

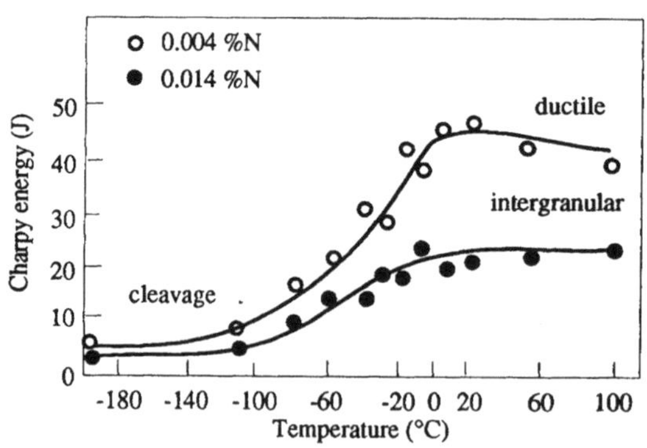

Fig. 2.94 Comparison of transition curves of Charpy energy for steels of high and low nitrogen content respectively, embrittled by intergranular aluminum nitrides.

The only way to remove this embrittlement is to heat the steel to a temperature high enough to dissolve the precipitates and follow this by a rapid cooling; the disadvantage is that this increases the grain size.

2.14 HYDROGEN EMBRITTLEMENT

A great deal of study has been given to this very important effect , but certain aspects are still not fully understood. The difficulty arises from the fact that although the hydrogen atom is very simple we do not have a good model for its effects on the cohesion energies of metals or on its interactions with the dislocations. Further, it is a very mobile atom, which can greatly complicate both experiments and their interpretation.

Introduction of hydrogen into a metal can be from a variety of causes: cathodic hydrogen, gaseous hydrogen , decomposition of water vapour, etc. Its solubility is fairly high in austenite and falls in ferrite so that on cooling it can precipitate out and form
flakes, small cracks that are very embrittling.

Hydrogen embrittling can be categorised by two main classes: one containing all those metals that can form hydrides, the other those that cannot.

2.14.1 Embrittlement of hydride-forming metals

These are the metals of columns IVB and VB of the periodic table: Ti, Zr, V, Nb, Ta and also Mg and Al. They include the case in which the hydride is not thermodynamically stable under ambient conditions but in which the energy of formation is affected by the stress field - this is the case for vanadium. Embrittlement of these metals is important and has the characteristic of first increasing with increasing temperature and then vanishing at high temperature. The deformation rate has the opposite of its usual effect: increasing this *lowers* the temperature of the transition to ductility; both this and the increase in ductility at low temperature are easily explained in terms of the hydrogen mobility. In the embrittled state, fracture has the character of cleavage, but it can occur after non-negligible extensions (Fig. 2.95)

In this case the model is based on the formation of hydrides under the influence of the stresses, which change the chemical potential of the hydrogen (see I -§3.3.8.5). In the very inhomogeneous stress field at the tip of a crack there is a high gradient of chemical potential, causing hydrogen to diffuse towards this region. Further, the stress field changes the Gibbs free energy of formation of hydrides, with the result that these form at the tip; then, because of the extreme brittleness of the hydrides, the tip progresses, and the process continues.

2.14.2 Metals and alloys that do not form hydrides

The general character of embrittlement is the same here as before - return to ductility at low temperature, lowering of the brittle-ductile transition temperature by increased rate of deformation - but the morphology of the fracture surface is very varied. It is often intergranular but can be transgranular cleavage or even dimpled fracture. The phenomena that come into play are more complex than in the case of hydride-forming metals.

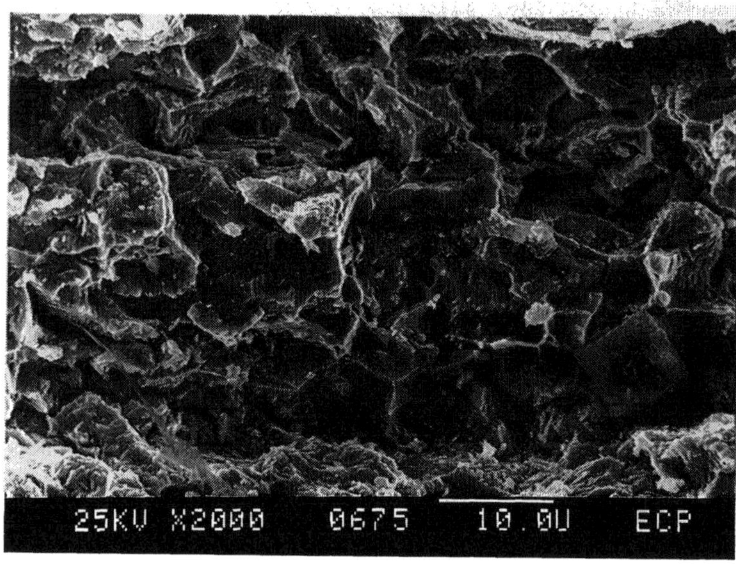

Fig. 2.95 Zirconium alloy (Zircaloy 4, stressed at *475°C*) containing *1055 ppm* hydrogen, showing fracture across the hydrides.

In a homogeneous solid solution hydrogen does not produce embrittlement; expressed otherwise, it does not seem to reduce the cohesion energy of the matrix. But it can concentrate in various regions, in particular in the grain boundaries. There it brings about an embrittlement which can be interpreted by the same model as that for embrittlement by tempering, except that here the high mobility of the hydrogen atom keeps the chemical potential constant - there is time for the redistributions to achieve this. Effects of co-segregation with hydrogen can also be seen, for example with sulfur in the case of nickel.

Here, as in the case of hydride-forming metals, hydrogen can concentrate in the plastic region at the head of a crack. The effects of this can be very complex because it can affect the mobility of the dislocations by forming atmospheres which these must then carry along, or by changing the dimensions of the core and so the Peierls force and the line tension. The result is not necessarily a hardening, and in certain cases can be a softening.

2.15 STRESS CORROSION CRACKING AND CORROSION FATIGUE

2.15.1 General character

Stress corrosion cracking (SCC) and corrosion fatigue are damage phenomena resulting from the combined actions of an aggressive environment and either steady or cyclic stress. They can have very varied manifestations of which it can be difficult to give any detailed description. The factors affecting either form are external - the environment - and internal - the metallurgical structure. In general, the lifetime of a component subjected to either will be shorter, the greater the stress; however, a small stress can suffice to cause the effects to appear, even though there is often a threshold below which cracking does not occur.

The residual stresses must not be forgotten, nor those that arise from the corrosion itself because of the difference in volume between the corrosion products and the original material. The dangerous stresses are the tensions. Creation of compressive surface stresses, for example by shot-peening, can be especially beneficial in combating SCC; it can be used also to show the existence of residual tensile stresses, possibly even to enable their intensity and direction to be found since a certain threshold has to be exceeded and cracks tend to form at right angles to the maximum principal stress.

The effect of SCC is accelerated at higher temperatures, but it can appear even below *0°C*.

Stress corrosion cracking is a feature of the couple consisting of the alloy and the environment; the latter, affecting different alloys differently, can be very varied: in pure water *1* to *2* parts per million of a particular ion can suffice to start SCC, or in a gaseous environment a very low partial pressure of water vapour or hydrogen sulfide, for example. These media, aggressive for the alloys concerned when under stress, are completely without effect in the absence of stress. Table 2.9 gives a few typical examples.

All alloys, and even non-materials, can be susceptible to SCC in certain media; resistance to attack increases with the mechanical resistance of the material. One has to be particularly on guard against this in the case of alloys that can develop passivity, that is, that can become coated with a fine protective layer, as can stainless steels. The chemical composition, in alloying elements or in impurities, is critical, and a very small variation can increase the susceptibility greatly.

The morphology of stress-corrosion cracks is brittle fracture, both inter- and transgranular; but the fracture surfaces differ from those of cleavages, which tend more to follow crystallographic planes. As a general rule they are characterised by much branching; often a small variation in some parameter - such as stress, oxydation potential, pH, heat treatment, temperature - will suffice to change the mode of cracking. In corrosion fatigue, however, fractures are transgranular.

SCC and corrosion fatigue cover different but juxtaposed fields (see also Chapter 3, §3.6.3). If the loading is cyclic and the frequency is reduced to the point where the period is equal to the time to fracture under SCC, the two effects coincide. The same happens if the ratio R of the minimum stress to the maximum in fatigue is close to *1*. However, in certain cases the environment can accelerate the fatigue, even though it seems to have no effect when the load is steady - an example is carbon steel in seawater. Finally, it should be kept in mind that the stress and the environment can act in combination in operations such as machining, polishing, drilling, erosion, grinding, friction, wear, fretting.

2.15.2 Mechanisms

Attempts to explain SCC and corrosion fatigue can call on a variety of mechanisms.

(a) *Anodic dissolution.* If the surface of a metal is raised to a positive potential it becomes susceptible to attack: this is *anodic dissolution* (Fig. 2.96); local variations in the composition can bring about the formation of *galvanic piles.* The fluctuations of the potential can be on a very small scale, and the anodic zones can concern very localised regions. Very often there will be micro-segregations of impurities. The grain boundaries will be attacked preferentially either because impurities concentrate there or because an exposed surface appears in their neighbourhood. Why then should any stress be needed in these circumstances? The answer is that it enables the aggressive medium to continue gaining access to the anodic zones which in the absence of stress would be protected by the corrosion products themselves. Further, the steps that form where the slip lines reach the surface can constitute anodic zones with respect to the passive film which they destroy; thus the rate of deformation is important, since if it is fairly slow the passive layer can reform faster than bare metal is exposed, and so halt the cracking. One method for countering attack by anodic solution is cathodic protection, which consists of giving the item a negative potential.

(b) *Hydrogen embrittlement.* Hydrogen is released in cathodic zones. The protons can embrittle the metal, either because they reduce the cohesion energy in being simply adsorbed on to the surface or because they penetrate the metal and diffuse towards the regions of high hydrostatic tension. They can also cluster and form micro-cavities, conducive to propagation of micro-cracks. These regions of high level of stress triaxiality are found at the tip of dislocation pile-ups or, once the cracks have formed, in the plastified zones at their tips. Propagation is often discontinuous, the cracks propagating only after a sufficient amount of hydrogen has diffused.

(c) *Surface tension.* Hydrogen is not alone in being able to reduce the surface energy: we have already met this effect in embrittlement by liquid metals. It occurs also in glass and concrete under the action of water molecules.

(d) *Fracture of the oxide layer.* The oxide layers which form on the surface are mostly brittle and break under stress. The resulting fractures provide an entry for corrosion, particulary next to pile-ups, which give rise to high stress concentrations. Fracture occurs when the thickness of the oxide layer reaches a critical value, as a result of the steady increase of the epitaxial stresses which are superimposed on the applied stress.

It is found that in many cases the more-or-less planarity of the slips plays a fundamental role. Planar slips favour the localisation of anodic attacks, hydrogen penetration and fracture of oxide films; thus a low stacking-fault energy is an unfavourable factor so far as stress-corrosion cracking is concerned.

Material	Corrosive agent	Comments
Carbon steels		
mild steel	molten (caustic) soda, NO_3^-, OH^-	
medium strength	HCN	
high strength	H_2O, H_2	
Stainless steels		
austenitic	Cl^- (with O_2), Br^-, OH^-	no effect if Ni > *50%*
ferritic	H_2	Cl^-, OH^-, NO_3
martensitic	H_2, H_2O	no effect
α *brasses*	cuprammonium compounds	
	NH_3, H_2O atmospheres	
	mercury salts	
β, α+β *brasses*	H_2O (soft, seawater)	
Aluminum alloys	NaCl,/H_2O or NaCl/HCl	
+ Zn, *4* to *20%*	H_2O	
+ Cu and Mg	NaCl, KCl, $MgCl_2$ solutions	
Magnesium alloys	distilled water, $K_2Cr_2O_7$	
Titanium alloys		
*6%*Al, *4%*V	N_2O_4 liquid	H_2O, NO inhibitors
*8%*Al, *1%*Mo, *1%*V	Cl^-, Br^-, I^-, H_2O, CH_3OH, CCl_4	
Nickel alloys	NaOH, KOH solutions	
	molten (caustic) soda	
	HF vapour	
Glass, concrete	H_2O	

Table 2.9 Combinations likely to lead to stress-corrosion cracking.

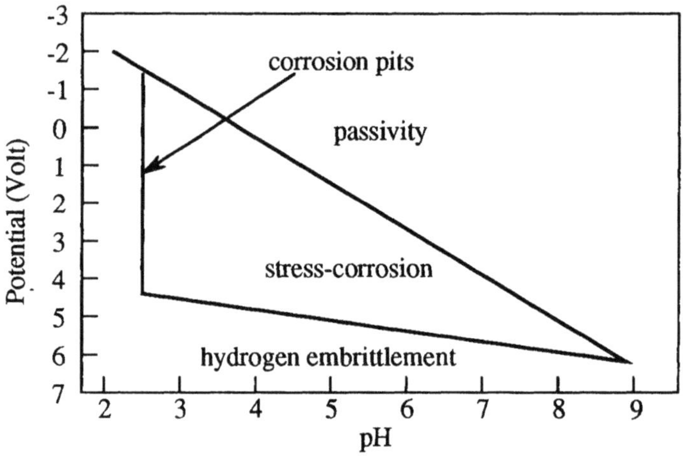

Fig. 2.96 Potential-pH graph showing domain of stress corrosion cracking (SCC) for carbon-manganese steels in phosphoric acid solution. (Parkins, 1979).

It should always be remembered that the very important phenomena of anodic solution and embrittlement by cathodic hydrogen are both controlled by the electrical potential of the item.

2.15.3 Propagation of cracks

It is usual in studying cracking to make use of the data of fracture mechanics, which we develop in the next chapter. It is sufficient for our present purposes to know that in conditions of linear elasticity the stress and deformation fields at the tip of a crack are characterised by the single parameter K, the stress intensity factor. For SCC and corrosion fatigue the assumption of linearity is well supported, since a low level of stress is sufficient to initiate the phenomenon. The velocity of the cracking is thus a function of K and experimental results are conveniently given on a log-log plot (Fig. 2.97). More directly, if measurements are made of the changing length of a crack under constant stress the velocity can be expressed as a function of length; this gives a graph similar to the above but does not require the intermediate calculation involving K.

Experiments have shown that for SCC there is a threshold, denoted by K_{ISCC}, below which cracking does not occur. Knowledge of its value is important in practice, because for a structure that contains a planar defect, whether real or postulated, theoretically there is no risk of fracture below this threshold.

As K_I increases, the cracking velocity increases very quickly, following a power law with a high index; quite often the line corresponding to this Stage I on a plot of log da/dt against log K_I is almost vertical. This will be followed on the same graph by a velocity plateau, showing that the velocity has become independent of K_I over a whole domain called Stage II.

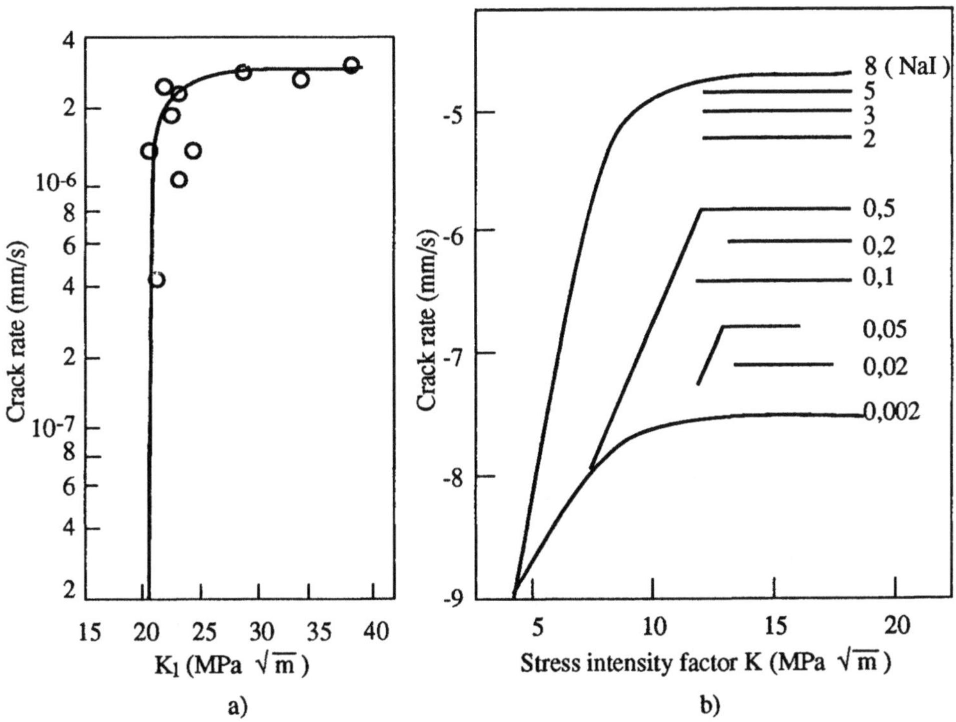

Fig. 2.97 Intergranular cracking velocity in SCC as a function of stress intensity factor. (a) mild steel in CO_3/HCO_3 at *-650 mV* and *75°C* (b) aluminum alloy 7079 T 651 at *-0.7V*, pH *6*, *23°C*, showing effect of NaI (Parkins (1979), Parkins and Greenwell (1977)).

Finally, as K_I approaches K_c, the toughness of the material, the velocity again increases very rapidly.

These various stages can be explained by recalling that whilst there is a certain chemical kinetics for the phenomenon at the tip of the crack, the aggressive medium has to penetrate as far as this by flowing along the crack. It is this mechanism of flow that determines the velocity with which the crack progresses when that is already high enough - that is, if the conditions are those of Stage II. Although the width of the crack is a function of K_I the flow velocity varies very little with this, and is so low that the chemical or electrochemical reactions can be regarded as instantaneous. The opposite is the case if K_I is small, that is, in Stage I. In certain systems Stage II does not occur at all.

As we have said previously, SCC often displays multiple ramifications, which complicate the analysis in terms of fracture mechanics.

Corrosion fatigue can be treated as a simple superposition of pure fatigue cracking and SCC, so that with an obvious notation the velocity can be written

$$da/dN = (da/dN)_{SCC} + (da/dN)_F \tag{2.235}$$

This superposition implies that fatigue cracking is not accelerated by the environment when K_I is below K_{ISCC}. The equation can be transformed by dividing through by the period T of the load cycle, giving

$$da/dt = (da/dt)_{SCC} + (1/T)(da/dN)_F \tag{2.236}$$

which highlights the effect of frequency. We can expect both the form of the cycle and the ratio of the maximum to the minimum stress also to have an effect; such effects have in fact been observed in very many systems.

However, this simple linear superposition is very often invalidated below the SCC threshold, where it is found that the environment continues to have an effect, at least if the rate at which the load increases in the cycling is high enough. In these conditions the time at which the stress is held at its maximum value has no effect.

These observations seem to suggest that it is the increase in plastic deformation at the crack tip that enables the corrosion to develop; such an assumption would be consistent with the analyses of the various mechanisms which involve the activity of slips: anodic attack on the emerging slip lines, introduction of hydrogen, fracture of the passive layer.

2.15.4 Metallurgical factors

Broadly, two types of metallurgical factors can be distinguished: those which affect sensitivity to corrosion and those which change mechanical properties. Among the first the most effective are those which act at the level of the grain boundaries, when segregations there favour anodic solution. It can need only a very small change in composition to bring about a large change in the susceptibility to SCC. Stainless steels, for example, can become susceptible after carbides of chromium have been precipitated in the grain boundaries, creating a region of reduced chromium content in their neighbourhood.

If the cracking is transgranular a much greater change in composition will in general be needed to suppress SCC.

The factors which raise the yield strength can become increasingly favourable with increasing importance of plastic deformation in causing corrosion to develop. In fact, observations have shown that in brass the threshold below which there is no cracking varies like the reciprocal of the square root of the grain size (Fig. 2.98), as does the yield strength. Fig. 2.99 shows that the SCC plateau for cracking for steels in water at $100°C$ increases with the yield strength.

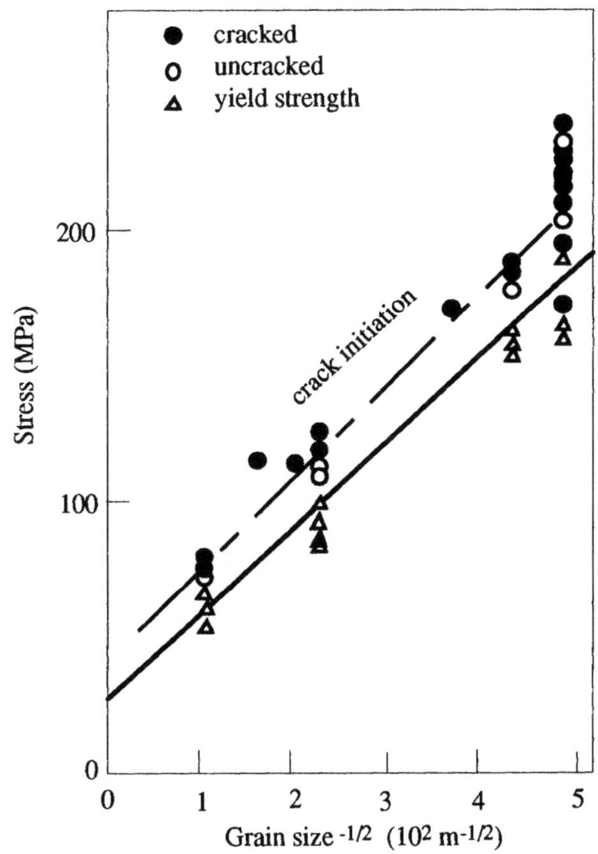

Fig. 2.98 Stress needed to initiate cracking in SCC, and yield strength, as functions of grain size. 70.30 brass in molar solution of $NaNO_2$, pH *9*, *-0.1V*, low strain-rate tensile test (Parkins 1979.).

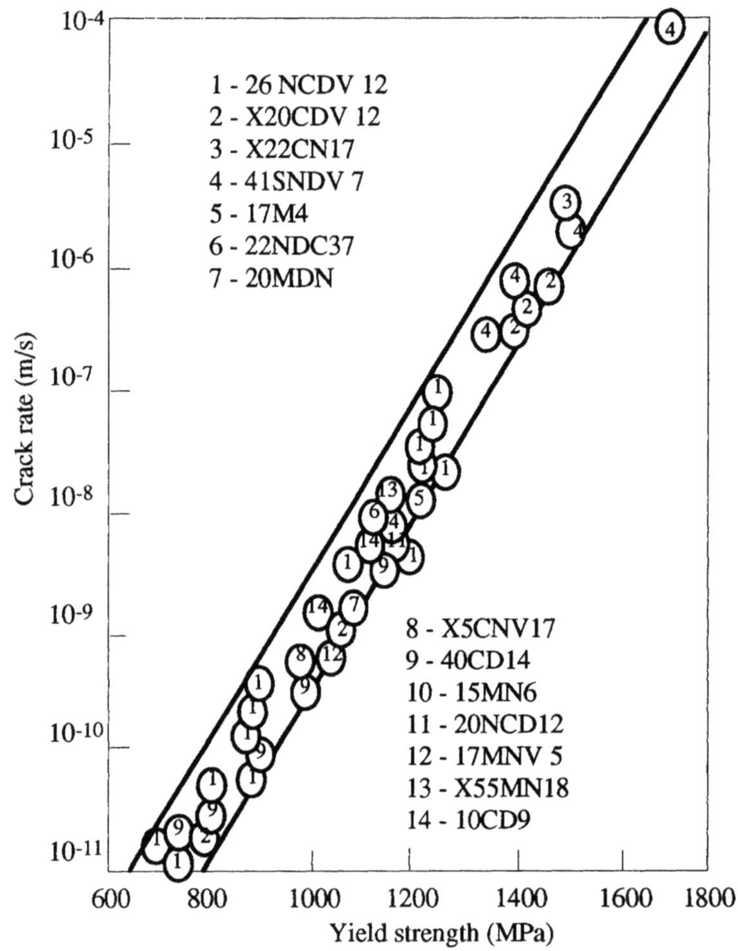

Fig. 2.99 SCC plateau in the crack velocity vs. stress intensity factor (Fig. 2.97) as a function of yield strength, for various steels in water at *100°C*. (Parkins 1979).

It is important to remember that it is the alloys with high yield strength that are most susceptible to stress-corrosion cracking. Thus the heat treatment T6, which gives aluminum alloys the highest strength, also introduces the risk of stress-corrosion cracking - which vanishes with over-ageing.

CHAPTER 2 EXERCISES

1. A single crystal of copper is oriented so that two slip systems are activated equally. Calculate the vertex angle at the point where necking occurs.

2. Investigate the conditions under which necking occurs in a material which follows a law of the type

$$\sigma = \sigma_0 \varepsilon^n \dot{\varepsilon}^m$$

Express the strain and the strain rate as a function of the cross-section of the sample and calculate the rate at which the section changes when a constant force is applied. Show that the time derivative of the expression found is positive if $\varepsilon > n/(1-m)$

3. Where a grain boundary is crossed steps are formed on the cleavage surface so as to accommodate the change of orientation; these steps are irregularly spaced and they oppose the propagation of the cleavage crack by exerting a retarding force on its front. Investigate the equilibrium of this front at the point where it encounters a step (a) when the two steps on either side are equidistant from the front (b) when these distances are different. In the second case show that the close steps converge.

4. In order to calculate the theoretical cleavage stress some assumption must be made concerning the variation of the binding energy with the distance between the lattice planes; possible forms are (a) sinusoidal (b) a Morse function $x^{-n} - x^{-m}$. The calculation involves finding the maximum stress when the lattice planes are separated, using the values of Young's modulus, the equilibrium inter-planar distance b and the surface energy γ_s corresponding to the breaking of the atomic bonds. Take into account the facts that the strain remains elastic for small separations and that $2\gamma_s$ represents the difference in energy between an intact crystal and one that has suffered cleavage.

5. (a) Calculate the stress needed to create an edge dislocation loop of radius a and with Burgers vector b. Given that the energy of such a loop is

$$(2\pi a)[\mu b^2/4\pi(1-\nu)]\log(R/b)$$

show that at equilibrium the energy change is balanced by the work done by the applied force on the dislocation, as given by the Peach-Köhler formula.

(b) Find the energy of a solid containing such a loop, considered as a crack of surface energy γ_s acted on by a stress σ normal to the loop. Investigate the variation of this energy as a increases, for different values of σ.

Find the stress needed to cause the cleavage to propagate (that is, to make the dislocation climb). What stress is needed to generate an unstable cleavage? Find the size of the "critical" nucleus, taking $\gamma_s \approx 0.1\ \mu b$.

6. Given that the energy of a cleavage formed by the Cottrell mechanism is

$$E_c = \frac{\mu(nb\sqrt{2})^2}{4\pi(1-\nu)}\log\frac{R}{a} + 4\gamma_s a - \frac{\pi\sigma^2 a^2(1-\nu)}{4\mu}$$

find how this varies as a increases, for different values of the stress σ. Show that the "critical nucleus" is reached when $\sigma = \sigma_f = 4\gamma_s/nb$, and calculate its size, taking

$\gamma_s \approx 0.1\ \mu b$

7. The problem is to calculate the energy γ_s dissipated by plastic strain when a crack propagates with velocity v_c; this requires the calculation of the power dissipated by dislocations gliding with velocity $v = v_0(\tau/\tau_0)^m$.

The power dissipated is $\dot{W} = \tau\dot{\gamma}$; $\dot{\gamma}$ is given by the standard formula as a function of the velocity and density (ρ_D) of the mobile dislocations, and at the crack tip, at a distance r, $\tau = \sigma\sqrt{a/r}\ \ \tau^*(\theta)$, where $a = v_c t$ is the length of the crack. The angular variation can be neglected, and the constant value $\tau^*(\theta) \approx 1$ taken. $\dot{W}$ is found by integration at the crack tip over $b \le r \le \infty$, $-\pi/2 \le \theta \le \pi/2$, and over the length of the crack front. This gives the energy dissipated per unit area, γ_p, when the crack propagates.

8. **The Rice-Thomson model** (Fig. 2.18)
The problem is to calculate the energy u_D corresponding to a dislocation loop of radius r and Burgers vector b at the end of a crack. The slip plane makes an angle Φ with the cleavage plane and the Burgers vector makes an angle Ψ with the projection of the direction of propagation on the slip plane. It can be shown that the energy of the loop is $\dfrac{2-\nu}{8(1-\nu)}\mu b^2 r \log\dfrac{8r}{e^2 r_0}$. The surface energy of the step created by the formation of the dislocation must be added to this, and the energy released by the formation of the loop must be subtracted; this latter can be found by calculating the work done by the local shear stress in causing the dislocation to slip. The stress field in the neighbourhood of the crack tip is given by

$$\begin{Bmatrix} \sigma_{11} \\ \sigma_{12} \\ \sigma_{13} \end{Bmatrix} = \frac{K_I}{\sqrt{2\pi r}} \cos\frac{\theta}{2} \begin{Bmatrix} 1 - \sin\frac{\theta}{2}\sin\frac{3\theta}{2} \\ \sin\frac{\theta}{2}\cos\frac{3\theta}{2} \\ 1 + \sin\frac{\theta}{2}\sin\frac{3\theta}{2} \end{Bmatrix}$$

where the reference system $Ox1$, $Ox2$, $Ox3$ is such that $Ox3$ is along the crack front and $Ox2$ is normal to the plane of the crack; r, θ are cylindrical coordinates in the $Ox1$, $Ox2$ plane and $K_I = [E\gamma_s/(1 - \nu^2)]^{1/2}$ when the crack propagates.

Find how the energy varies with the radius of the loop, and show that it has a maximum which ceases to occur when a certain relation involving μ, ν, b, γ_m, γ_s, β, and β' holds, with $\beta = \sin\Phi \cos \Phi/2 \cos\Psi$, $\beta' = \sin\Phi \cos\Psi$. It can be assumed that γ_m, and γ_s, are of the order of $0.1\mu b$.

9. **Fracture of aluminum-zirconia**
At *1100°C* tetragonal zirconia undergoes a martensitic transformation resulting in a monoclinic phase, with an increase in volume of *3%*; under a pressure of *3.8 GPa* the transformation temperature falls to *0°C*.

(1) For a spherical ZrO_2 particle in an Al matrix, calculate the pressure developed at the moment of the transformation: the values of Young's modulus are *390 GPa* for Al and *260 GPa* for ZrO_2, Poisson's ratio is *0.25* for each. Find the corresponding strain energy and hence the evolution of the transformation temperature, assuming that the latent heat and the entropy of the transformation remain constant.

(2) Intuitively, if the particles are of large size one can expect microcracking to occur at the interface with the matrix. By adding a fracture energy to the energy balance studied in (1), calculate the critical diameter of the particles, assuming that the size of the microcracks is proportional to the diameter of the particles.

10. Calculate the breaking load for a concrete beam acted on by a constant bending moment. Compare the result obtained when the concrete is assumed perfectly brittle with that obtained by the methods of damage mechanics. To simplify the calculation it can be assumed that the damage parameter is scalar-valued and is proportional to the strain when this is positive.

Find first the breaking load for a tensile test piece, assuming that the strain at fracture is 10^{-4} and Young's modulus is *30 GPa*. It can be shown that the stress-strain curve is parabolic. For the beam, use the Bernoulli hypothesis, giving the strain as varying linearly over the section. When there is no axial load the equilibrium conditions for the normal stresses enable the position of the neutral fiber to be determined as a function of the damage parameter for the fiber of maximum elongation. Show that when the maximum damage is 1 the ratio of the distances from the neutral axis of the extreme fibers is $1/\sqrt{3}$.

From the equilibrium conditions for moments in the section the applied moment M, and therefore the load, can be determined as a function of Young's modulus and the damage parameter for the most-elongated fiber. Show that if h is the height of the beam, then when this parameter = 1, $M/Eh^2 = 10^{-4}(2 - \sqrt{3})(15/4 + 1/3\sqrt{3})$.

Compare this result with the value obtained for the fracture moment when the concrete is perfectly brittle and the strain in the most-elongated fiber reaches 10^{-4}.

11. Growth of cylindrical cavities under axisymmetrical loading

We consider a cylindrical tube with internal and external radii R, R_e respectively, subjected to an axisymmetrical load. The velocity field has radial and axial components $\dot{u}_r$, $\dot{u}_z = z\,\dot{E}_z$ respectively.

Form the strain tensor and give the condition expressing the incompressibility of the matrix; deduce $\dot{u}_r$ as a function of the radius r and $\dot{E}_z$, together with an integration constant A, and determine the equivalent strain rate

$$\dot{\varepsilon}_{eq} = (2/3.\ \dot{\varepsilon}_{ij}\ \dot{\varepsilon}_{ij})^{1/2}$$

Assuming a constitutive equation $s_{ij} = (2/3)\sigma_0\,\dot{\varepsilon}_{eq}^{\,n-1}\,\dot{\varepsilon}_{ij}$, calculate $\sigma_r - \sigma_\theta$ and $\sigma_z - \frac{1}{2}(\sigma_r + \sigma_\theta)$; using the boundary conditions $\sigma_r\,(r = R) = 0$, $\sigma_r\,(r = R_e) = \Sigma_r$, $< \sigma_z > \Sigma_z$, together with the equilibrium equations, show that

$$\Sigma_r = \int_R^{R_e} \frac{\partial \sigma_r}{\partial r} dr = \frac{4}{3}\sigma_0 \int_R^{R_e} \left(\dot{E}_z^2 + \frac{4A^2}{3r^4} \right)^{\frac{1}{2}(n-1)} \frac{A}{r^3} dr$$

$$\Sigma_z - \Sigma_r = \frac{2}{R_e^2} \int_R^{R_e} \left(\sigma_z - \frac{\sigma_r + \sigma_\theta}{2} \right) r dr = \frac{2\sigma_0}{R_e^2} \int_R^{R_e} \left(\dot{E}_z^2 + \frac{4A^2}{3r^4} \right)^{\frac{1}{2}(n-1)} \dot{E}_z r dr$$

Transform these integrals by substituting $\omega = \dfrac{1}{\sqrt{3}} \dfrac{\dot{V}}{\dot{E}_z V}$, where V is the volume of the cavity, and $x = \omega(R/r)^2$. Note that $\dot{V}/V = 2\dot{R}/R + \dot{H}/H = 2A/R^2$, where H is the height of the cylinder.

Show that for an infinite medium

$$\frac{\Sigma_r}{\Sigma_z - \Sigma_r} = \frac{1}{\sqrt{3}} \int_0^{\omega} \left(1 + x^2\right)^{\frac{1}{2}(n-1)} dx,$$

that for an infinite rigid-perfectly plastic medium ($n = 0$)

$$\frac{\Sigma_r}{\Sigma_z - \Sigma_r} = \frac{1}{\sqrt{3}} \sinh^{-1} \omega, \quad \frac{\dot{V}}{V\dot{E}_z} = \sqrt{3} \sinh\left(\sqrt{3} \frac{\Sigma_r}{\Sigma_z - \Sigma_r}\right)$$

and that for a finite rigid-perfectly plastic medium this gives the Gurson criterion

$$\frac{\Sigma_{eq}^2}{\sigma_0^2} + 2f \cosh\left(\sqrt{3} \frac{\Sigma_r}{\sigma_0}\right) - 1 - f^2 = 0$$

where $f = (R/Re)^2$ is volume fraction of the cavity; this can be found by calculating first Σ_r and then $\Sigma_{eq} = \Sigma_z - \Sigma_r$, using the above integrals. Putting

$$A_1 = \sqrt{3} \frac{|\Sigma_r|}{\sigma_0}, \quad A_2 = \frac{\Sigma_{eq}}{\sigma_0}, \quad A_3 = \left(1 + \omega^2\right)^{1/2} + \omega, \quad A_4 = \left(1 + \omega^2 f^2\right)^{1/2} + \omega f$$

and using the relations

$$A_2 = 1 / A_4 - f / A_3, \quad \exp A_1 = A_3 / A_4, \quad A_4 = \frac{A_2}{1 - f \exp A_1},$$

$$A_3 = \frac{A_2 \exp A_1}{1 - f \exp A_1}, \quad 2\omega = \frac{A_3^2 - 1}{A_3} = \frac{A_4^2 - 1}{f A_4}$$

we obtain finally

$$A_2^2 = \left(1 - f \exp A_1\right)\left(1 - f \exp(-A_1)\right) = 1 + f^2 - 2f \cosh A_1$$

12. Calculate the growth rate of a spherical cavity in a medium subjected to a hydrostatic loading.

Consider a spherical shell of internal and external radii R, R_e respectively under hydrostatic loading Σ_m. Find the tensor for the strain rates as a function of the radial velocity; from the incompressibility condition for the matrix find the radial velocity as a function of the radius r and an integration constant A, and deduce the equivalent strain and the components of deviator of the stresses, given that the constitutive equation is $s_{ij} = (2/3)\sigma_0\, \dot{\varepsilon}_{ij}^{\,n-1}\, \dot{\varepsilon}_{ij}$, (as in Exercise 11.).

With the boundary conditions $\sigma_r\ (r = R) = 0$, $\sigma_r\ (r = R_e) = \Sigma_m$ show that

$$\Sigma_m = \frac{2}{3}\sigma_0 \frac{2^m}{m}|A|^m\left[\frac{1}{R^{3m}} - \frac{1}{R_e^{3m}}\right]$$

Find A as a function of Σ_m and the fractional cavity volume $f = (R/R_e)^3$, and hence the value of $\dot{V}/V$, where V is the volume of the cavity.

13. Calculate the characteristic time $\Delta\tau_p$ for establishment, by creep, of a uniform shear stress τ between intergranular particles of size p and distances apart L_p. This can be done by first calculating the elastic sliding of segments of length L_p; the rigidity $k = d\tau / du$ of the segment is $\pi\mu / L_p\beta$, where μ is the shear modulus and β is a parameter for the interaction between segments.

Boundary sliding by creep, on the other hand, is regarded as Newtonian, with viscosity η. Form the equation expressing the fact that elastic sliding is balanced by viscous creep over a boundary thickness δ_j. Show that the stress decreases exponentially with relaxation time $\eta L_p\beta/\pi\mu\delta_j$.

14. Calculate the potential g of vacancies in the neighbourhood of a cavity in equilibrium at a boundary.

This involves solving the differential equation for the flux $\vec{J}$ of vacancies, expressing the fact that the continuous creation of vacancies along the boundary is balanced by the diffusion of these into the cavity. Set up this conservation equation, taking the flux to be proportional to the gradient of the potential; and assuming appropriate boundary conditions show that this leads to equation 2.170.

15. Compare the fracturing pressures for thin-walled cylindrical and spherical containers, both made from the same aluminum alloy or from the same cast iron. Assume that the materials obey a Hollomon work-hardening law, that in the case of the aluminum alloy necking precedes fracture whilst for the cast iron ductile fracture occurs before necking.

16. Calculate the fracturing bending moment for a steel plate loaded at very low temperature, for a given probability of fracture, Assume that cleavage fracture follows a Weibull law, with $m = 22$, $\sigma_u = 1500\ MPa$, $V_0 = 30x\ 10^{-18}\ m^3$, and that the behaviour is perfectly plastic with yield strength $R_p = 400\ MPa$.

The steps in solving the problem are as follows:

(1) Use the Bernouilli hypothesis (that plane sections remain plane) to determine the thickness of the plastic zone as a function of the strain in the most-deformed fiber.

(2) Determine this thickness as a function of the applied moment.

(3) Construct the equation for the Weibull law, noting that cleavages can occur only in the plastic zone under tension; if M_L is the limit bending moment and M_0 the moment at which the most heavily-loaded fiber becomes plastic, this is

$$\log\left(\frac{1}{1-P_R}\right)=\frac{V}{2V_0}\left(\frac{R_p}{\sigma_u}\right)^m\left[1-\left(\frac{M_L-M_R}{M_0/2}\right)^{1/2}\right]$$

Assuming that the yield strength is inversely proportional to the absolute temperature, find the brittle-ductile temperature for the plate.

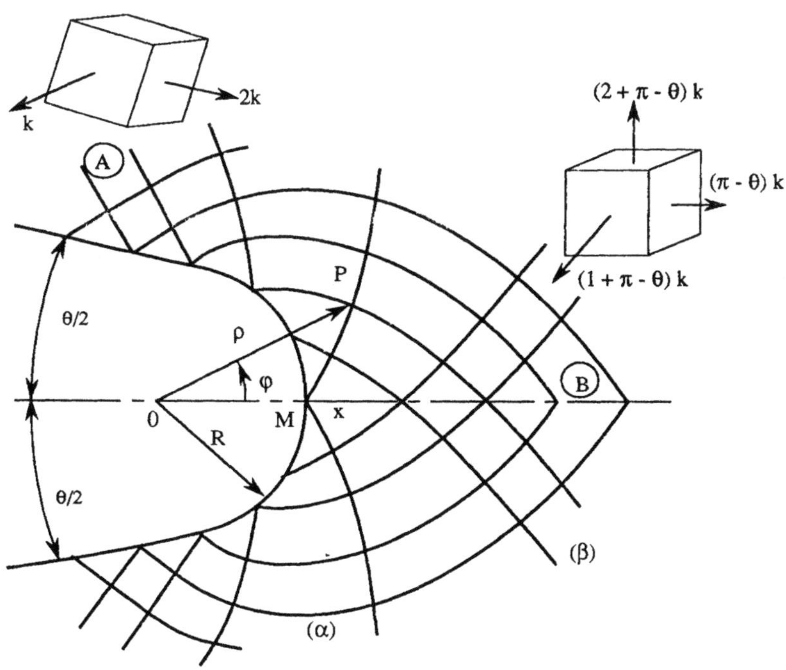

Fig. 2.100 Slip-line field in the neighbourhood of a notch.

17. Mechanics of notches; logarithmic spiral fields in the neighbourhood of the tip of a notch

We consider a notch as shown in Fig. 2.100: the angle is θ and the tip is blunted by a circle of radius R.

(1) Construct the field of slip lines, or characteristics.

Note that because the sides of the notch are free of stress, the slip lines make angles of $\pm \pi/4$ with the normals there; also that because of the axial symmetry a slip line passing through any point P (Fig. 2.100) will cut the radius OP at an angle $\pm \pi/4$. Show that this means that the slip lines must constitute two orthogonal families of logarithmic spirals α, β defined by

$$\varphi - \log(\rho/R) = \beta, \ \varphi + \log(\rho/R) = \alpha$$

(2) Show that in these cylindrical coordinates the stress state is given by

$$\sigma_{rr} = 2k \log(\rho/R), \ \sigma_{\theta\theta} = \sigma_{rr} + 2k = 2k[1 + \log(\rho/R)]$$

Check that these relations satisfy the plasticity criterion and the Hencky conditions for the slip lines.

If x is the distance from the point M to the crack tip, deduce that the opening stress $\sigma_{\theta\theta}$ is given by

$$\sigma_{\theta\theta} = 2k[1 + \log(1+x/R)]$$

(3)This slip line field is linked to two rigid zones: one (A) at the edges of the notch, the other (B) within the material. Starting from the edge of the notch and using the Hencky relations, show that the principal stresses are those acting on the small cubes of material shown in the diagram.

3a: Show that the pressure on the lips of the notch at A is k, and on the material at B is

$$k(1 + \pi - \theta).$$

3b: Deduce that the maximum stress in the interior is

$$(\sigma_{\theta\theta})_{max} = 2k(1 + \pi/2 - \theta/2).$$

3c: Show that the distance x from M to the point on the notch at which the stresses are maximum is given by $x/R = \exp(\pi/2 - \theta/2) - 1$. Calculate this distance for a Charpy test piece for which $\theta = \pi/4$, $R = 0.25$ mm.

18 Notched axisymmetrical test pieces

1. In the minimum section (Bridgman's hypothesis).

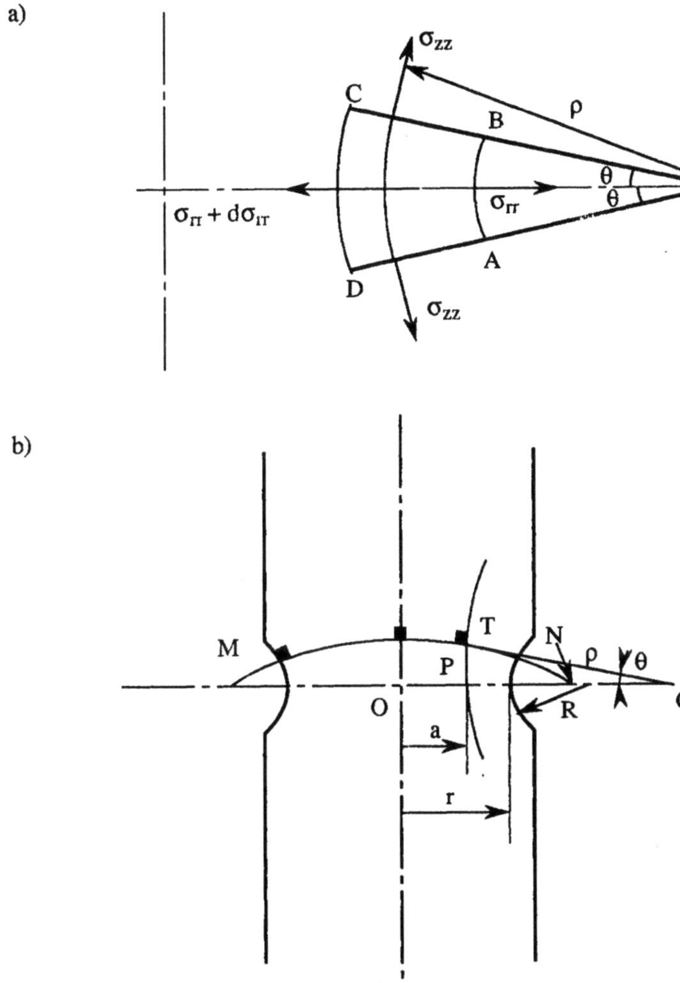

Fig. 2.101 Minimum section of the notch: (a) equilibrium of a sector centered on the minimum section and an isostatic line (b) orthogonal trajectories to the principal directions.

1a. Find the equation for the equilibrium of a sector ABCD (Fig. 2.101a), of angle 2θ, centered on the minimum section and an isostatic line, and show that

$$\frac{d\sigma_{rr}}{dr} = \frac{\sigma_{rr} - \sigma_{zz}}{\rho} = -\frac{\overline{\sigma}}{\rho} \qquad \text{(where } \overline{\sigma} \text{ is the yield strength)} \qquad (1)$$

1b. Assuming that the orthogonal trajectories to the principal directions are circular arcs (Fig. 2.101b), show that if P is a point on the axis distant r from the axis of symmetry then $ON^2 = r^2 + 2\rho r$. From the fact that this holds for the point A show that

$$\rho = \frac{a^2 + 2aR - r^2}{2r} \tag{2}$$

1c: Substituting from (2) into (1), and integrating, show that

$$\sigma_{rr} = \sigma_{\theta\theta} = \bar{\sigma} \log\left(1 + \frac{a^2 - r^2}{2aR}\right), \quad \sigma_{zz} = \bar{\sigma}\left[1 + \log\left(1 + \frac{a^2 - r^2}{2aR}\right)\right] \tag{3}$$

Hence show that the mean stress is

$$< \sigma_{zz} > = \frac{1}{\pi a^2} \int_0^a \sigma_{zz} . 2\pi r dr = \bar{\sigma}\left(1 + \frac{2R}{a}\right) \log\left(1 + \frac{a}{2R}\right) \tag{4}$$

2. The neighbourhood of the minimum section

The profiles of σ_{zz} and σ_{rr} are approximately parabolas, so we can put

$$\sigma_{zz} / \bar{\sigma} = 1 + 2\alpha\left(1 - r^2 / a^2\right), \quad \sigma_{rr} / \bar{\sigma} = \sigma_{\theta\theta} / \bar{\sigma} = 2\alpha\left(1 - r^2 / a^2\right) \tag{5}$$

where

$$\alpha = \frac{1}{2} \log\left(1 + \frac{a}{2R}\right) \approx \frac{a}{4R} \quad \text{if } a << R$$

2a From the first equilibrium equation $\partial\sigma_{rr} / \partial r + \partial\sigma_{rz} / \partial z = 0$

deduce that $\sigma_{rz} = 4\alpha\bar{\sigma}\left(rz / a^2\right)$

2b: From the second equilibrium equation $\partial\sigma_{zz} / \partial z + \partial\sigma_{rz} / \partial r + \sigma_{rz} / r = 0$

deduce $\qquad \sigma_{zz} / \bar{\sigma} = 1 + 2\alpha - 2\alpha(r / a)^2 - 4\alpha(z / a)^2$

Show that the stress tensor in the neighbourhood of the minimum section is approximated by the following expression:

$$\frac{\sigma_{ij}(r,z)}{\bar{\sigma}} = \left[\frac{1}{3} + 2\alpha - 4\alpha(z / a)^2 - 2\alpha(r / a)^2\right]\delta_{ij} + \begin{pmatrix} -1/3 & 0 & 4\alpha rz / a^2 \\ 0 & -1/3 & 0 \\ 4\alpha rz / a^2 & 0 & 2/3 \end{pmatrix}$$

2c: Similarly, an expression can be found for the strain field. Since the strain is axisymmetrical, $\varepsilon_{rr} = \varepsilon_{\theta\theta}$ and $\sigma_{rr} = \sigma_{\theta\theta}$; the displacement u in the r direction is therefore a linear function of r, say $u = -(r/2) f(z)$.

If v is the displacement in the z direction, show that the incompressibility condition gives

$$\varepsilon_{rr} = \varepsilon_{\theta\theta} = -(1/2) f(z), \quad \varepsilon_{zz} = f(z)$$

so that the equivalent strain is $\bar{\varepsilon} = f(z)$. Deduce that

$$\varepsilon_{rz} = \frac{3}{2} \bar{\varepsilon} \frac{\sigma_{rz}}{\bar{\sigma}} = 6\alpha f(z) \frac{rz}{a^2}$$

Further, we know that

$$\varepsilon_{rz} = \frac{1}{2}\left(\frac{\partial v}{\partial r} + \frac{\partial u}{\partial z}\right) = \frac{1}{2}\left(\frac{\partial v}{\partial r} - \frac{r}{2}\frac{df}{dz}\right)$$

Equating the two expressions for ε_{rz} and assuming that v does not depend on r; derive the differential equation $df/f = -24\,(\alpha z/a^2)\,dz$, and integrate this to obtain

$$f(z) = \bar{\varepsilon}_0 \exp\left(-12\frac{\alpha z^2}{a^2}\right)$$

From this, show that the strain tensor can be written

$$\frac{\varepsilon_{ij}(r,z)}{\bar{\varepsilon}_0} = \exp\left(-12\frac{\alpha z^2}{a^2}\right)\begin{pmatrix} -1/2 & 0 & 6\alpha rz/a^2 \\ 0 & -1/2 & 0 \\ 6\alpha rz/a^2 & 0 & 1 \end{pmatrix}$$

Recall that this assumes that v does not depend on r.

19. **Limit load and stress field for a notched test piece under bending**

This is an exercise in using the theorems of limit-load analysis to determine the limiting moment M_L for a notched test piece acted on by bending load; a kinematic approach is used to find an upper limit M^+_L and a static approach to find a lower limit M^-_L.

Kinematic approach. Referring to Fig. 2.44a, the deformation of the test piece under the limiting bending load can be considered as consisting of a rotation around two circular arcs of angle 2α and length l, with angular velocity ω.

(1) By equating the work done by the applied couple, $2M\,\omega$, to the plastic power dissipated per unit of thickness ($2klr\omega$), show that $M = klr$.

Show that the power dissipation is minimum when the angle satisfies $\tan\alpha = 2\alpha$, giving $\alpha = 66°\,50'$; and hence $M_L^+ = 0.69k\,(w - a)^2$.

(2) Compare this value with that of the moment M_0 necessary to reach the yield limit in the remaining section of the beam below the notch, the height of which is w-a. Recall that with Tresca's criterion the limit bending load is reached when the maximum stress at the surface reaches the value $1.5\ R_p$, that is, $3k$; deduce that $M_0 = 0.50k(w - a)^2$.

Static approach. The slip line field in the neighbourhood of the notch tip was studied in Exercise 17. On the lower face, loaded in compression, this field is that for a rigid block, and the lines are straight; the stress field here is uniform with $\sigma_{\theta\theta} = -2k$ (see Fig. 2.44b).

Under limit load the zones AAC, BBC meet at C, the apex of the triangle; the position of this point is determined by the condition that there is no resultant stress across the section OO', that is

$$\int_R^{w-a+R} \sigma_{\theta\theta} d\rho = 0$$

(3) Taking the values found in exercise 17 -

$$\sigma_{\theta\theta} = 2k\left(1 + \log\frac{\rho}{R}\right), \quad R < \rho < R + h, \quad = -2k, \quad R + h < \rho < R + w - a$$

show that if $\rho_1 = 1 + h/R$, $\rho_2 = 1 + (w-a)/R$ then $\rho_2 = \rho_1(1 + \log \rho_1)$

Show that for a Charpy test piece with $w = 10\ mm$, $a = 2\ mm$ and $R = 0.25\ mm$, h is approximately $2.25\ mm$.

(4) Figure 2.45 gives the distribution of the stress $\sigma_{\theta\theta}$ over the section OO'. This stress is discontinuous at the point C. The lower bound M_L^- for the limit load can be found from the equilibrium condition, thus

$$M_L^- = -\int_R^{R+w-a} \sigma_{\theta\theta} \rho d\rho$$

Show that the result can be expressed as

$$\frac{M_L^-}{k(w-a)^2} = \left(1 + \frac{R}{w-a}\right)^2 - \left(\frac{h+R}{w-a}\right)^2 + \frac{1}{2}\left(\frac{R}{w-a}\right)^2 - \left(\frac{h+R}{w-a}\right)^2\left(\frac{1}{2} + \log\frac{h+R}{R}\right)$$

where the value of h (or ρ_1) is found as before.

Show that for a Charpy test piece $M_L^- = 0.69\ (w - a)^2$, so the two bounds are identical.

20 **Brittle fracture by cleavage, Weibull statistics**

Tensile tests have been performed on notched axisymmetrical samples (Fig. 2.41) of a steel of type 15 MND 6, of the following composition (percentage by weight):

C	S	P	Si	Mn	Ni	Cr	Mo	Cu
0.15	*0.009*	*0.009*	*0.24*	*1.36*	*0.68*	*0.24*	*0.46*	*0.063*

This steel is used for the pressure vessels for pressurised-water nuclear reactors, therefore it is important to investigate its brittle fracture not only in the state in which it is produced (by bainitic quenching followed by annealing) but also after irradiation. The effect of irradiation on the yield strength R_p has been investigated by means of tensile tests at various temperatures (Fig. 2.102).

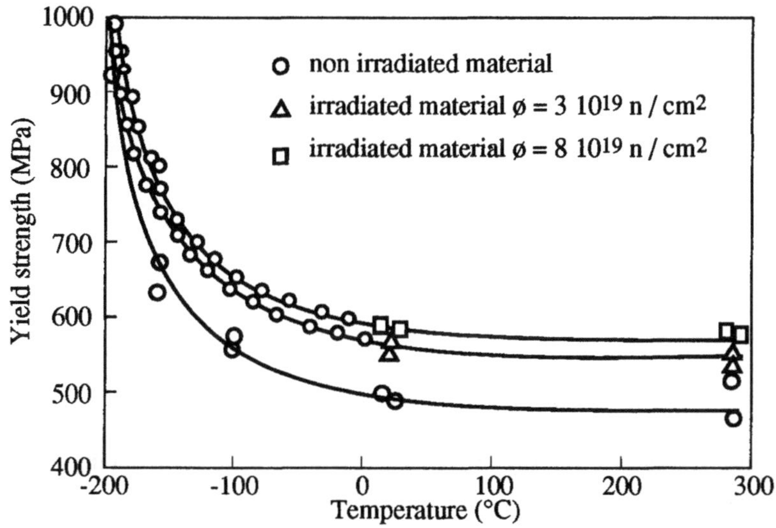

Fig. 2.102 Variation of yield strength with temperature for steel 15 MND 6, unirradiated and irradiated at two neutron fluxes.

(1) Comment briefly on the composition of this steel, bearing in mind its intended application.

(2) How do you explain the increase in yield strength brought about by irradiation?

(3) The tests enable the Weibull stress σ_W to be determined, that is

$$\sigma_W = \frac{1}{V_0} \int \Sigma_I^m \, dV \quad \text{where } \Sigma_I = \sigma_{zz} \text{ is the maximum principal stress.}$$

Fig. 2.103 gives the results found by taking $V_0 = (50\ \mu m)^3$.

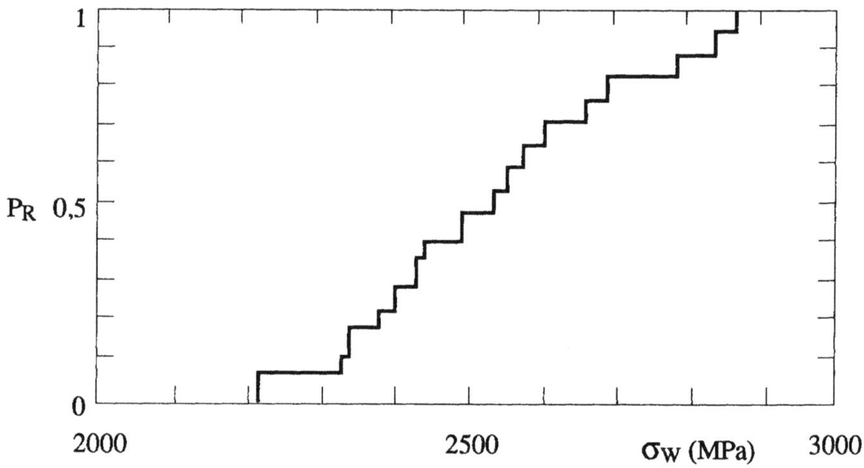

Fig. 2.103 Cumulative probability of fracture as a function of the Weibull stress.

3a. In the light of what you know about the mechanism of fracture by cleavage, comment of the choice of value for V_0.

3b. With the cumulative probability of fracture expressed as

$$P_R = 1 - \exp[-(\sigma_W/\sigma_u)^m]$$

show that the values given in Fig. 2.103 are well accounted for by taking $m = 22$ and $\sigma_u = 2575\ MPa$.

3c. The test results show that the stress for fracture by cleavage is little affected by irradiation; what would you expect to be the effect on the brittle-ductile transition temperature?

21. **Brittle fracture by cleavage**: use of the Cottrell formula $\sigma_f = \dfrac{E\gamma_s}{\pi(1-\nu^2)k_y}$

We consider a medium-strength steel with bainitic structure. The grain size is 10µm, the yield strength is *800 MPa* at *20°C* and *2000 MPa* at *-196°C*. The yield strength follows the Hall-Petch law: $R_p = \sigma_i + k_y\, d^{-1/2}$, $k_y = 0.74\ MPa\sqrt{m}$ and the cleavage stress follows the Cottrell law $\sigma_f = k_y'\, d^{-1/2}$. This last is independent of temperature and has the value *1500 MPa*.

(1) Using Cottrell's relation with $k_y' = E\gamma_s / \pi(1 - v^2)k_y$ calculate the corresponding surface energy γ_s; knowing that in the iron this is of the order of $1\ J\ m^{-2}$, what conclusion do you draw?

(2) Assuming for simplicity that the friction stress σ_i varies linearly with temperature in the range [*-196°C, 20°*], and that above *20°C* it remains constant,

(a) find the value of σ_i at *-196°C* and at *20°C*.

(b) find the transition temperature, defined as the temperature at which the yield strength R_p becomes equal to the cleavage strength σ_f.

(c) To what must the grain size be reduced so that this transition temperature is reduced to *-196°C*?

CHAPTER 3 FRACTURE MECHANICS

3.1 THE VARIOUS MODES OF DAMAGE AND FAILURE: IMPORTANCE OF FRACTURE MECHANICS

A piece of material or a structure can have demands of various sorts made on it - mechanical, chemical (corrosion), thermal (fatigue, creep, creep-fatigue) - which can lead to failure. The main ways in which this can occur are:

- elastic instability, or buckling,
- plastic instability, or excessive deformation: an example is progressive deformation, the *ratchet* effect (Chapter 1, §1.4.4.4),
- damage and cracking.

In general, damage manifests itself by the appearance of discontinuities (microcavities, microcracks), either throughout the body or only on the surface, as we saw in Chapter 2. Table 2.I summarises the various kinds of damage that are encountered; the development of these leads more or less rapidly to the creation of macroscopic cracks, of sufficient size for the surrounding material to satisfy the assumptions of the mechanics of continuous media. We express this by saying that we are then in the domain of *fracture mechanics*.

Very generally, two stages can be distinguished, of initiation and of propagation respectively of cracks, ending with fracture when the velocity of propagation is very high. This leads to the distinguishing between sudden or unstable propagation on the one hand and stable propagation on the other. For the latter, a dominant role is played by the time factor in the case of creep and stress corrosion, and by the number of load cycles in that of fatigue. In certain cases the initiation stage can be greatly reduced or even non-existent; there is often a risk of this happening in welded structures or when there are manufacturing defects such as quench cracks, or presence of inclusions.

Fracture mechanics is concerned with the study of macroscopic cracks. By this is meant the determining of the stress field and the corresponding deformations in the neighborhood of such a crack and the obtaining of relevant experimental data, with the aim of establishing a body of knowledge from which to predict the kinetics of crack propagation and critical loadings beyond which a given loading will initiate fracture. Fig. 3.1 illustrates the underlying philosophy. The subject has become important since the 1960s; its theoretical development owes a great deal to advances made in non-destructive testing.

In this chapter we describe the tools needed to attack the problems of fracture mechanics, limiting the treatment to two-dimensional defects and in particular to linear elasticity. The study of these problems in non-linear behaviour is being pursued vigorously but is outside the scope of this book; simple accounts of some recent advances are given in Gdoutos (1990, 1993).

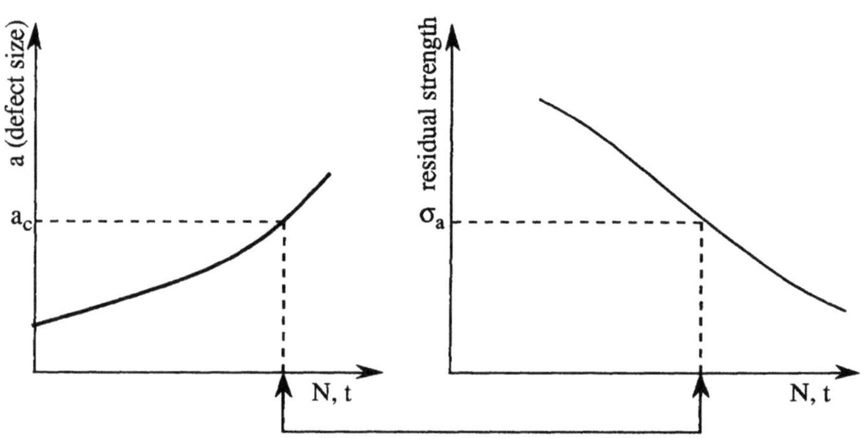

Fig. 3.1 Diagram illustrating the change in size of a crack with the number of load cycles N (case of fatigue) or time t (case of stress corrosion or creep), and the critical crack size a_c at fracture, for a given loading.

3.2 ANALYSIS OF 2-DIMENSIONAL CRACKS, IN LINEAR ELASTICITY

3.2.1 Notation and assumptions

We consider a planar crack whose tip is a straight line (Fig. 3.2); this corresponds to a discontinuity such that the normal stresses are zero and there is a discontinuity in the displacement across the two sides.

The points of the plane can be referenced in either cartesian (x,y) or cylindrical (r,θ) coordinates. The following important assumptions will hold throughout the treatment unless anything to the contrary is stated:

1. The material is homogeneous and isotropic and its behaviour is linear elastic.

2. Inertial forces and accelerations can be neglected, consequently dynamic fracture is not considered.

3. All the problems are 2-dimensional.

A 2-dimensional representation will correspond to different problems according to the assumptions made concerning the third dimension: these relate to plane stress, plane strain, Mode III deformation and axi-symmetry, which we consider in turn.

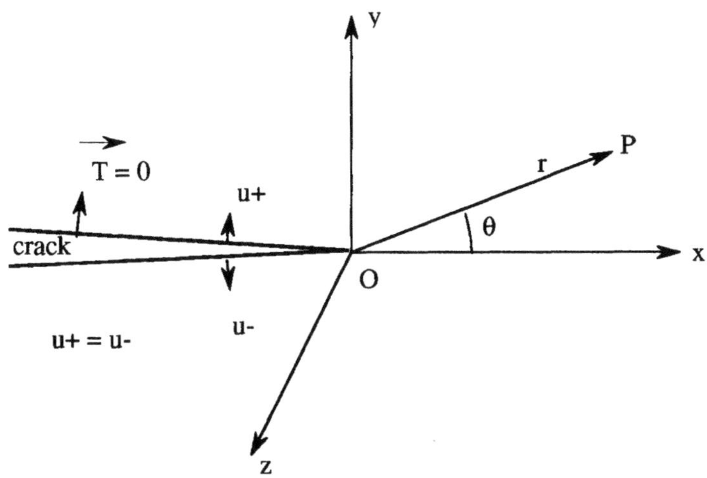

Fig. 3.2 Notation used for a plane crack

Plane stress

This assumption, which holds most closely for thin sheets, corresponds to stress and strain tensors of the form

$$\underline{\underline{\sigma}} = \begin{pmatrix} \sigma_{xx} & \sigma_{xy} & 0 \\ \sigma_{xy} & \sigma_{yy} & 0 \\ 0 & 0 & 0 \end{pmatrix} \qquad \underline{\underline{\varepsilon}} = \begin{pmatrix} \varepsilon_{xx} & \varepsilon_{xy} & 0 \\ \varepsilon_{xy} & \varepsilon_{yy} & 0 \\ 0 & 0 & \varepsilon_{zz} \end{pmatrix}$$

It is important to remember that here $\varepsilon_{zz} \neq 0$

Plane strain

This applies to thick plates. It is expressed by the fact that $\varepsilon_{zz} = 0$ and that $\sigma_{zz} = \nu(\sigma_{xx} + \sigma_{yy}) \neq 0$. The tensors are now

$$\underline{\underline{\sigma}} = \begin{pmatrix} \sigma_{xx} & \sigma_{xy} & 0 \\ \sigma_{xy} & \sigma_{yy} & 0 \\ 0 & 0 & \sigma_{zz} \end{pmatrix} \qquad \underline{\underline{\varepsilon}} = \begin{pmatrix} \varepsilon_{xx} & \varepsilon_{xy} & o \\ \varepsilon_{xy} & \varepsilon_{yy} & 0 \\ 0 & 0 & 0 \end{pmatrix}$$

Mode III deformation

This refers to a displacement in the third dimension $u_z(r,\theta)$ which depends only on the coordinates in the plane and is therefore associated with what is called Mode III of the opening of a crack (see Fig. 3.4 below). Consequently all the in-plane elements of the tensors vanish and the forms are as follow:

$$\underline{\underline{\sigma}} = \begin{pmatrix} 0 & 0 & \sigma_{xz} \\ 0 & 0 & \sigma_{yz} \\ \sigma_{xz} & \sigma_{yz} & 0 \end{pmatrix} \qquad \underline{\underline{\varepsilon}} = \begin{pmatrix} 0 & 0 & \varepsilon_{xz} \\ 0 & 0 & \varepsilon_{yz} \\ \varepsilon_{xz} & \varepsilon_{yz} & 0 \end{pmatrix}$$

Axi-symmetry

Problems here involve a rotation about an axis. We use a cylindrical coordinate system (R,φ,z) (Fig. 3.3), reserving (r,θ) for the region ahead of the crack. By symmetry, displacements normal to the radius, u_φ, are zero and u_R, u_z are functions of r and z only; problems of this type are solved in a meridional plane. The stress and strain tensors are:

$$\underline{\underline{\sigma}} = \begin{pmatrix} \sigma_{RR} & \sigma_{R\varphi} & 0 \\ \sigma_{R\varphi} & \sigma_{\varphi\varphi} & 0 \\ 0 & 0 & \sigma_{zz} \end{pmatrix} \qquad \underline{\underline{\varepsilon}} = \begin{pmatrix} \varepsilon_{RR} & \varepsilon_{R\varphi} & 0 \\ \varepsilon_{R\varphi} & \varepsilon_{\varphi\varphi} & 0 \\ 0 & 0 & \varepsilon_{zz} \end{pmatrix}$$

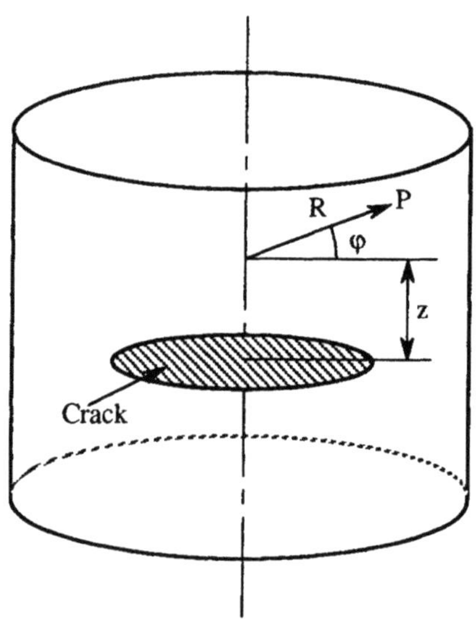

Fig. 3.3 Notation for axi-symmetric problems.

Using a *2D* representation makes great simplifications possible, but when a real-life problem has to be treated it is not always clear which is the most appropriate of these assumptions. We shall return to this important point later.

3.2.2 Resume of the general methods for solving problems of linear elasticity.
A reason for the importance of *2D* representation is that it enables known methods to be used, in particular those based on functions of a complex variable. Other works Labbens (1980), Bui (1978), Liebowitz (1968) should be consulted for a full treatment of the application of these methods in fracture mechanics; here we shall give only the general forms for the equations in each of the above cases, and of the solutions in the case of linear elasticity.

3.2.2.1 Mode III deformation
There is only one displacement $u_z(x,y)$, and hence only two non-zero components:

$$\varepsilon_{xz} = (1/2)\partial u_z/\partial x,\ \varepsilon_{yz} = (1/2)\partial u_z/\partial y$$

Correspondingly there is a stress vector (σ_{xz}, σ_{yz}) which satisfies the equilibrium equations

$$\vec{div}\,\underline{\underline{\sigma}} = 0 \text{, that is } \partial\sigma_{xz}/\partial x + \partial\sigma_{yz}/\partial y = 0$$

In linear elasticity we have also $\underline{\underline{\sigma}} = 2\mu\underline{\underline{\varepsilon}}$, whence $\vec{div}\left(2\mu\,\vec{grad}\,\vec{u}\right) = 0$, leading to

$$\nabla^2 u_z = 0, \text{ where } \nabla^2 \text{ is the Laplacian operator } \partial^2/\partial x^2 + \partial^2/\partial y^2 \qquad (3.1)$$

The solution has the form $u_z = (1/\mu)\,\mathrm{Im}[f(\zeta)]$ (3.2)
(where $\zeta = x + iy$)
and the stresses satisfy $\sigma_{yz} + i\sigma_{xz} = f'(\zeta)$ (3.3)
(see Exercise 1)

3.2.2.2 Plane stress and plane strain
In plane strain the components u_x, u_y are independent of z; the equilibrium conditions are $\partial\sigma_{xx}/\partial x + \partial\sigma_{xy}/\partial y = 0,\ \partial\sigma_{xy}/\partial x + \partial\sigma_{yy}/\partial y = 0$

which are satisfied if the stresses are expressed in terms of the Airy function U such that

$$\sigma_{xx} = \partial^2 U/\partial y^2,\ \sigma_{xy} = -\partial^2 U/\partial x\partial y,\ \sigma_{yy} = \partial^2 U/\partial x^2 \qquad (3.4)$$

The strain compatibility equations are

$$2\partial^2\varepsilon_{xy}/\partial x\partial y = \partial^2\varepsilon_{xx}/\partial x^2 + \partial^2\varepsilon_{yy}/\partial y^2 \qquad (3.5)$$

Assuming linear elasticity, substitution of (3.4) into (3.5) gives a biharmonic equation for U:

$$\nabla^2(\nabla^2 U) = \nabla^4 U = 0 \qquad (3.6)$$

U can be expressed in terms of functions ϕ and ψ of the complex variable $\zeta = x + iy$ $(= re^{i\theta})$ by

$$U = \mathrm{Re}\left[\bar{\varsigma}\phi(\varsigma) + \int \Psi(\varsigma)\, d\varsigma\right] \qquad \text{(where } \Psi'(\zeta) = \psi(\zeta)\text{)} \tag{3.7}$$

when the stresses satisfy the relations

$$\sigma_{xx} + \sigma_{yy} = 4\,\mathrm{Re}(\phi'(\varsigma)), \quad \sigma_{yy} - \sigma_{xx} + 2i\sigma_{xy} = 2[\bar{\varsigma}\phi''(\varsigma) + \psi'(\varsigma)] \tag{3.8}$$

The difference between plane stress ($\sigma_{zz} = 0$) and plane strain ($\sigma_{zz} = \nu(\sigma_{xx} + \sigma_{yy})$) conditions is seen at the level of the displacements, which can be put in the form

$$u_x + iu_y = \frac{1}{2\mu}\left[\kappa\phi(\varsigma) - \varsigma\bar{\phi}'(\varsigma) - \bar{\Psi}'(\varsigma)\right] \tag{3.9}$$

where $\kappa = 3 - 4\nu$ for plane strain and $(3 - \nu)/(1 + \nu)$ for plane stress (3.10)

3.2.2.3 Axisymmetric problems

A problem of revolution is a plane problem in a meridional plane (Fig. 3.3), which can be solved in terms of the biharmonic stress function in cylindrical coordinates:

$$\nabla_r^2 \nabla^2 U = 0 \tag{3.11}$$

where $\nabla_r^2 = \partial^2/\partial r^2 + (1/r)\partial/\partial r + (1/r^2)\partial^2/\partial\theta^2$

The difference from the plane problem is that the Laplacian operator now contains the term $(1/r)\partial/\partial r$; solution in terms of two analytic functions is no longer possible. There is no simple general relation between the solution of a problem of revolution and that of the analogous plane problem in the meridian plane, it is intuitively clear that the two will not differ greatly if both the thickness and the curvature are small.

3.2.3 Singularities in the stress field

3.2.3.1 The three modes of crack opening.

Any displacement of the faces of a crack can be expressed as a combination of the three principal modes shown in Fig. 3.4.

Mode I. The displacement is wholly in the direction of *Oy*. As we shall see later, this mode is often the most dangerous of the three. It is similar to the displacement produced by a rectilinear edge dislocation parallel to the tip of the crack with Burgers vector parallel to *Oy*.

Mode II. This is brought about by a shear in the plane of the crack, parallel to *Ox*. It is similar to the displacement produced by a rectilinear edge dislocation parallel to the tip of the crack with Burgers vector parallel to *Ox*.

Mode III.. This is due to anti-planar shear in the plane *Oxz* of the crack and parallel to *Oz*. It is similar to the displacement produced by a screw dislocation parallel to the tip of the crack with Burgers vector in the direction of *Oz*.

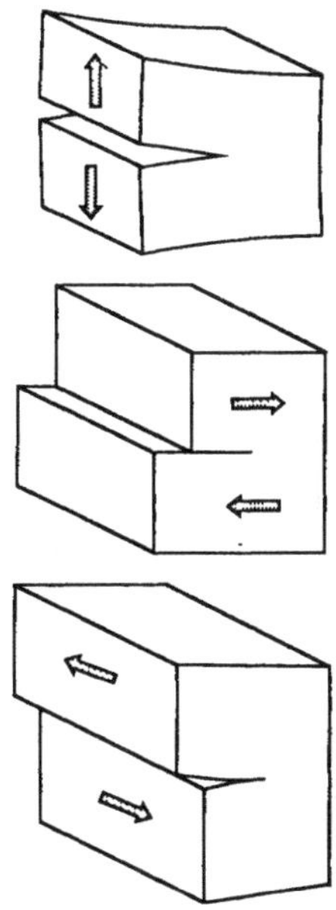

Fig. 3.4 The three modes of crack opening.

Problems concerning the Mode III are solved in terms of a *2D* anti-planar representation , those concerning Modes I and II in terms of *2D* plane stress or strain.

3.2.3.2 Stress intensity factor *K*

In the cylindrical coordinate system described, for all plane problems involving a crack the stresses and strains tend to infinity as the radius r tends to zero: that is, there is a singularity at the tip of the crack. At first sight this may seem surprising but it is easily understood when the crack is regarded as the limit of a notch (Fig. 3.5). For a notch in the form of a very narrow ellipse ($b << a$) the stress concentration factor is $K_T \cong 2\sqrt{a/\rho}$ (see §2.2.5), assuming that $a >> \rho$; for a crack $\rho \rightarrow 0$ and therefore $K_T \rightarrow \infty$. This is the case for a very sharp crack such as that resulting from fatigue.

In all problems the dominant term in the singularity is $r^{-1/2}$, and it is only for this form that the elastic energy can remain finite. This can be expressed as

$$W = \int_V \frac{1}{2} \underline{\underline{\sigma}} : \underline{\underline{\varepsilon}} \, dV = \int_V \frac{1}{2} \underline{\underline{\sigma}} : \underline{\underline{\varepsilon}} \, B \, dr d\theta \qquad \text{for a plane problem}$$

$$= \int_V \frac{1}{2} \underline{\underline{\sigma}} : \underline{\underline{\varepsilon}} . 2\pi R \, dRdz \qquad \text{for an axisymmetric problem}$$

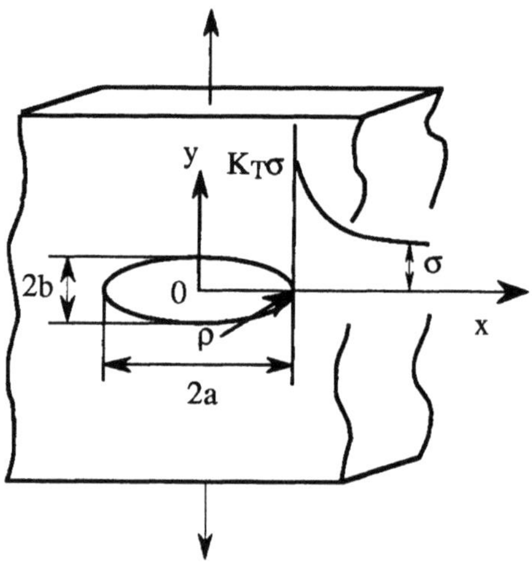

Fig. 3.5 Plate with an elliptical-section hole

where, for a plane problem, B is the thickness of the plate. For the energy W to remain bounded when V becomes small the singularity in the product $\sigma \, \varepsilon$ must be of the order of $1/r$ and therefore, in linear elasticity, the stress and strain must be each be of order $1/\sqrt{r}$. A justification of this argument is given in Volume I, Annex 3 (Exercise 3.2.1).

A result of fundamental importance in linear elasticity has been obtained by Irwin: that the leading term in the series expansion of the stress at the tip of a crack is the same for all problems concerning a given mode of opening. For Modes I, II and III this term is known to within a numerical factor written K_I, K_{II}, K_{III} respectively and called the *stress intensity factor* for the mode. Thus the solution of a crack problem is reduced to the problem of determining the relevant K.

At this stage it becomes important to avoid confusing the *stress concentration factor* K_T, which carries only local information about the conditions around the tip of the notch, with the *stress intensity factor K* which relates to the spatial stress field as a whole. K_T is dimensionless but K is not: the latter is the product of a stress and the square root of a length and therefore has the dimensions

$$[K] = FL^{-2}L^{1/2} \tag{3.12}$$

Thus K is expressed in $Nm^{-3/2}$, or $Pa\sqrt{m}$, and the unit most often used is $MPa\sqrt{m}$; another that is used is $ksi\sqrt{inch}$, which is about $1.12\ MPa\sqrt{m}$.

3.2.3.3 Stress and strain fields

We now give the results for the three modes of opening; proofs are given in specialised works, e.g. Bui (1978), Labbens (1980), Liebowitz (1968).

Mode I.

Stresses

$$\sigma_{xx} = \frac{K_I}{\sqrt{2\pi r}} \cos\frac{\theta}{2}\left(1 - \sin\frac{\theta}{2}\sin\frac{3\theta}{2}\right)$$

$$\sigma_{yy} = \frac{K_I}{\sqrt{2\pi r}} \cos\frac{\theta}{2}\left(1 + \sin\frac{\theta}{2}\sin\frac{3\theta}{2}\right)$$

$$\sigma_{xy} = \frac{K_I}{\sqrt{2\pi r}} \cos\frac{\theta}{2}\left(\sin\frac{\theta}{2}\cos\frac{3\theta}{2}\right)$$

$$\sigma_{zz} = \nu(\sigma_{xx} + \sigma_{yy}) \quad \text{plane strain}$$

$$\sigma_{zz} = 0 \quad \text{plane stress}$$

Displacements

$$u_x = \frac{K_I}{2\mu}\sqrt{\frac{r}{2\pi}} \cos\frac{\theta}{2}\left(\kappa - 1 + 2\sin^2\frac{\theta}{2}\right)$$

$$u_y = \frac{K_I}{2\mu}\sqrt{\frac{r}{2\pi}} \sin\frac{\theta}{2}\left(\kappa + 1 - 2\cos^2\frac{\theta}{2}\right)$$

$$\kappa = 3 - 4\nu \quad \text{plane strain}$$

$$\kappa = \frac{3-\nu}{1+\nu} \quad \text{plane stress}$$

Mode II

Stresses

$$\sigma_{xx} = \frac{K_{II}}{\sqrt{2\pi r}}(-\sin\frac{\theta}{2})\left(2+\cos\frac{\theta}{2}\cos\frac{3\theta}{2}\right)$$

$$\sigma_{yy} = \frac{K_{II}}{\sqrt{2\pi r}}\sin\frac{\theta}{2}\left(\cos\frac{\theta}{2}\cos\frac{3\theta}{2}\right)$$

$$\sigma_{xy} = \frac{K_{II}}{\sqrt{2\pi r}}\cos\frac{\theta}{2}\left(1-\sin\frac{\theta}{2}\sin\frac{3\theta}{2}\right)$$

$$\sigma_{zz} = \nu(\sigma_{xx}+\sigma_{yy}) \quad \text{plane strain}$$

$$\sigma_{zz} = 0 \quad \text{plane stress}$$

Displacements

$$u_x = \frac{K_{II}}{2\mu}\sqrt{\frac{r}{2\pi}}\sin\frac{\theta}{2}\left(\kappa+1+2\cos^2\frac{\theta}{2}\right)$$

$$u_y = \frac{K_{II}}{2\mu}\sqrt{\frac{r}{2\pi}}(-\cos\frac{\theta}{2})\left(\kappa-1-2\sin^2\frac{\theta}{2}\right)$$

Mode III

Stresses

$$\sigma_{xz} = \frac{K_{III}}{\sqrt{2\pi r}}(-\sin\frac{\theta}{2})$$

$$\sigma_{yz} = \frac{K_{III}}{\sqrt{2\pi r}}\cos\frac{\theta}{2}$$

Displacements

$$u_z = \frac{2K_{III}}{\mu}\sqrt{\frac{r}{2\pi}}\sin\frac{\theta}{2}$$

NOTES

1. These relations show that we need to know only the stress intensity factor K, which is a function of load and geometry, in order to give a detailed description of the stress and strain fields around the tip of the crack. This enables results derived from tests to be compared with the behaviour of structures containing cracks. Provided that the conditions are those of linear elasticity the equality of the stress intensity factors guarantees that the corresponding stress and strain fields will be identical.

However, it must not be forgotten that the expressions involve only the leading terms in the expansions. In general the relations are of the form

$$\sigma_{ij} = \frac{K_\alpha}{\sqrt{2\pi r}} f_{ij}^{\alpha}(\theta) + o\left(\frac{1}{\sqrt{r}}\right) \tag{3.13}$$

As the distance from the tip of the crack increases "regular" - that is, non-singular - terms become increasingly important; the notation $o\left(1/\sqrt{r}\right)$ means that these are negligible compared with $1/\sqrt{r}$ when r is small.

2. In a real-life problem the solution will often be a linear combination of the separate solutions for the three modes (see Exercise 2): this is justified by the assumption of linear elasticity.

3. We have already emphasised the fundamental difference between K and K_T; however, the argument used to establish the presence of the singularity suggests that the two are related, and Irwin has shown that K_I can be expressed as a limit, thus:

$$K_I = \lim_{\rho \to 0} \frac{\sqrt{\pi}}{2} \sigma_{\max} \sqrt{\rho}$$

where ρ is the radius of curvature of the notch (Fig. 3.5). For the geometry of this figure this gives, for a crack of length $2a$ in an infinite plate.

$$K_I = \sigma \sqrt{\pi a}\ \mathrm{K} \tag{3.15}$$

4. Mode I is the most important mode physically, since this form of opening can lead easily to fracture. Further, the other modes are not always easy to handle; in particular, problems in these other modes concerning friction between the faces of a crack are often difficult to tackle. Lastly, a Mode I crack propagates in its own plane, by symmetry, which is not the case for the others. For all of these reasons we shall confine our treatment mainly to pure Mode I cracks; in Exercise 3 a criterion (that of Erdogan and Sih (1963)) is suggested for the bifurcation of a crack under mixed-mode loading.

3.2.3.4 Stress intensity factors for some particular cases

Table 3.1, at the end of this chapter, gives a number of useful results; the following four books give a large number of solutions - see also Exercises 3,5,6.

Tada et al (1985); Rooke and Cartwright (1976); Sih (1973); Murakami (1987).

However, practical problems often do not conform to published solutions, so we shall give later an indication of general methods of attack.

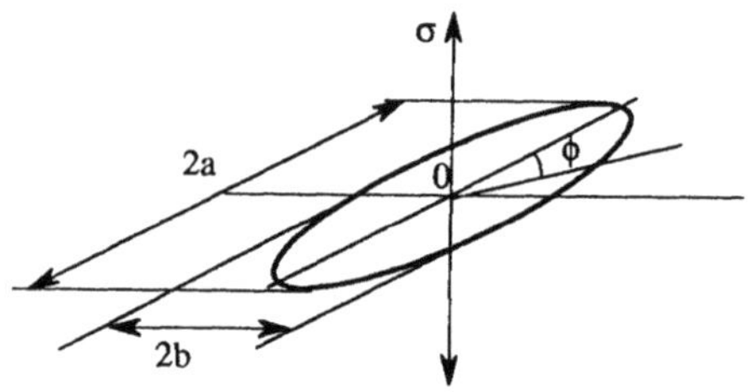

Fig. 3.6 Elliptical crack in an infinite medium.

We have treated only 2-dimensional cracks, but real cracks can have a 3-dimensional character. It is often found that a crack has an elliptical or semi-elliptical contour where it meets the surface, in which case the factor K will vary along the tip. Thus for a plane elliptical crack with axes a, b ($a < b$), loaded in Mode I, in an infinite medium (Fig. 3.6)

$$K_I(\Phi) = \frac{\sigma\sqrt{\pi a}}{E(k)}\left(\sin^2\Phi + \frac{a^2}{b^2}\cos^2\Phi\right)^{1/4}$$

where $E(k)$ is the elliptic integral $E(k) = \int_0^{\pi/2}\sqrt{1-k^2\cos^2\Phi}\,d\Phi$, $k^2 = 1\text{-}a^2/b^2$

K_I is maximum at the end of the minor axis, where it has the value $K_{I\max} = \dfrac{\sigma\sqrt{\pi a}}{E(k)}$

and if $a = b$, when the crack is circular ("penny-shaped") this is $K_I = \dfrac{2}{\pi}\sigma\sqrt{\pi a}$

3.2.4 Rate of energy release, *G*

3.2.4.1 Griffith's theory of brittle fracture

The first theories of fracture dealt with very brittle materials such as glass. Griffith (1920) was the first to propose a theory based on the competition between the release of elastic energy with the advance of the crack and the absorption of energy in the creation of new surfaces. His theory is based on the elastic solution around an elliptical cavity; since then it has been extended.

Let Π denote the potential energy stored in the structure, ∂A the increment of surface resulting from the growth of the crack and G the *crack extension force* or *rate of release of elastic energy*. G is defined by

$$G = -\partial \Pi / \partial A \tag{3.16}$$

Consider the system of Fig. 3.7. This is a closed system, exchanging no energy of any form with the exterior, the latter consisting of the material containing the crack and the forces acting on it; the material is assumed to be linear elastic.

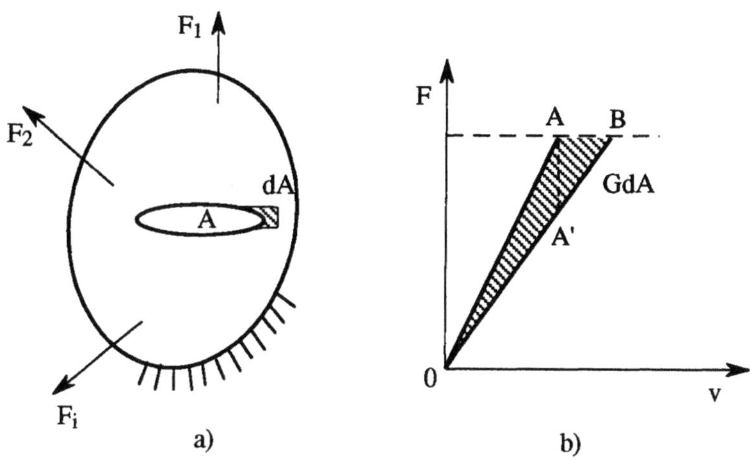

Fig. 3.7 Energy release rate.

The energy of this system, which remains constant, consists of the potential energy U of the applied forces, the stored elastic energy E, the surface energy $(S + 2A)\gamma_S$ where S is the area of the external surface and A that of the crack, which has two faces, and possibly a kinetic energy W_k. The conservation principle gives

$$dU + dE + d(S + 2A)\gamma_S + dW_k = 0$$

We can assume $dS = 0$, and putting $\Pi = U + E$, the total potential energy, we have

$$d\Pi + 2\gamma_S dA + dW_k = 0$$

that is $G = 2\gamma_S + dW_k/dA$ (3.17)

At equilibrium, without motion, $dW_k = 0$, and therefore

$$G = G_c = 2\gamma_S \tag{3.18}$$

If G exceeds this critical value G_c the crack will propagate; if it is less, the crack should reclose ($dA < 0$ since $dW_k > 0$), but this does not in fact happen because various irreversible processes intervene, in particular the oxidation of the crack surfaces.

3.2.4.2 Relation between G and the compliance C of the structure

The behaviour of the cracked structure is shown diagrammatically in Fig. 3.7b. If P denotes the applied forces, v the displacements at the points of application and C the compliance (the inverse of the stiffness) of the structure, then

$$v = CP \tag{3.19}$$

G is represented by the area of triangle OAB if a test is made at constant load, or by OAA' if at constant displacement

Constant displacement:

$$dU = 0$$

$$E = \frac{1}{2}Pv = \frac{1}{2}\frac{v^2}{C}$$

$$G = -\frac{\partial \Pi}{\partial A} = -\frac{\partial E}{\partial A} = \frac{1}{2}\left(\frac{v}{C}\right)^2\frac{\partial C}{\partial A} = \frac{1}{2}P^2\frac{\partial C}{\partial A}$$

Constant load:

$$U = -Pv = -CP^2$$

$$E = \frac{1}{2}Pv = \frac{1}{2}CP^2$$

$$G = -\frac{\partial \Pi}{\partial A} = -\left(\frac{\partial U}{\partial A} + \frac{\partial E}{\partial A}\right) = \frac{P^2}{2}\frac{\partial C}{\partial A}$$

It is important to note that in the case of prescribed load the amount of energy stored increases as the crack propagates ($\partial E > 0$), but the work done by the external forces compensates for this increase. In both cases we have the general relation

$$G = \tfrac{1}{2}P^2\,\partial C/\partial A \tag{3.20}$$

which, as we shall see later, can be useful in certain cases for determining the stress intensity factor.

3.2.4.3 Relation between K and G

The stress intensity factor K describes the distribution of stress and strain in the neighbourhood of a crack. Intuitively, one would expect the energy release rate G to be very sensitive to the stress singularity; we shall show that K and G are in fact related.

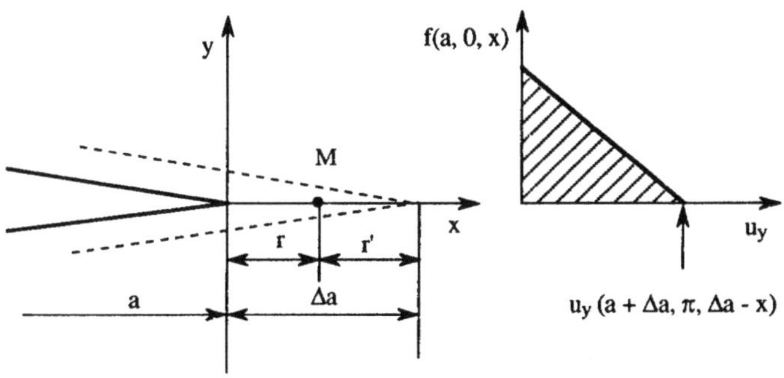

Fig. 3.8 Advance and postulated closing of a crack.

Consider the plane case represented by Fig. 3.8, with Mode I opening. We want to find the amount of energy released when the crack advances from length a to length $a + \Delta a$; in absolute value this is equal to the work ΔW required to reclose the crack over the length Δa. When the length of the crack is $a + \Delta a$ the displacement of the points on its sides between a and $a + \Delta a$ is that given by the Mode I expressions of §3.2.3.3 for a crack of length $a + \Delta a$, with angle $\theta = \pi$ and distance $r' = \Delta a - x$.

When the length Δa of the crack is reclosed its length returns to a. The "cohesion forces" (not strictly the physical forces) at a distance x ahead are given by $df = B\sigma_{yy}\, dx$, where B is the thickness of the plate. σ_{yy} can be calculated from the expressions given in§3.2.3.3 for a crack length a, angle $\theta = 0$ and a distance x. In the reclosing, the displacement at each point M changes from $u_y(a + \Delta a, \pi, \Delta a - x)$ to 0 while the force per unit of thickness increases from 0 to $\sigma_{yy}(a, 0, x)dy$. The corresponding work is

$$dW = 2(B/2)\sigma_{yy}(a, 0, x)u_y(a + \Delta a, \pi, \Delta a - x)dx$$

and the total work for the two sides of the crack is

$$\Delta W = B\int_0^{\Delta a} \sigma_{yy}u_y\, dx = G.B\Delta a$$

$$= B\int_0^{\Delta a} \frac{K_I}{\sqrt{2\pi}}(x-a)^{-1/2}\frac{\kappa+1}{2\mu}\frac{K_I}{\sqrt{2\pi}}(a+\Delta a-x)^{1/2}dx$$

Evaluation of the integral gives $G = \frac{\kappa + 1}{2\mu}\frac{K_I^2}{4}$ so with $\kappa = 3 - 4\nu$ for plane strain and $(3 - \nu)/(1 + \nu)$ for plane stress we have finally

$$G = K_I^2(1 - \nu^2)/E \qquad \text{(plane strain)} \qquad (3.21a)$$

$$G = K_I^2/E \qquad \text{(plane stress)} \qquad (3.21b)$$

A similar calculation can be carried through for the shear modes II and III but it must then be assumed that the crack propagates in its own plane (coplanar extension); the general result is

$$G = (1/E)[(K_I^2 + K_{II}^2)(1 - \nu^2) + (1 + \nu)K_{III}^2] \qquad (3.22)$$

These relations are useful in particular for calculating the value of the factor K (see Exercises 7 and 8).

As a general rule, the extension of the crack does not remain in the original plane; we have then to treat the propagation as that of a virtual crack at an angle α and find the work done by the "cohesion forces" of the *initial* crack in the displacement field of the *branched* crack. A simple argument would assume that the stress intensity factors for the branched crack tend towards those for the initial crack as Δa tends to zero, but this would be false.

3.2.5 Methods for determining the stress intensity factors

Here we take a strictly practical view, keeping always to the 2-dimensional scene; at the end we shall discuss some aspects of 3-dimensional problems.

3.2.5.1 Use of known solutions: principle of superposition

As we said earlier, solutions to many problems have been published in the literature; some of these are given in Table 3.I at the end of this chapter. The commonest practice in attacking a new problem is to attempt to reduce it to a set of problems for which the solutions are known and appeal to the *principle of superposition* to construct the solution as a combination of these known solutions: clearly, this requires the conditions to be those of linear elasticity (see Exercises 9 and 10).

Suppose our problem concerns a 2-dimensional solid containing a crack and acted on by various forces and moments (Fig. 3.9), and that the solution is not already known. It can be split into two parts, Problems 1 and 2:

We first try to solve the problem with the same boundary conditions but with no crack in the material - Problem 1. The solution will give the stress and strain distributions at all points of the solid. With a general point M on the side[s] of the crack we associate a force $\vec{F}(M)$ such that $\vec{F}(M)=\sum_j \sigma_{ij} n_j$ where $\vec{n}$ is the normal to the plane of the crack and is directed towards this; $\vec{F}$ can be split into three parts:

$F_I(M)$ normal to the plane of the crack,

$F_{II}(M)$ in this plane and parallel to it,

$F_{III}(M)$ normal to the plane of the problem.

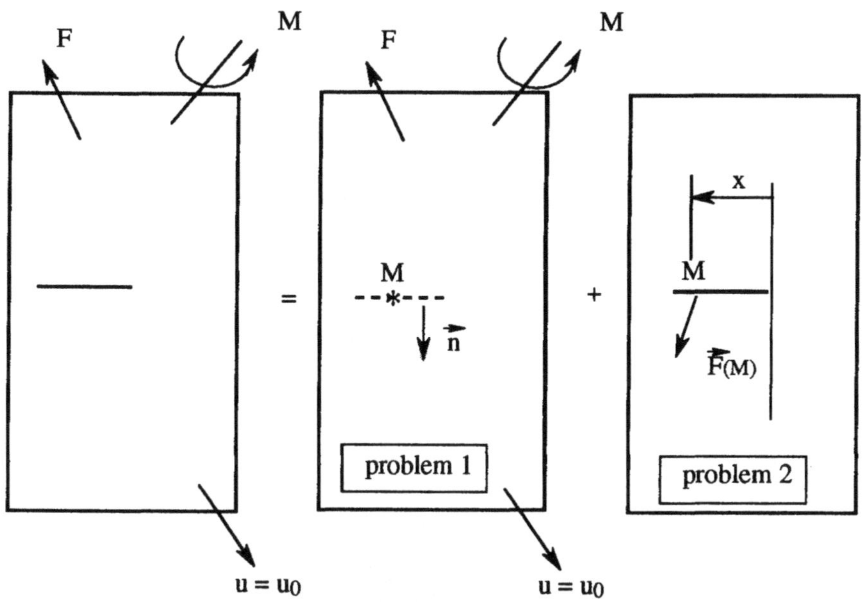

Fig. 3.9 Application of the method of superposition.

Problem 2 is as follows. With boundary conditions of zero forces or displacements at the external surface, find the stress intensity factor for a crack subjected to the load $\vec{F}(M)$.

The solution to Problem 2 may be known for certain forms for $F(M)$, for example for constant value or for linear or parabolic variation with distance; we can then approximate it by a polynomial $F(M) = A_0 + A_1x + A_2x^2 + \ldots$, where x is the distance from M to the tip of the crack and obtain a solution in the form $A_0K_{I0} + A_1K_{I1} + A_2K_{I2} + \ldots$ where K_{I0} etc are the solutions corresponding to the elementary distributions $F(M)= 1, x, x^2$, etc. Fig. 3.10 gives the solutions for various geometries; these are taken from Buchalet and Bamford (1974).

Finally, the solution to the original problem is expressed as a linear combination of the solutions to Problems 1 and 2.

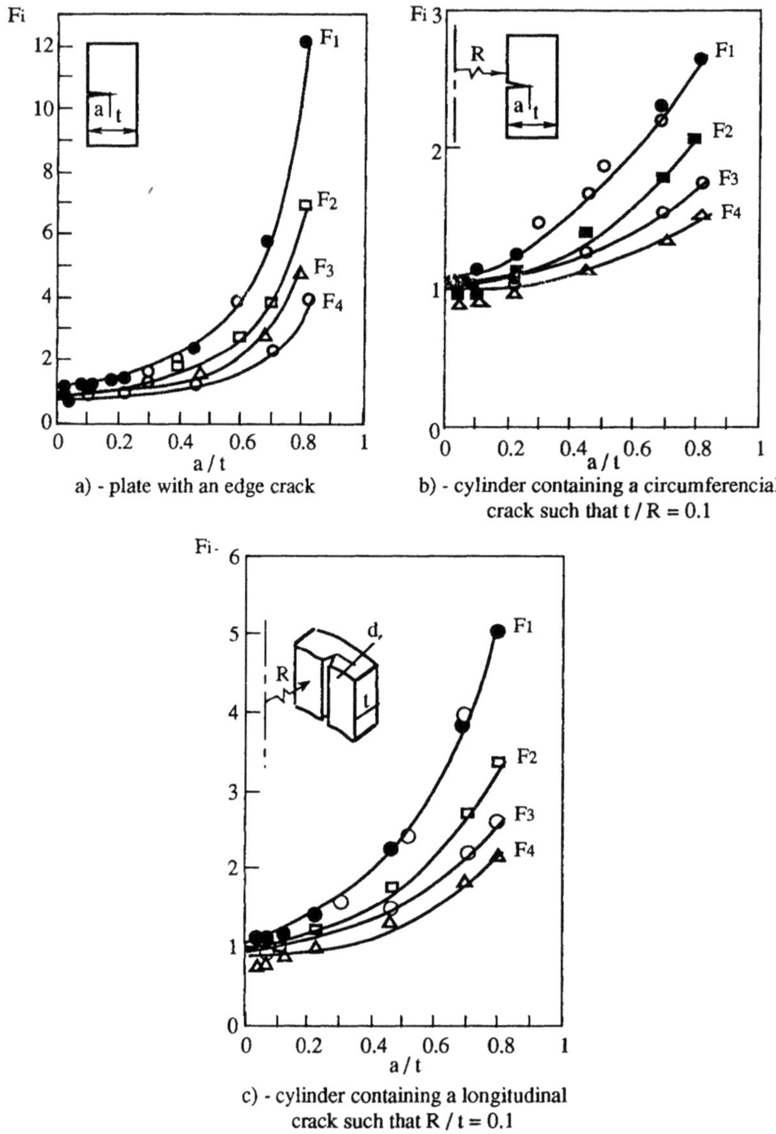

Fig. 3.10 Application of the method of superposition to determine the stress intensity factor K_I, in three geometries: (a) Plate with lateral crack
(b) Cylinder with circumferential crack, $t/R = 0.1$
(c) Cylinder with longitudinal crack, $t/R = 0.1$

To *3rd* order, K is expressed as

$$K = \sqrt{\pi a}\ [A_0F_1 + (2a/\pi)A_1F_2 + (a^2/2)A_2F_3 + (4a^3/3\pi)A_3F_4]$$

The values of the factors F_i are given for each of the three geometries.

An alternative method for attacking the same problem is to attempt to find the value of $K_I(x)$ due to a unit force acting at a distance x from the tip of the crack; given this the value K_I corresponding to a distribution $F(x)$ is

$$K_I = \int_{crack} F(x)K(x)\,dx \tag{3.23}$$

The function $K(x)$, called the *weight function*, has been determined for various particular cases; for example, for a crack of length a in an infinite plate (Fig. 3.11)

$$K(x) = \sqrt{(2a - x) / (\pi ax)} \tag{3.24a}$$

which can be written alternatively as

$$K(x) = \sqrt{(a + t) / (\pi a(a - t))} \qquad (3,24b)$$

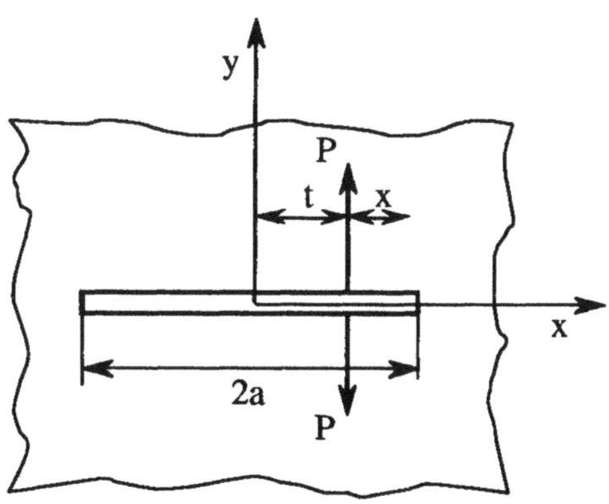

Fig. 3.11 Weight function for a point load

More results for weight functions are given in Labbens et al (1976).

NOTES

1. Before attempting to use these methods it is important to check that the boundary conditions are fulfilled. In any practical problem the dimensions are necessarily finite and the true solution may be very different from that obtained with the mathematically convenient but physically unrealistic assumption of infinite size. Similarly for problems of revolution, where it is important to ensure that the values of the thickness-to-radius ratio B/R for the problem are in the range for which the given solution has been tabulated.

2. The influence of the boundary conditions is seen in the expression for K, as is shown diagramatically in Fig. 3.12: the value increases along the length of the crack when the stress is prescribed and decreases when the strain is prescribed.

3. The weighting-function method assumes that the stress is prescribed. If what is available is the solution corresponding to these conditions but what is wanted is the solution for prescribed displacements, then displacements can be transformed into forces by use of the compliance of the sample (Equation 3.19). Further, equations (3.20) and (3.22) make it possible to pass from one problem to the other, obtaining, for plane strain,

$$G = \frac{P^2}{2}\frac{\partial C}{\partial A} = \frac{K_I^2(1-\nu^2)}{E}$$

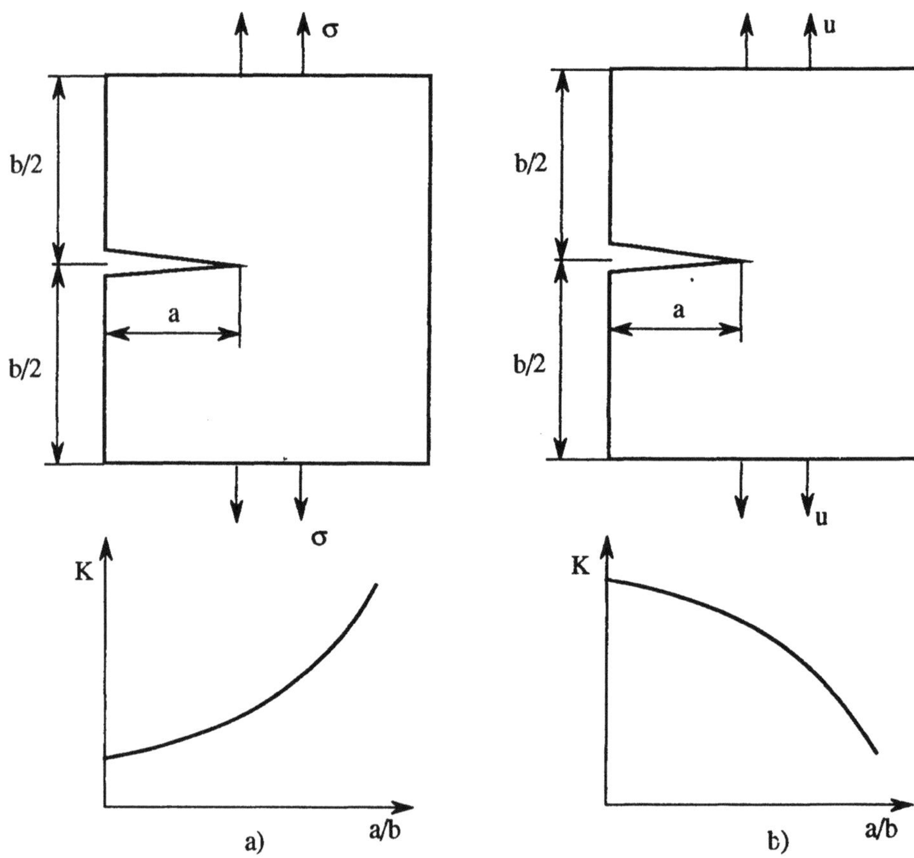

Fig. 3.12 Variation of K with boundary conditions of (a) prescribed stress (b) prescribed displacement, for the same geometry.

3.2.5.2 Experimental methods

The first method uses the relation between K and G (§§3.2.4.2-3): the variation of compliance with crack length is determined experimentally, and the values of K follow immediately.

Other methods are based on the experimental determination of the stress and strain fields in the region of the crack tip. The measured distributions are studied in as great detail as possible in the light of the facts that the stresses vary like $r^{-1/2}$ and the strains like $r^{1/2}$, and that the constant of proportionality is in each case directly related to K (see §3.2.3.3). Since the singularity occurs only at the crack tip the measurements have to be very precise and allow high gradients to be studied: photo-elasticity has been used to achieve this. There are possibilities of using this method for *3*-dimensional problems.

Finally, there are experimental methods based on analogies between the equations of mechanics on the one hand and on different physical phenomena on the other, but these are little used in practice. The interested reader will find details in Bui (1978).

3.2.5.3 Numerical methods

Numerical methods are now the ones most commonly used.

The first we shall describe is based on the known form of the solutions of the equations of elasticity, involving one or more unknown functions of a complex variable. Singularities are accounted for by including a term in $\zeta^{1/2}$ and it is assumed that all other terms are regular and can be expressed as power series in ζ, with coefficients determined by the boundary conditions.

More precisely, we know (§3.2.2.2) that in plane stress and plane strain the solution involves two functions $\phi(\zeta)$ and $\psi(\zeta)$. In order to satisfy the boundary conditions on the faces of the crack these are put in the form

$$\phi(\varsigma)=\varsigma^{1/2}F(\varsigma),\quad \psi(\varsigma)=\varsigma^{1/2}\overline{F}(\varsigma)-\varsigma\phi'(\varsigma),\quad \overline{F}(\varsigma)=F(\overline{\varsigma}) \qquad (3.25)$$

and we recall that

$$\sigma_{xx}+\sigma_{yy}=4\,\mathrm{Re}\,\phi'(\varsigma),\quad \sigma_{yy}-\sigma_{xx}+2i\sigma_{xy}=2\left[\overline{\varsigma}\phi''(\varsigma)+\psi'(\varsigma)\right]$$

We can check (see Exercise 11) that these forms for ϕ and ψ give a singularity $\zeta^{1/2}$ in the stress, as required; consequently we have to determine the function $F(\zeta)$ so that the other boundary conditions for the problem are satisfied. For this we can express it either as

$$F(\zeta) = a_0 + a_1\,\zeta + a_2\,\zeta^2 + \ldots.$$

or as

$$F(\zeta) = a_0 + a_1\,\zeta + a_2\,\zeta^2 + \ldots + b_1/(\zeta - C_1) + b_2/(\zeta - C_2)$$

The second expression enables certain geometries to be handled better, with the proviso that the points C_i all lie outside the external boundary.

The process reduces the problem to that of minimising, in the least-squares sense, the difference between the actual boundary values and those given by the approximation chosen for $F(\zeta)$.

The second method, the *finite element method* , is the one most widely used and is the only practical method for any but simple geometries. There are several forms.

Extrapolation of the stress and strain fields This is related to the experimental method described in §3.2.5.2. It requires a knowledge of the stress and strain fields, for the determination of which very fine meshes must be used so the high gradients at the crack tip are represented accurately. Most often special elements are used, which give the $r^{-1/2}$ singularities automatically.

Energy method. This also is based on the known relation between K and G (§3.2.4.3) and the determination of the compliance (§3.2.4.2). Requiring two meshes it is relatively costly; the meshes must have comparable finenesses, otherwise numerical errors will be introduced. A further disadvantage is that what the method gives is the value of G and that in the case of a mixture of Modes I and II the separate values of K_I and K_{II} cannot be determined directly.

Perturbation method. This method, proposed by Parks (1974) involves calculating the change in energy when the node representing the crack is moved No special elements are needed at the tip of the crack, and the method is less costly than the energy method since it requires only the solution corresponding to a given crack length. Further, it lends itself reasonably well to generalisation to *3*-dimensional problems.

Fig. 3.13 illustrates the method. The mesh in full lines is that for the initial position of the crack and that in dotted lines is for the displaced position. The contour Γ_1 remains unchanged while Γ_0 , in the immediate neighbourhood of the crack, moves when the crack advances a distance dl.

Parks showed that if $[K]$ is the stiffness matrix the rate of release of energy can be determined with the aid of the expression

$$G = -\frac{1}{2}{}^t\vec{u}\frac{d[K]}{dl}\vec{u} \qquad (3.26)$$

where ${}^t\vec{u}$ is the transpose of the displacement vector $\vec{u}$; $[K] = \sum_{i=1}^{N_c}[k_i^c]$.

$[k_i^c]$ is the stiffness matrix for an element between the contours Γ_0 and Γ_1 and N_c is the number of elements between these contours. Thus in contrast to the previous methods this requires only a few nodes of the stiffness matrix to be modified.

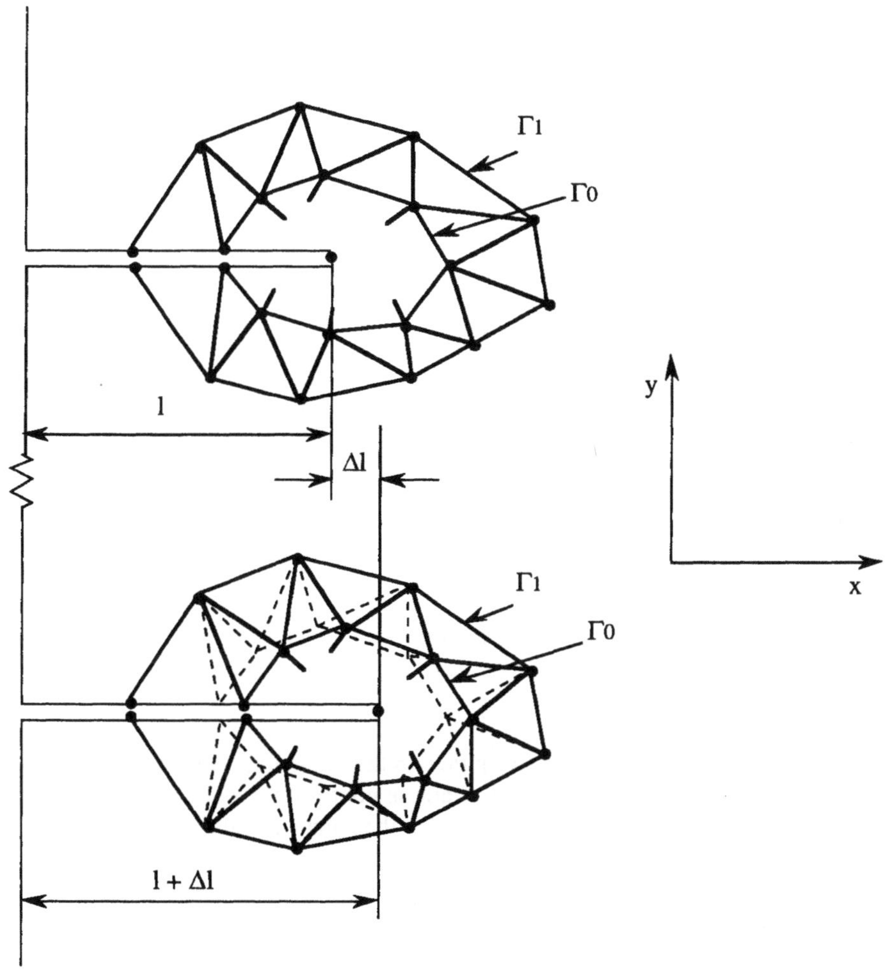

Fig. 3.13 Parks's method: advancing a crack by displacing nodes in its immediate neighbourhood.

This method has advantages, but being based on the calculation of G it suffers from the same disadvantages as the previous method in the case of mixed-mode opening of the crack.

Method of the integral J. This is a contour integral which we introduce later. It can be shown that, given certain conditions, it represents the elastic energy released in an infinitesimal virtual advance of the crack, so that in linear elasticity $J = G$. From this it becomes possible to calculate G and therefore K_I for Mode I.

3.2.6 Notes on *3*-dimensional cracks

As we said at the start of this chapter, we shall not be dealing with *3*-dimensional cracks; however, at this stage we can make a few comments that may be helpful in drawing attention to certain specific points.

In three dimensions, at all points along the tip of a crack there are three stress intensity factors, $K_I(s)$, $K_{II}(s)$, $K_{III}(s)$ (Fig. 3.14).

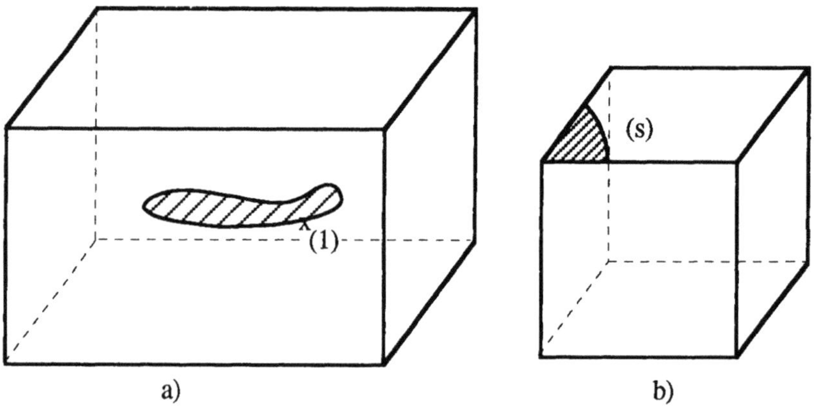

Fig. 3.14 *3*-dimensional cracks: (a) general form (b) crack very often found at the edge of a hole in a plate. s is measured along the crack front.

Theory shows that the form of the stress field in the immediate neighourhood of a point on the crack tip is the same as in plane strain; in contrast, the points at which the crack reaches the surface are special in the sense that the singularity there is weaker than $r^{-1/2}$. Fig, 3.15 shows this edge effect for a crack in the form of a circular arc of radius a, emerging at a free surface; note that values of K_I for θ close to zero are not given when c is much greater than a.

In two dimensions, writing the fracture conditions in the form of $G = G_c$ or of $K_I = K_{Ic}$ comes to the same thing, but this is not so in three dimensions since then K_I varies along the crack tip. The theory of *3D* brittle fracture is still under discussion: some authors prefer to use an energy criterion, and therefore the form $G = G_c$, whilst others take the view that fracture results when $\max_S K_I(s) = K_{Ic}$ along the crack front.

Finally, local approach to brittle fracture in steels by cleavage (see Chapter 2, §2.6.5) leads one to think that the fracture condition would be better expressed as

$$\int_S K_I^4(s)\, ds = K_{Ic}^4 B \qquad (3.27)$$

where B is the thickness of a plate for which an experiment on fracture with a 2-dimensional crack has given the value K_{Ic}.

More information on *3D* cracks is given in Labbens (1980) and Bui (1978).

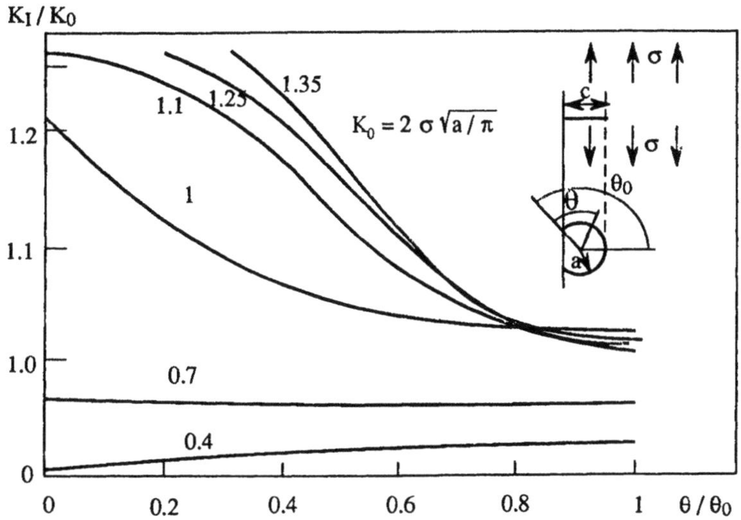

Fig. 3.15 Circular crack emerging from a semi-infinite body acted on by uniform tension σ.

3.3 CRACKS WHEN THE CONDITIONS ARE ELASTO-PLASTIC; CASE OF SMALL-SCALE YIELDING

3.3.1 Introduction

It is clear that the solution based on the assumption that conditions are elastic is not physically realistic, especially in the case of metals; qualitatively, one can see that peaks in the stresses at the tip of the crack will be smoothed out by plasticity. Two important questions arise:

First, it is reasonable to ask to what extent does plasticity affect the elasticity solution. The meaning of the energy relations for the fracture, derived from Griffith's energy balance (§3.2.4) is less certain here, and in fact, since plasticity is by definition irreversible, the meaning of energy releases rate is less clear.

In this section we first study the case in which the plastic region is confined to the immediate neighbourhood of the tip of the crack, as illustrated in the diagram of Fig. 3.16: the assumption is that the extent of this region is small compared with the rest of the structure, that is, with the length *a* of the crack and the width *b* of material ahead of the crack. We shall show that under certain conditions concerning the dimensions of the sample the problem can still be treated within the framework of linear fracture mechanics.

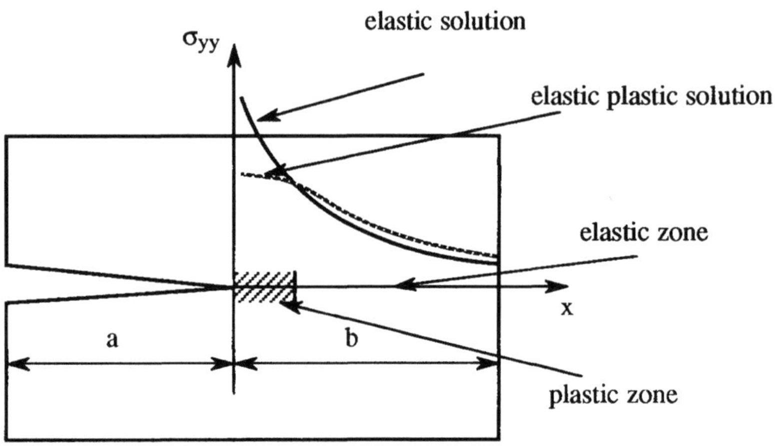

Fig. 3.16 Effect of plasticity in reducing the stresses around the tip of a crack.

3.3.2 Qualitative analysis of plastic flow at the crack tip: case of plane stress and of plane strain

We consider plane stress and plane strain separately (see also Exercise 12).

Plane stress (Fig. 3.17) In the plastically-deformed zone and in the prolongation of the plane of the crack we take axes Ox, Oy, Oz in the directions of the principal stresses. The Mohr circle representing the stress state in this zone has radius k, the yield strength of the material in shear, and passes through the origin since $\sigma_{zz} = 0$. Since the maximum principal stress is σ_{yy} the maximum shear acts on planes at angles of $\pm\pi/4$ to the directions Oy, Oz. It can be seen, as is shown in Fig. 3.17, that the plastic deformation results from slip along these directions, producing local necking of an extent depending on the thickness of the plate. Fig. 3.17 shows the trace of the slip lines on the rear surface of the plate.

Plane strain (Fig. 3.18) As before, the principal stresses are in the directions of the axes. It is easily seen that $\sigma_{xx} < \sigma_{yy}$ - because of the blunting of the crack, however small this effect may be. In plasticity σ_{zz} is half the sum of σ_{xx} and σ_{yy}, since the relation $\sigma_{zz} = \nu(\sigma_{xx} + \sigma_{yy})$ always holds and $\nu = 0.50$. Thus the intermediate stress is now σ_{zz} instead of σ_{xx} as before . The slip planes contain Oz and are at angles $\pm\pi/4$ with Ox and Oy, as shown in Fig, 3.18b. With the plasticity confined to a small volume the plastic zones have the form of butterfly wings (Fig. 3.18a). At the surface, where flow occurs much more easily, approaching that of the plane stress case, the plastic zone is more extended. The Mohr circle (Fig. 3.18c) is now displaced a long way on the tensile axis. Thus in the plastic zone there is a *high level of triaxiality* of the stresses, characterized by the ratio $\sigma_m/2k$ where $\sigma_m = (\sigma_{xx} + \sigma_{yy} + \sigma_{zz})/3$.

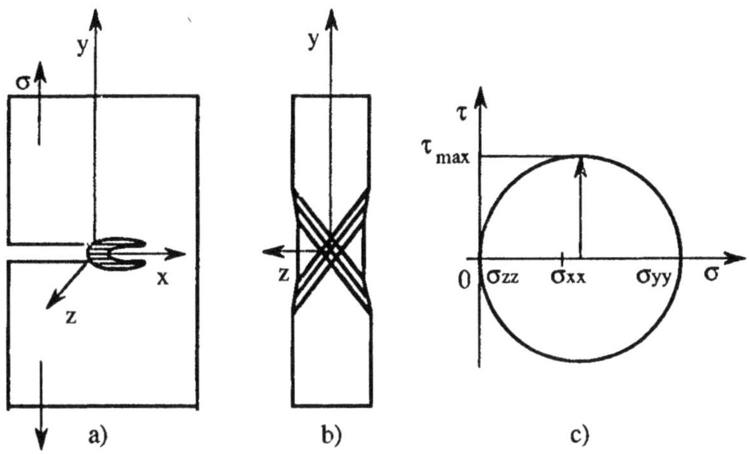

Fig. 3.17 Plastic deformation in plane stress. (a) View in the plane of the plate (b) View of the rear surface of the plate (c) Mohr circle.

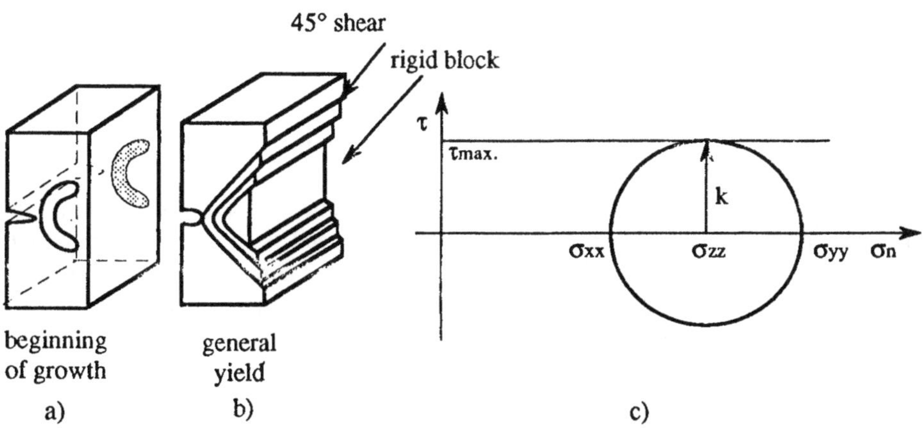

Fig. 3.18 Plastic deformation in plane strain. (a) View in the plane of the plate. (b) View on the rear surface (c) Mohr circle.

3.3.3 Solutions in confined plasticity small-scale yielding

We give two solutions, one proposed by Irwin (1958) and the other based on the Dugdale-Barenblatt model; both are important for understanding the effect of plasticity.

3.3.3.1 Irwin's model (Fig. 3.19)

Irwin's model is applied with conditions of plane stress. The material is assumed to be perfectly plastic and to fulfill Tresca's criterion $2k = R_p$, where R_p is the yield strength.

As we have just seen, the stress σ_{yy} on the plane prolonging that of the crack is a principal stress and in the plastic zone it is constant with the value R_p.

With the stresses previously calculated for elastic conditions, in the plane $\theta=0$ $\sigma_{yy} = K_I / \sqrt{2\pi r}$ and therefore reaches the limit R_p at a distance R_y given by

$$R_y = (1/2\pi)(K_I/R_p)^2 \tag{3.28}$$

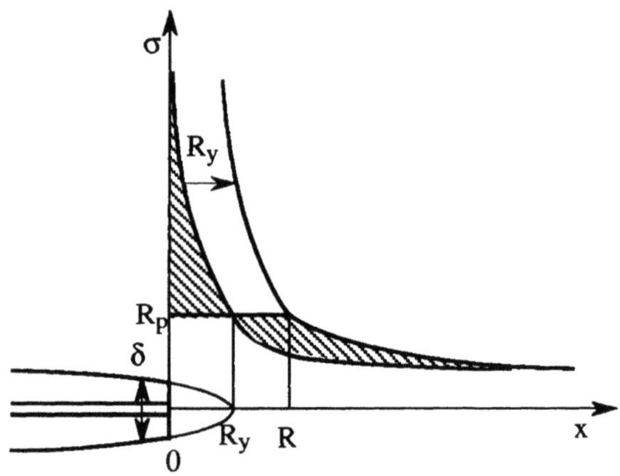

Fig. 3.19 Calculation of the extent of the plastic zone at the crack tip, using Irwin's model in plane strain.

In reality, the plastically-deformed zone is more extended, because the equilibrium conditions for the stresses have to be satisfied. The need is to compensate for the truncation of the elastic distribution for $x \le R_y$; with this as aim a simple assumption is made, that the distribution of σ_{yy} is shifted a distance X determined by equating the areas of the shaded regions in Fig. 3.19. Thus

$$\int_0^{Ry}\left(\frac{K}{\sqrt{2\pi x}} - R_p\right)dx = \int_{R_y}^{R_y+X}\left(R_p - \frac{K}{\sqrt{2\pi x}}\right)dx + \int_{R_y}^{\infty}\left(\frac{K}{\sqrt{2\pi(x+X)}} - \frac{K}{\sqrt{2\pi x}}\right)dx \quad \text{and}$$

hence

$$X = \frac{1}{2\pi}\left(\frac{K_I}{R_p}\right)^2 = R_y$$

The size of the plastic zone is thus

$$R_{ZP}^{CP} = \frac{1}{\pi}\left(\frac{K_I}{R_p}\right)^2 \tag{3.29}$$

This result is clearly only approximate, since all that the calculation ensures is that the stresses are in equilibrium in the prolongation of the plane of the crack; but the same result is given by a more rigorous calculation, using Mode III. It is valid only so long as the extent of the plastic zone remains small - the assumption of confined plasticity.

For a strain-hardened material the constitutive equation is $\varepsilon/\varepsilon_0 = \alpha(\sigma/\sigma_0)^N$, where ε_0 is the deformation at the elastic limit and $\sigma_0 = E\varepsilon_0$; the size of the plastic zone in the plane $\theta = 0$ is

$$R_{ZP}^{CP} = \frac{1}{\pi}\frac{N-1}{N+1}\left(\frac{K_I}{\sigma_0}\right)^2$$

Since for most metallic materials N is large, between 5 and 20, we can see that work-hardening has only a small effect on the result (3.29). An approximation for the size of the plastic zone is often used in which the yield strength R_p is replaced by a flow stress equal to mean of R_p and the ultimate strength R_m.

The model of Fig. 3.19 shows also that the profile of the crack is blunted by the plasticity; at the tip the faces are separated by a quantity δ known as the Crack Tip Opening Displacement (CTOD), whose approximate value is

$$\delta = 2u_y(\theta = \pi, r = R_y) = 2\frac{K_I}{2\mu}\sqrt{\frac{R_y}{2\pi}}\frac{4}{1+\nu} = \frac{4}{\pi}\frac{K_I^2}{ER_p}$$

(E is Young's modulus)
Various more rigorous calculations have been made, giving in all cases

$$\delta \approx K_I^2/ER_p \tag{3.30}$$

This result is easily arrived at by recalling that K_I^2/E is the rate of release of elastic energy and equating this to the work required in deforming, by a distance δ, a virtual tensile test piece of yield strength R_p situated at the tip of the crack.

A reasoning by analogy can be used to lift the restriction to plane stress. In this a fictitious elastic yield strength is introduced so as to take account of the triaxiality, as we noted previously (§3.3.2); the calculation, which is not completely rigorous, gives the result

$$R_{ZP}^{CP} = \frac{1}{3\pi}\left(\frac{K_I}{R_p}\right)^2 \tag{3.31}$$

NOTES

1. The model just described does not claim to represent the plastic zone correctly over the whole of the plane. More elaborate analytical and numerical calculations have been made, some of the results of which are given in Fig. 3.20, together with references sources of further information. The estimates of size given by Irwin's simple model are

always of the right order of magnitude, but those of the shape depend strongly on the problem treated, as we should expect from the discussion on the slip planes in the previous section. The treatment of Mode III is the subject of Exercise 13.

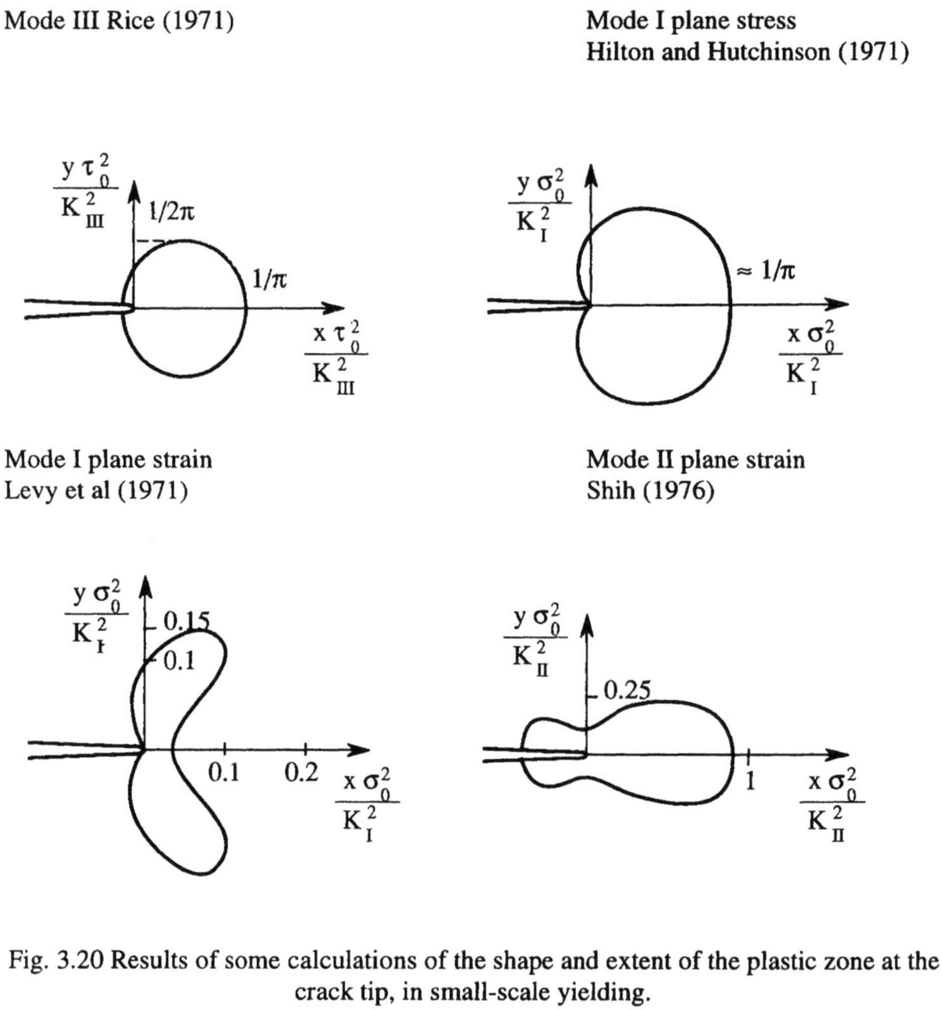

Fig. 3.20 Results of some calculations of the shape and extent of the plastic zone at the crack tip, in small-scale yielding.

2. Some experimental investigations have been made of the extent of the plastic zone, and in some cases of the form also; most of these have concerned fatigue cracks. Various techniques have been used, including:
- measurement of micro-hardness,
- examination of surface relief, for example by holography,
- metallographic attack.

Fig. 21. gives an example.

3.3.3.2 Dugdale-Barenblatt and Bilby-Cottrell-Swinden (1963) models

These models enable us to relax requirement that the plastic zone is confined to a small region in the immediate neighbourhood of a singularity; they provide an illustration of the extension of the principle of superposition of solutions.

Fig. 3.22 gives the principle. We consider a crack of length $2a$ in an infinite plate acted on by a uniform stress σ_∞ at infinity.

The plasticity is modeled by constant forces equal to the yield strength of the material; these are exerted over lengths R_D in the material just before each extremity of the crack, where R_D is the size of the plastic zone, the quantity to be determined.

Thus the model concerns a plastified zone in an perfectly plastic material under conditions of plane stress.

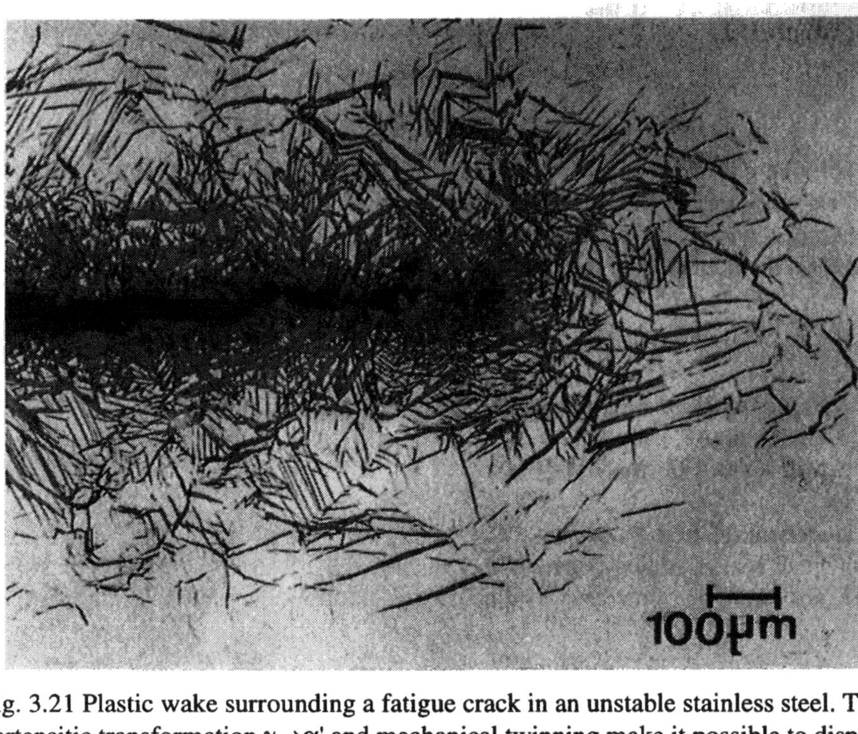

Fig. 3.21 Plastic wake surrounding a fatigue crack in an unstable stainless steel. The martensitic transformation $\gamma \rightarrow \alpha'$ and mechanical twinning make it possible to display the plastic zone.

By superposition the boundary conditions for the problem, transferred to the crack, are

$$\sigma_{yy} = -\sigma_\infty \quad if \ |x| \leq a,$$
$$= -\sigma_\infty + R_p \ if \ a \leq |x| \leq a + R_D$$

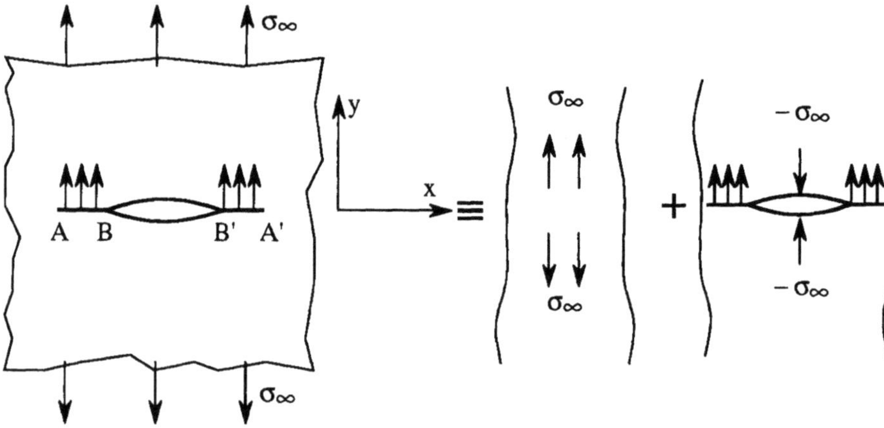

Fig. 3.22 Dugdale-Barenblatt model

The solution can be obtained by applying the weight function method (§3.2.5.1) to a fictitious crack of length $2(a + R_D)$ which includes the plastic zones. R_D is found as a function of a and σ_∞ by requiring that stress intensity factor is zero, so as to give finite stresses on the boundary of the plastic zone (see Exercise 14); the result is

$$R_D = a\left[\sec(\pi\sigma_\infty/2R_p) - 1\right] \tag{3.32}$$

If $\sigma_\infty << R_p$ the power-series expansion of this can be truncated to give

$$R_D \approx (\pi/8)\,(K_I/R_p)^2 \tag{3.33}$$

The numerical factor $\pi/8$ does not differ greatly from the $1/\pi$ given by Irwin's model in which small-scale yielding is assumed. The model due to Bilby, Cottrell and Swinden (1963) arrives at the same result by employing a fictitious distribution of dislocations.

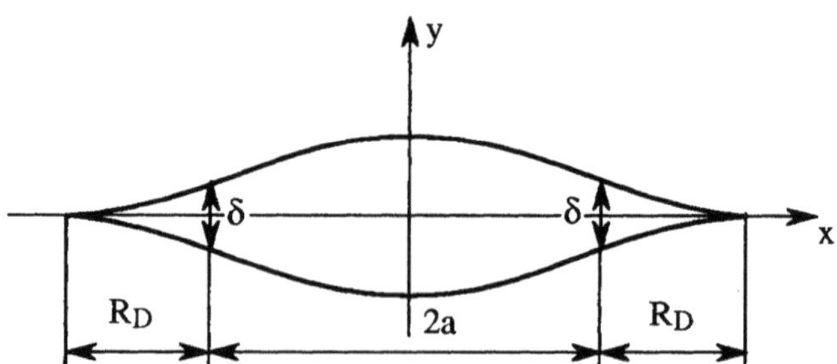

Fig. 3.23 Form of an open crack in the Dugdale-Barenblatt model.

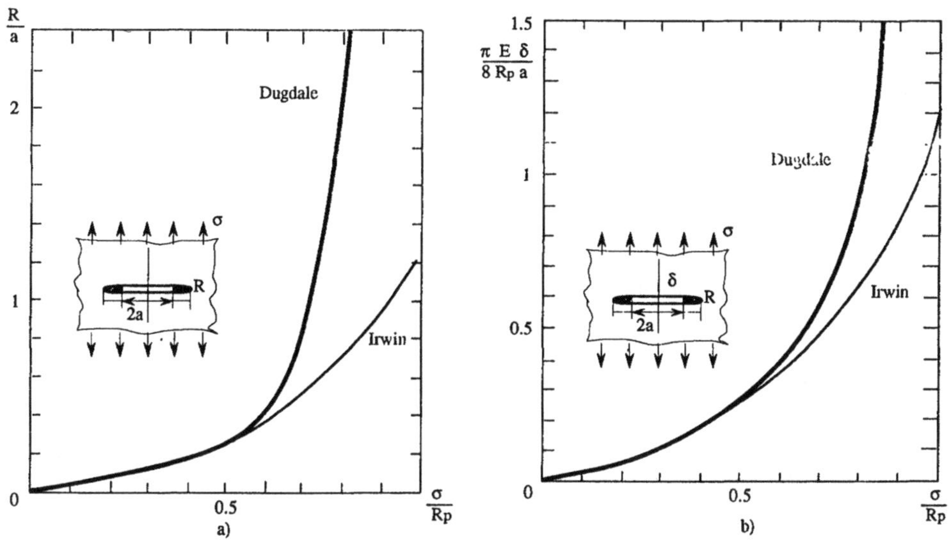

Fig. 3.24 Comparison of Dugdale-Barenblatt and Irwin models, for plane stress. (a) Size of plastic zone (b) CTOD δ

These two models correspond quite well with reality in the case of thin plates; further, they enable the CTOD, δ (§3.3.3.1) to be determined (see Exercise 13). Fig. 3.23 shows the shape taken by the crack in response to the cohesion forces due to the plasticity; the separation at the tips is given by

$$\delta = (8R_p/\pi E)\, a\, \mathrm{logsec}(\pi\sigma_\infty/2R_p) \tag{3.34}$$

It is interesting that experimental measurements made with a polymer confirm this result quite well: the "crazing" (see Chapter 2, §2.10) found in such material represents a system that is doubtless well adapted to the cohesion forces associated with the plasticity.

The Dugdale-Barenblatt model also enables sharper bounds to be given for the validity of the small-scale yielding solutions. For this, the size R of the plastic zone and the CTOD δ, calculated from equations 3.32 and 3.34 respectively, are represented as non-dimensional numbers and these are compared with the values found from equations 3.29 and 3.30, plane stress being assumed in all cases. Fig. 3.24 gives this comparison, showing that the previous solutions cease to be reasonably valid when the applied stress exceeds one-third, perhaps one-half, of the yield strength.

It is only recently that crack phenomena in conditions of *large scale plasticity* - that is, when the size of the plastic zone is no longer negligible on the scale of the crack length or the dimensions of the crack-free material - has been studied. This constitutes the subject of *fracture mechanics in non-linear behaviour*; in this section we give some of the results that have been obtained.

3.4.1 Limit load

If a component made from a perfectly plastic material is acted on by a steadily increasing load a stage is reached at which there is full-scale plastic yielding, as shown in Fig. 3.25. Since there is no work-hardening the load cannot then be increased further, and this is called the *limit load.*

The limit load concept applies to any component; a crack always reduces its value below that for the uncracked component. Calculation of the value involves using tools developed in the theory of plasticity, in particular the extremal theorems of Volume I, §3.4; thus in the simple case corresponding to Fig. 3.25 it can be shown (Exercise 15) that

$$P_L = P_y(1 - a/W) \qquad (3.35)$$

where $P_y = WBR_p$ and B, W $(= 2b)$ are the thickness and width of the plate respectively.

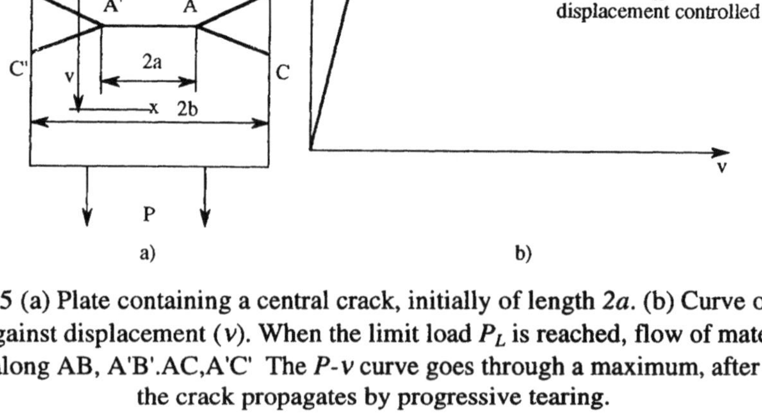

Fig. 3.25 (a) Plate containing a central crack, initially of length $2a$. (b) Curve of load (P) against displacement (v). When the limit load P_L is reached, flow of material occurs along AB, A'B'.AC,A'C' The P-v curve goes through a maximum, after which the crack propagates by progressive tearing.

The limit load represents a distinctive mode of failure when the structure is subjected to a steadily increasing load; but this is not necessarily the case when the displacement is prescribed. Further, when the crack propagates stably the progressive tearing entails a decreasing load.

Fig 3.26 gives a simple illustration of the regions of linear fracture mechanics in linear and non-linear behaviour respectively. We can imagine experiments made with materials of very different types, one very brittle (such as glass), one very ductile (such as mild steel), with equation 3.35 used to represent the results as a graph of P_R/P_L against a/W as in Fig. 3.26.

So far we have treated only the case of fracture mechanics in linear elasticity, corresponding to the loading ab and without stable propagation; whilst with very ductile materials the stable propagation, with displacement prescribed, corresponds to the trajectory cd.

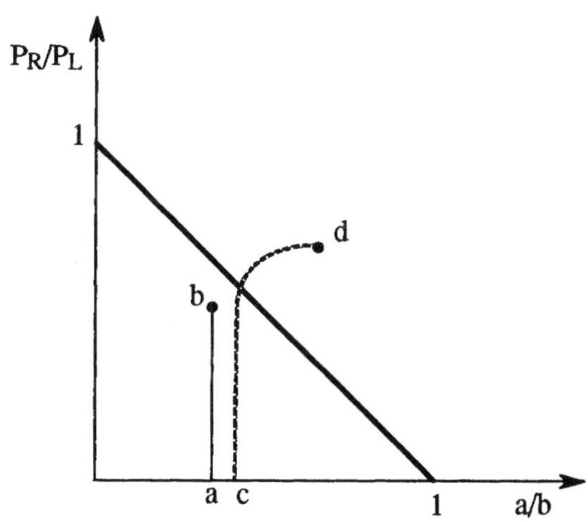

Fig. 3.26 Diagrammatic representation of results of fracture tests on a plate containing a central crack, for very brittle (AB) and very ductile (CD) materials. Points *b* and *d* are the loads at fracture.

Fig. 3.27 shows an alternative representation, in which K/K_{Ic} is plotted against P/P_L.

For a given geometry and a given material there is a curve corresponding to fracture at a prescribed load; even if this curve is not fully determined there is certainly, whatever the case, a minimal envelope for the fracture, taking a pessimistic view of the possible geometric effects.

If on the other hand the displacement is prescribed it is difficult to take so simple an approach, since then the stable propagation modifies the geometry and it becomes necessary to find an accurate solution to the problem at the tip of the crack. In perfect plasticity there are only partial solutions. There are however solutions taking work-hardening into account. Here we describe only few results, limited to the case of stationary cracks.

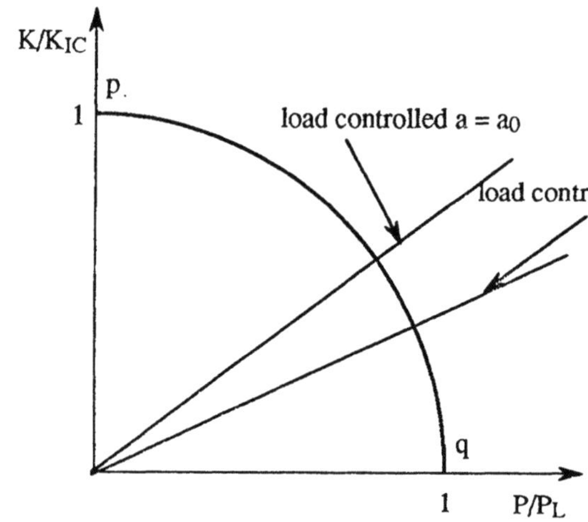

Fig. 3.27 Envelope (*pq*) in the K/K_{Ic}-P/P_L plane for a structure with a crack.

3.4.2 Perfect plasticity; the Prandtl field

We consider here a Mode I crack in a perfectly plastic material in conditions of plane strain. The stress distribution ahead of the crack can be obtained from the equilibrium equations and the plasticity criterion, for which we can use the slip-lines model (see I-§3.4, and Exercise 16 of this chapter); this enables it to be shown, analogously to the case of a punch (cf. I-§3.4.3.5), that the stresses are as shown in Fig. 3.28.

In this figure, in the first "diamond" region $0 < \theta,\ \pi/4$

$$\sigma_{xx} = \pi k,\ \sigma_{xy} = 0,\ \sigma_{yy} = (2 + \pi)k$$

In the "fan" region $\pi/4 < \theta < 3\pi/4$

$$\sigma_{\theta\theta} = (1 + 3\pi/2 - 2\theta)k,\ \sigma_{r\theta} = k,\ \sigma_{rr} = (1 + 3\pi/2 - 2\theta)k$$

In the second "diamond" region $3\pi/4 < \theta < \pi$

$$\sigma_{xx} = k,\ \sigma_{xy} = o,\ \sigma_{yy} = 0.$$

where k, the yield strength in shear, is $R_p\sqrt{3}$ for a material to which the von Mises criterion applies and $R_p/2$ for the Tresca criterion.

It follows that there is no singularity at the tip of the crack and that the stresses near there depend only on θ; this is true also in the case of small-scale yielding. This distribution does not apply until after a short distance from the tip ($r > 2\delta$), as would be suggested by the slip lines in Fig. 3.28; it can be shown that all the strains are regular with the exception of the $\varepsilon_{r\theta}$ component in the fan, which has the form $(k/2\mu r)R(\theta)$. This strain field thus has a strong singularity and, unlike the stress field, is dominant at long distances.

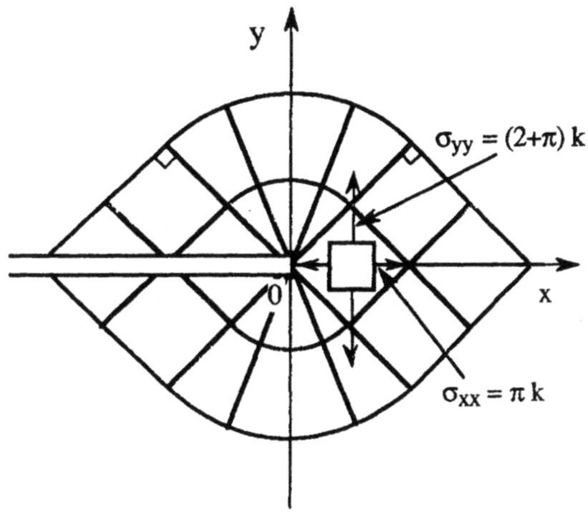

Fig. 3.28 Slip lines and Prandtl field. *k* is the yield strength in shear.

3.4.3 The contour integral J

To lift the restriction of perfect plasticity and so to approximate more closely to the behaviour of real materials we introduce a non-linear elastic material that gives the same strain curve as that for perfect plasticity. The possibility of this is reasonable provided that the loading remains radial, which suggests that there must be no sharp changes in direction of the applied load and therefore no unloading. The assumption of non-linear elasticity implies, in fact, that any unloading follows the same curve as the loading, which clearly is not the case for a real elasto-plastic material.

We know that in the general framework of elasticity there is an elastic potential W whose derivatives are the stresses: $\sigma_{ij} = \partial W/\partial \varepsilon_{ij}$. In such conditions, as in linear elasticity, it is possible to calculate the amount of elastic energy released during the virtual advance of a crack, under the assumptions that the crack propagates without changes of direction and that there are no forces acting on its sides. It can be shown that the energy change is given by the value of a certain contour integral, unaffected by the particular contour chosen. This integral, introduced by Rice (1968) and Cherepanov, is

$$J = \int_{\Gamma} \left(W dy - \vec{T} \frac{\partial \vec{u}}{\partial x} ds \right) \tag{3.36}$$

where W is the elastic density, so that $\sigma_{ij} = \partial W/\partial \varepsilon_{ij}$, $\vec{T}$ and $\vec{u}$ are the stress and displacement vectors at a general point M on the contour Γ. The contour is oriented as in Fig. 3.29, with the normal directed outwards .

This integral has three important properties; we give the first two here and the third in the paragraph that follows.

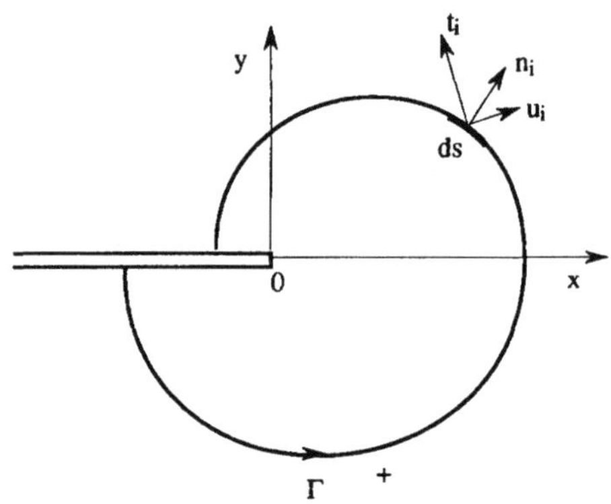

Fig. 3.29 Notation used at the crack tip, to define the contour integral *J*.

(i) *J is independent of the contour Γ*. This interesting and important result can be proved without great difficulty. A contour that keeps very close to the tip of the crack will enclose only the singularity there, whilst one that is situated at a great distance will bring in the boundary conditions for the problem. This gives a convenient way to relate the behaviour at the crack tip to that at a distance.

(ii) In linear elasticity *J* represents the energy release rate for an infinitesimal advance of the crack $J = -(1/B)d\Pi/da$ where Π is the total potential energy and *a* is the length of the crack. For a proof of this see Rice (1971). The property is illustrated in Fig. 3.30, where the two fictitious loading (*P*, *v*) curves correspond to two crack lengths differing by *da*.

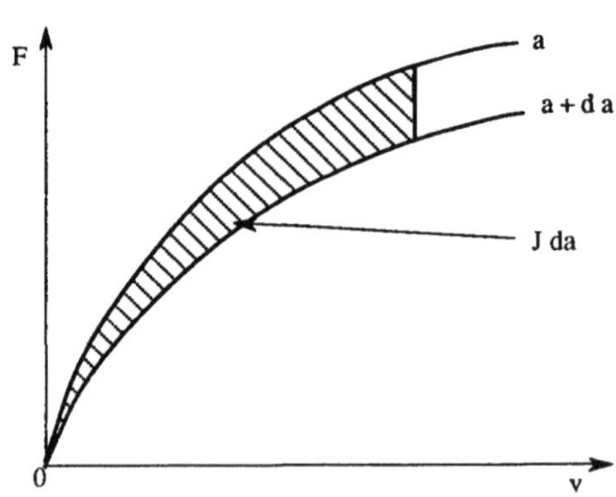

Fig. 3.30 Determination of *J* by the method of compliance.

It follows that $J = - \int [\partial P/B\partial a] dv$ (3.37)

or the dual

$$J = \int [\partial v/B\partial a] dP \tag{3.38}$$

This property is very useful when conditions are in the neighbourhood of the limit load.

The form of the loading curve can always be determined if the geometry is known; in particular, it can be put in the form $P/P_L = f(v/l)$ where P is the load and l a length, for example that of the crack-free material. The form of the function f can be determined by a single experiment or a single calculation.

It can be shown (see Exercise 17) that in general J can be expressed as

$$J = -\frac{1}{P_L}\frac{\partial P_L}{\partial a}\int P dv + \frac{1}{l}\frac{\partial l}{\partial a} P v - \frac{1}{l}\frac{\partial l}{\partial a}\int P dv$$

If U is the area under the $(P,\ v)$ curve this can be written

$$J = U\left[-\frac{\partial P_L}{P_L \partial a} - \frac{\partial l}{l\partial a}\right] + Pv\frac{\partial l}{l\partial a} \tag{3.39}$$

3.4.4 Asymptotic solution. HRR field

This solution aims to represent the stress and strain fields at a moderate distance from the tip. At the tip the blunting and the Prandtl field (§3.4.2) lead to finite stresses, and damage phenomena also can perturb the solutions we shall describe. At greater distances the distributions can be approximated by an asymptotic solution proposed by Hutchinson (1968) and by Rice and Rosengren (1968) (the HRR solution). Here we shall give only some results. The constitutive equation for the material is written

$$\varepsilon/\varepsilon_0 = (\sigma/\sigma_0)^N \tag{3.40}$$

where ε_0, σ_0 and N are constants for the material. $N = 1$ corresponds to linear elasticity and $N \rightarrow \infty$ to the approach to perfect plasticity.

The asymptotic solution can be written as follows.

$$\sigma_{ij} = \sigma_0 \left(\frac{J}{\sigma_0 \varepsilon_0 I_N r} \right)^{1/(N+1)} \tilde{\sigma}_{ij}(\theta, N)$$
$$\varepsilon_{ij} = \varepsilon_0 \left(\frac{J}{\sigma_0 \varepsilon_0 I_N r} \right)^{N/(N+1)} \tilde{\varepsilon}_{ij}(\theta, N) \tag{3.41}$$

For $N = 1$ this corresponds to that found for linear elasticity, since then, in plane strain,

$$J \equiv G = \frac{K_I^2}{E}(1 - \nu^2) \tag{3.42}$$

The functions $\tilde{\sigma}_{ij}(\theta, N)$, $\tilde{\varepsilon}_{ij}(\theta, N)$ are solutions of a differential equation and have been tabulated for various values of N; Fig. 3.31 shows them for $N = 13$. I_N is a function of N such that $(I_N/\pi)^{1/(N+1)}$ is approximately equal to 1

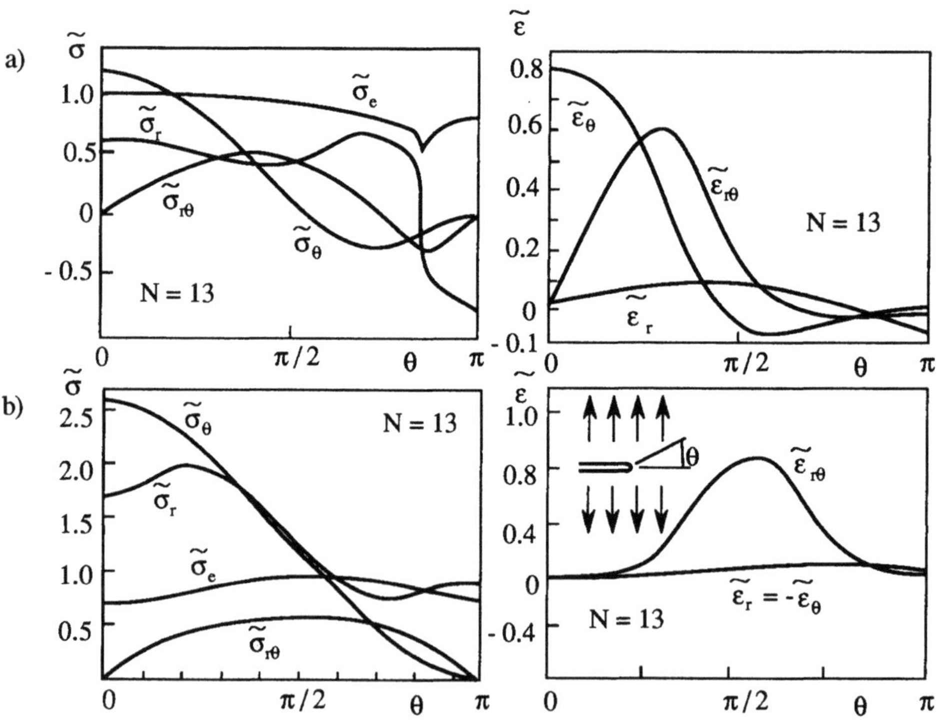

Fig. 3.31 Values of $\tilde{\sigma}$ and $\tilde{\varepsilon}$ as functions of θ for $N = 13$
(a) plane stress (b) plane strain.

As in linear elasticity, in the neighbourhood of the tip of a crack - but not too close - the stress and strain fields should be asymptotically functions of a single parameter, J in this case: this property is essential for ensuring that results obtained with test pieces can be transferred to structures containing cracks.

For non-linear behaviour, however, these fields depend on the constitutive equations, which is not the case for the stress field in linear elasticity.

This third property of the integral J, meaning the role it plays in the form of the distributions of stress and strain, has led to important applications, the bases of which are as follows.

Particular test pieces are studied, for which, using the arguments indicated in §3.4.3, the value of J can be found experimentally by simply measuring the area under the (P,δ) loading curve. From observation of the initiation and propagation of cracks in such pieces it is possible in principle to construct a *tearing resistance curve* $(P, \Delta a)$, where Δa represents the advance of the crack. Fig. 3.32 shows such a curve.

We postulate that even in conditions of large scale yielding the stress field is given by equations (3.41), from which it follows in particular that the initiation phase is fully defined by the critical value of J, that is, J_{Ic}. The crack growth $(\Delta a)_c$ associated with this corresponds to the blunting of the crack, which, as in small-scale yielding, is directly related to J: $\delta = J/mR_p$ (m a numerical constant) (3.43)

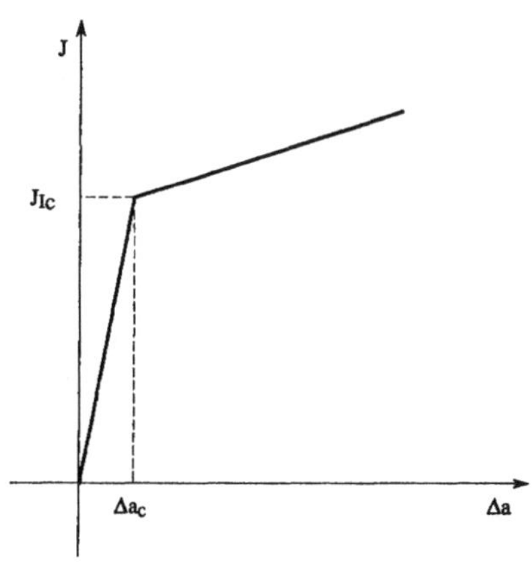

Fig. 3.32 Schematic representation of the fracture resistance curve.

We know that in these conditions the value of m for plane stress is approximately *1*; otherwise, m is a function of the geometry, the yield strength and the work-hardening coefficient, and values observed up to now lie between *2* and *4*.

The assumption of a stress field independent of geometry, and completely characterized by J, can be questioned. It holds quite well for numerical calculations concerning test geometries in which there is a strong bending component, but much less so when the plasticity results from tensile loading. Any use of the J integral for predicting the fracture resistance of materials of under large scale yielding conditions should therefore be treated with great reserve.

3.5 SUMMARY OF LINEAR AND NON-LINEAR FRACTURE MECHANICS

3.5.1 Strictly linear elasticity. In the region in which the elasticity is strictly linear all the results we have given are derived ultimately from solutions of crack problems in plane elasticity; further, they are concerned mainly with Mode I opening. In these conditions a single parameter suffices to specify the complete set of important values, including the energy release rate G; this leads naturally to the defining of the *fracture toughness* K_{Ic}.

As soon as we abandon the assumptions on which Griffith's theory of brittle fracture is based we enter much less well-charted territory, where we meet problems such as those of anisotropy, of mixed modes and of 3-dimensional systems.

3.5.2 Confined plasticity In small-scale yielding, provided that the extent of the plastically-deformed zone around the tip of the crack is small enough, the concepts of linear elastic fracture mechanics can still be applied. Whilst the parameter K may not have the significance for energy considerations that it has in linear elasticity, it still describes the stress and strain fields and consequently gives the critical loading conditions that lead to fracture. We thus have every right to speak of toughness; the important need is that the conditions for small-scale yielding are satisfied. The Irwin and Dugdale-Barenblatt models in principle provide means for evaluating the extent of the plastic zone at the crack tip, and from the results of a large number of experiments and calculations it has become usual to accept that the size b of the uncracked material satisfies $b > 2.5(K_I/R_p)^2$.

3.5.3 Use of the contour integral *J*. If the assumptions of small-scale yielding do not hold then the form of the singularity in the stresses is greatly modified. Whilst in linear elasticity this is like $r^{-1/2}$, it is much weaker in plasticity where it is like $r^{-n/(n+1)}$, n being the work-hardening exponent, values for which lie between 0 and 0.2. In certain cases the integral J gives a good value for the plastic singularity.

The significance of J for energy considerations rests on the schematic representation of true elasto-plastic behaviour, with work-hardening, by non-linear elastic behaviour; consequently J represents the rate of energy release to only a limited extent. Nevertheless the asymptotic forms for the stress and strain distributions in terms of J are still very useful when combined with a more realistic solution in the region of the crack tip that gives the blunting there. Further, the approximate methods for evaluating J based on the limit load make calculation of its value reasonably easy.
For all these reasons, most of the recent work on non-linear fracture mechanics makes use of J, even if this use is not always fully justified.

3.6 APPLICATIONS OF FRACTURE MECHANICS

To end the chapter we give some applications of fracture mechanics, mainly restricted, for the reasons given, to linear elasticity; other applications are given in the form of exercises (numbers 18, 19, 21). We deal first with the concept of fracture toughness K_{Ic}, concentrating here on the determination of its value; later we consider applications concerning stable crack propagation in fatigue and in stress-corrosion cracking.

3.6.1 Determination of fracture toughness K_{Ic}: the *R*-curve

For test pieces experimentalists use most often one or other of three types of geometry: these are:

- the *compact tensile test piece* (CT) shown in Fig. 3.33,

- the *bending* (3PB),

- the *centered crack panel* (CCP).

The two first are usually preferred since they require lower applied forces for a given *K*. With these geometries the value of *K* is given by

$$K = PY/BW^{1/2} \tag{3.44}$$

where *P* is the applied load, *B* and *W* are the thickness and width of the piece respectively and *Y* is a known (tabulated) function of the length of the crack.

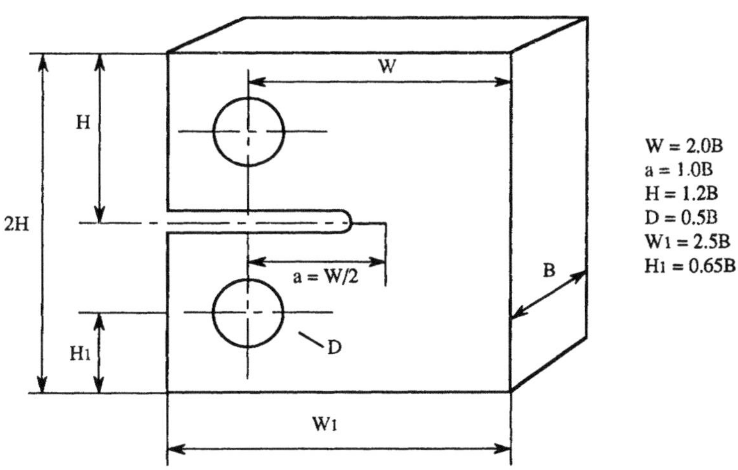

Fig. 3 33 Test piece of type CT

If the behaviour of the material remains perfectly elastic throughout the test (3.44) enables the fracture toughness K_{Ic} to be determined from the load P_R at fracture. For most of the usual metallic materials there is in fact a departure from linearity, due either to a greater or less development of plasticity or to stable propagation of the crack (Fig. 3.34)

As a result of many experiments it has become possible to lay down standards for the preparation and the dimensions of test pieces, and these have been made the subject of the standards ASTM E 399-74 and AFNOR A03-80. Without going into detail we give the principles here.

Effect of thickness

Over the sides of the test piece, where the conditions are those of plane stress, fracture occurs not in pure Mode I but by shear, as indicated in Fig. 3.35; when measuring fracture toughness the aim is always to limit the importance of these *shear lips.*

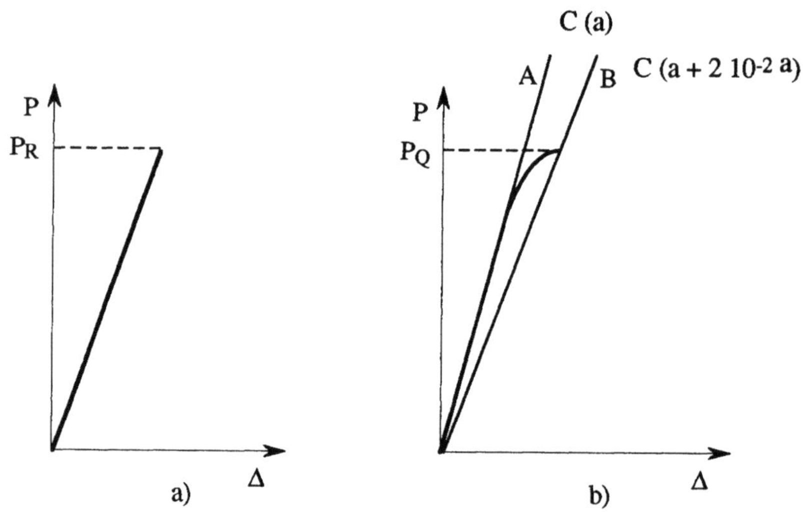

Fig. 3.34 Load-displacement diagram recorded in a test with (a) perfectly brittle (b) more ductile material.

The fracture energy in the plastic zone is much lower for plane strain than for plane stress; this is due to the difference in size of the two zones, since a large part of the fracture energy comes from plastic deformation, and also to the difference in the mean stresses.

In fact, as we saw in Chapter 2, when considering the mechanisms of fracture (§2.8), the elongation at fracture is less, the higher the mean stress. If G_{Ic} and G_{Cmax} are the fracture energies for plane strain and plane stress respectively and c is the width of the shear lips corresponding to plane stress, G_C is approximated by the weighted mean of these energies, thus:

$$G_C = \frac{2c}{B} G_{C_{\max}} + \frac{B-2c}{B} G_{Ic}\text{, i.e. } \frac{G_C}{G_{Ic}} = 1 + \frac{2c}{B}\left[\frac{G_{C_{\max}}}{G_{Ic}} - 1\right] \tag{3.45}$$

Here c is of the order of the extent of the plastic zone in plane stress, that is

$$c \approx R_P^{CP} \approx \frac{1}{\pi}\left(\frac{K_C}{R_p}\right)^2 = \frac{1}{\pi}\frac{EG_C}{R_p^2}$$

Fig. 3.35 gives the variation of G_C with thickness B. The thickness B_C corresponds to the vanishing of the plane fracture zone; the condition that this occurs is approximately

$$B = 2c \approx \frac{2}{\pi}\left(\frac{K_C}{R_p}\right)^2 \tag{3.46}$$

If B is less than B_C the size of the plastic zone is close to B; this is why the fracture energy decreases.

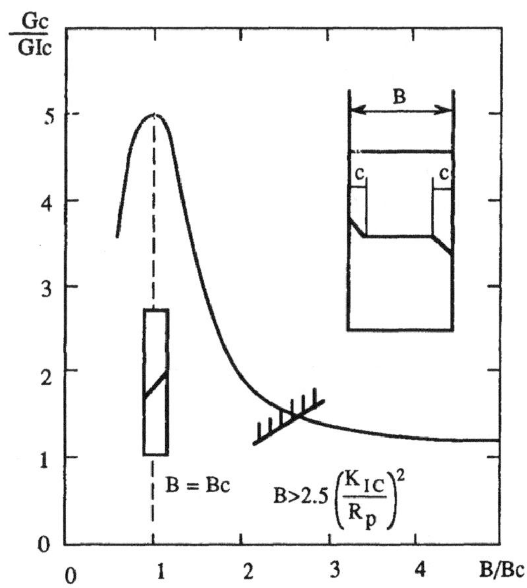

Fig. 3.35 Variation of G_C with thickness and the schematic fracture aspect.

If we wish to find a value for the fracture energy, and therefore for the fracture toughness, that is independent of thickness and is therefore an intrinsic property of the material, we must ensure that the thickness of the test piece is great enough. Further, we should wish the plasticity to remain confined, both in front of and behind the crack. Thus the practical conditions are

$$a, B, b > 2.5\,(K_{Ic}/R_p)^2 \tag{3.47}$$

When the asymptotic value G_{Ic} is reached the toughness K_{Ic} is given by

$$K_{Ic} = \sqrt{\frac{EG_{IC}}{(1-\nu^2)}}$$

The R-curve

Since the formation of the shear lips is progressive, the fracture energy, according to the above approximation, increases from G_{Ic} when the crack starts to propagate to G_C when these are fully developed. This provides a possible explanation for the curve R already introduced, the $(J, \Delta a)$ curve of Fig. 3.32, §3.4.4, which gives the fracture energy in the course of the growth Δa of the crack (Fig. 3.36).

Another possible explanation is that the plastic zone becomes very different when the crack propagates, as indicated in Fig. 3.37. Whatever the reason, it seems that an R curve has to be introduced, as in Fig. 3.36; the slope of the $(G_C, \Delta a)$ or the $(J, \Delta a)$ curve will increase with increasing ductility of the material.

In Fig. 3.38 the evolution of the crack driving force G is shown, for two different loading conditions, as a function of crack length for constant values of applied stress σ and applied load F respectively; the R curve relates G_c to an initial length a_0.

Crack propagation requires $G > G_C$. If $\partial G/\partial a > dG_C/da$: it occurs only if the stress or the load is increased, that is, there is stable crack growth; but if $\partial G/\partial a < dG_C/da$ then keeping the stress or the load constant will provide more energy than is needed and consequently the growth becomes unstable.

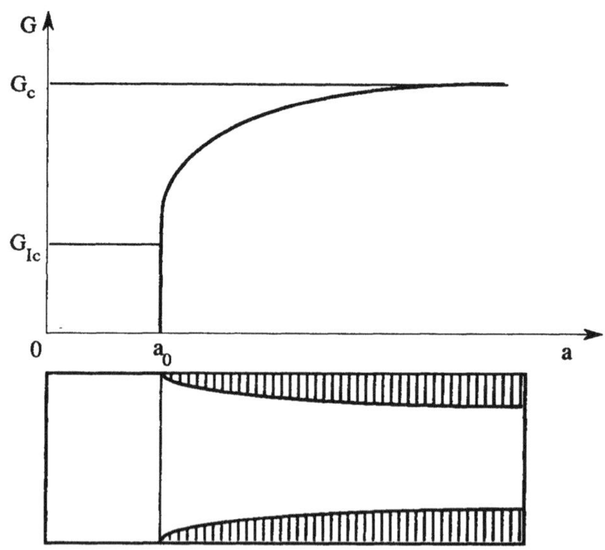

Fig. 3.36 Fracture resistance curve and development of shear lips.

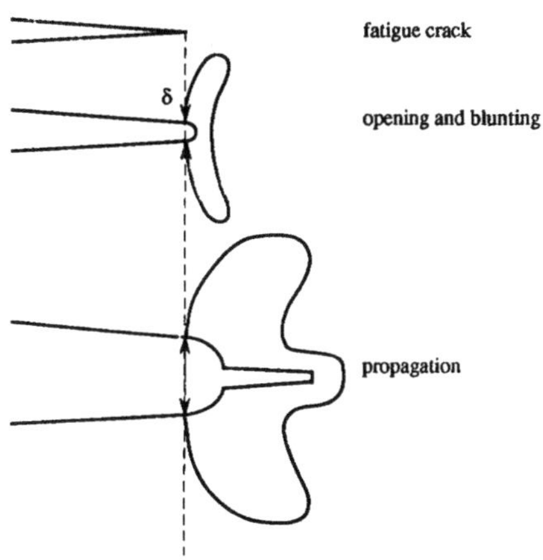

Fig. 3.37 Shape of plastic zones at crack initiation and during propagation

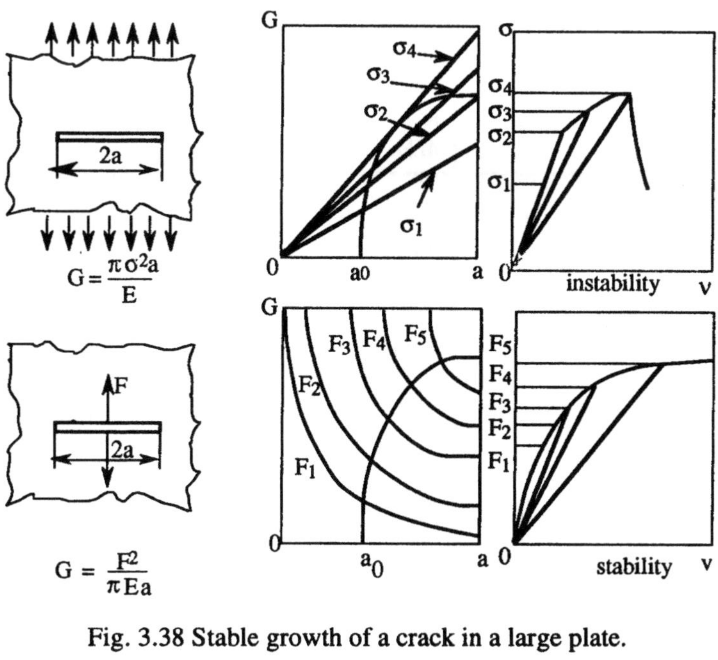

Fig. 3.38 Stable growth of a crack in a large plate.

Limitation of the curvature of the load / displacement curve so as to restrict the stable advance of a crack

In measuring toughness we are looking primarily for criteria for the occurrence of unstable fracture; for this reason, when in practice the experimental curve shows a departure from linear elasticity (Fig. 3.34b) we seek to limit this departure. Since an advance Δa_i of the crack is expressed by a change in the compliance $C(a)$, we construct practical rules based on the measurement of this latter: this leads to the condition

$$\Delta a_i < \frac{1}{2} R_{ZP}^{DP} \approx \frac{1}{6\pi}\left(\frac{K_{Ic}}{R_p}\right)^2 \qquad (3.48)$$

which with the condition (3.47) for the length of the crack becomes $\Delta a_i/a < 2.10^{-2}$

This relative change of length is easily translated into a change in compliance, as indicated in Fig. 3.34b: in practice, $\Delta C/C \approx 5.10^{-2}$ for a crack length such that $a/W \approx 0.5$. From this, if we know the value of the corresponding load (P_Q in Fig. 3.34b) and can be confident that the condition (3.47) on the thickness is satisfied, we can determine the fracture toughness. Other requirements are laid down in the test procedures, and these must be fulfilled; the interested reader should consult the documentation.

Table 3.2 gives the values of fracture toughness measured for several families of structural materials. For low-strength materials the thickness condition (3.41) gives large values for the dimensions of the test piece, in some cases so large as to be unattainable in practice when one thinks of the quenching of such steels

Material	R_p MPa	K_{Ic} MPa $\sqrt{m}$	$2.5(K_{Ic}/R_p)^2$ mm
Steels, very high resistance e.g. 35 NCD 16	*1500*	*60*	*4*
Steels, medium resistance very low temperature	*1000*	*40*	*4*
ductile e.g. 15 MND 6	*500*	*200*	*400*
Aluminium alloys, high resistance e.g 7075	*500*	*30*	*9*
Titanium alloys, high resistance e.g. TA6V	*800*	*80*	*25*

Table 3.2 Values for yield strength R_p and fracture toughness K_{Ic} for various materials.

Since toughness controls the critical size of allowable defects (see for example Exercises 18 and 19) it will be readily agreed that a compromise has to be sought between resistance to deformation R_p and fracture toughness K_{Ic}. This need is illustrated in Fig. 3.39, which shows how, with suitable heat treatment, these two properties can be made to vary in the case of two aluminium alloys. Aiming for high yield strength at all costs can result in a bad choice of material in certain cases (see Exercise 18), especially when over-riding safety considerations limit the critical size of admissible defects - the "leak before break" problem for pressure vessels, for example.

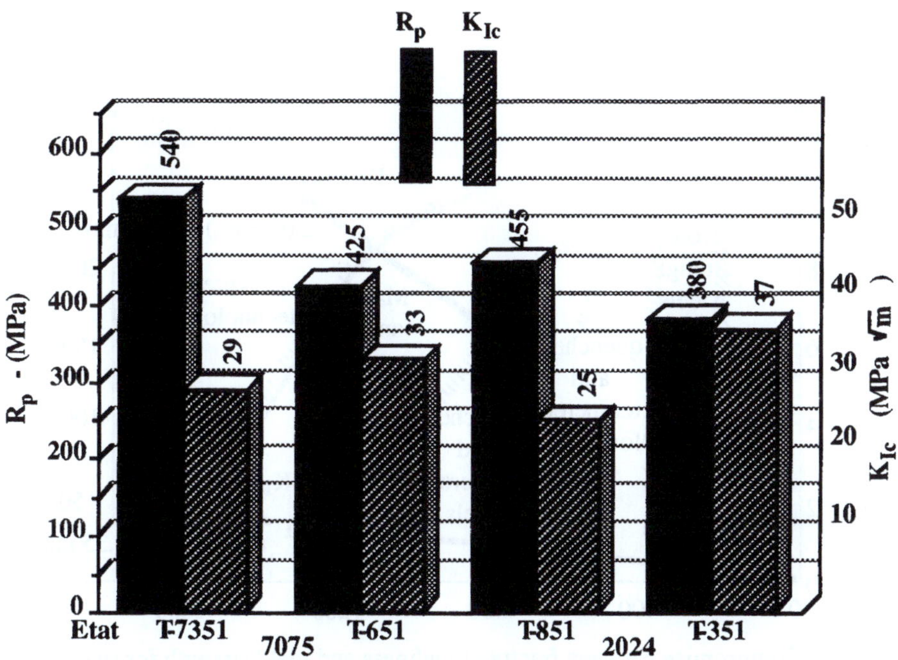

Fig. 3.39 Variation of yield strength and fracture toughness in two high-strength aluminium alloys

Fig. 3.40 illustrates this compromise for the case of steels.

Here, in order to characterise the toughness, we show both K_{Ic} and the fracture energy K_{CV} measured in Charpy tests. A diagram of this type shows the risks that can be run when the value of the ratio of toughness to yield strength is small.

In fact, for any given working stress, yield strength for example, the critical defect size, which is proportional to $(K_{Ic} / R_p)^2$, can be very small, necessitating methods for non-destructive testing which have to be very powerful and very reliable, as we stated at the beginning of the chapter.

3.6.2 Crack propagation by fatigue

We consider only strictly periodic loads, with no overloading and the overall behaviour of the structure elastic but with the possibility of small plastic deformations in restricted areas. This is the case, for example, for a structure with a geometrical discontinuity, in which, after initiation, the crack propagates in an elastic stress field - as in Fig. 3.41. To the extent that the plasticity at the crack tip is governed by the loading parameter K we can expect the crack propagation rate to be correlated with the variation of stress intensity factor $\Delta K = K_{max} - K_{min}$.

Experiments have shown that there is indeed such a correlation, at least for sufficiently long cracks, over a few millimeters. The basic experiments are simple, as illustrated in Fig. 3.42.

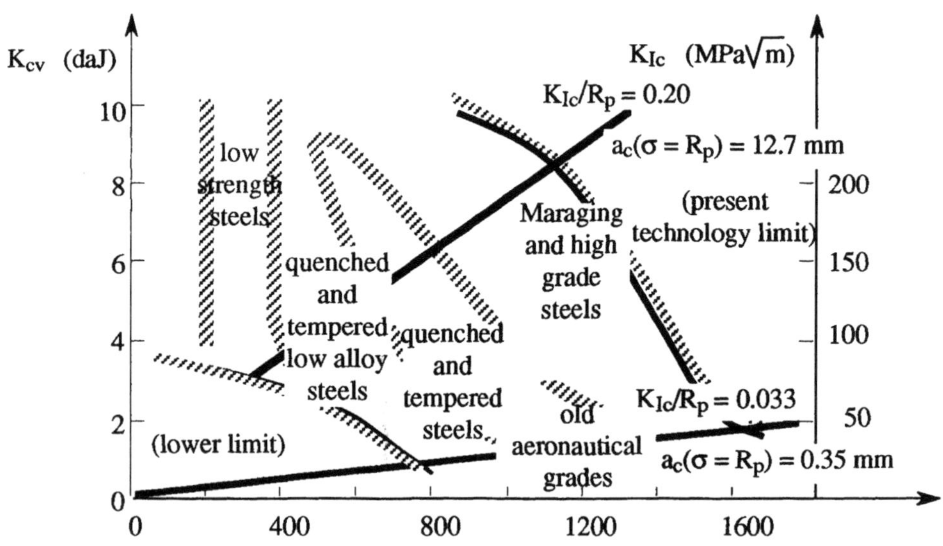

Fig. 3.40 Compromise between fracture toughness and yield strength for steels. The «correlation» between K_{Ic} and K_{CV} suggested by this diagram must not be interpreted too strictly.

A load varying periodically between P_{min} and P_{max} is applied to a test piece for which the stress intensity is known (one of type CT, for example); the crack is allowed to propagate and the change in da/dN corresponding to a change in ΔK is recorded. Fig. 3.42 shows the type of record obtained; this agrees with the results derived in the theoretical treatment of fatigue damage in Chapter 2 (§ 2.11).

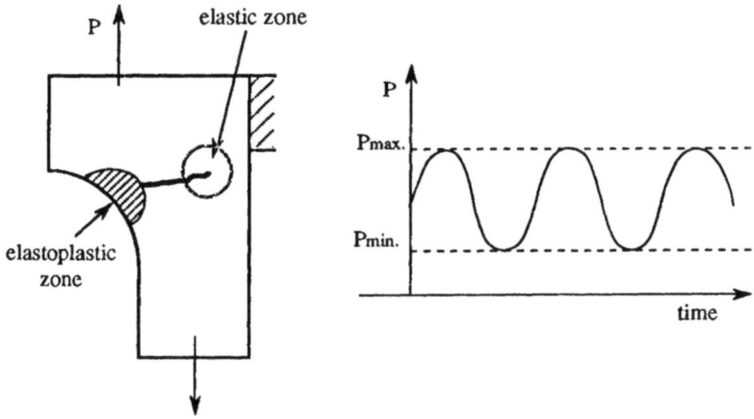

Fig. 3.41 Fatigue crack developing in a structure

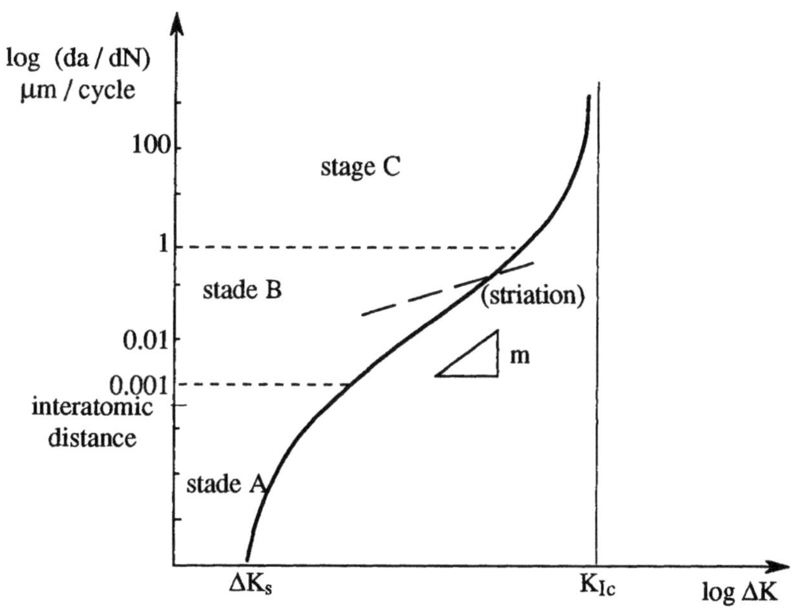

Fig. 3.42 Propagation of fatigue cracks.

It is now standard practice to present these (*da/dN, ΔK*) relations as log-log plots; Fig. 3.43 shows some actual results, for a high-strength steel of type 40 CD 3. Three stages can be distinguished in the schematic Fig. 3.42, as indicated there.

Stage A. The propagation rate is very low and there is a threshold ΔK_s more or less well-defined according to the material, below which there is no crack growth. Its value is much less than that of K_{Ic}; to show the order of magnitude, some typical values are given in Table 3.3. The value is strongly influenced by a number of metallurgical parameters: thus as a general rule it increases with grain size and falls with increasing yield strength of the material. It can also be strongly affected by the environment, for example by the humidity of the air, as is shown in Fig. 3.43. These important effects are seen very clearly in steels and in aluminium alloys. Another strong influence is the load ratio $R = K_{min}/K_{max}$, the value of ΔK_s falling as this increases.

Material	Steels high resistance	Steels low resistance	Aluminium alloys	Titanium alloys
ΔK_S $MPa\sqrt{m}$	*1 - 4*	*3 - 6*	*1.5 - 4*	*2 - 8*

Table 3.3 Order of magnitude of ΔK_s for various materials

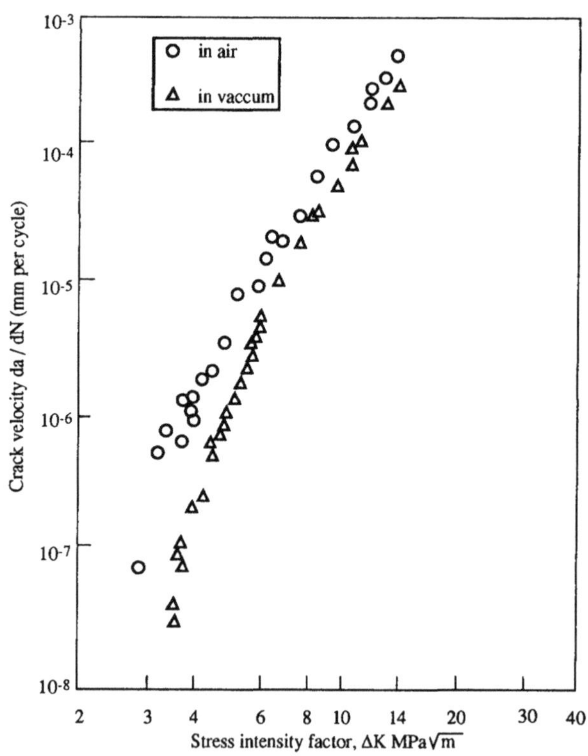

Fig. 3.43 Experimental data for propagation of fatigue cracks in a high-strength steel (Ritchie (1977).

Stage B. This corresponds to the propagation rates that are most commonly observed. We recall that the propagation law is

$$da/dN = C(\Delta K)^m \tag{3.40}$$

with m usually between 2 and 6. With some materials, particularly FCC alloys, the fatigue striations discussed in Chapter 2, §2.11.5 are often seen on the fracture surfaces. The problem of the formation of these striations has been the subject of a great deal of study; it has been found, for example, that there is a range of rates within which one striation is formed in each cycle, a result of great practical importance for failure analysis. The dashed line on Fig. 3.42 indicates the variation of distance between successive striations with ΔK; to a first approximation this is like $(\Delta K)^2$.

Stage C. This is characterised by an increasing propagation rate, ending in sudden fracture when $K_{max} = \Delta K/(1 - R)$ reaches the critical value K_{Ic}. As in Stage A, it is strongly influenced by the load ratio R; this is only what one should expect, since here conditions are approaching those of fracture under steadily increasing load. Several empirical rules have been suggested, among which that due to Foreman is often used; this takes the form

$$da/dN = A(\Delta K)^m/[(1 - R)K_{Ic} - \Delta K] \tag{3.50}$$

An accurate knowledge of the relation between da/dN and ΔK (the propagation law) enables the harmfulness of defects, whether already in the structure or appearing in service, to be assessed. It also provides a sound foundation on which to base the inspection program, as is done, for example, for airframes (Exercise 20)

3.6.5 Crack propagation in stress corrosion and corrosion fatigue

As we said in the introduction to this chapter, these modes of damage result from the combined effects of mechanical loading and chemical attack of the material by a corrosive gaseous or liquid environment. One example that has been much studied because of its practical consequences is that of aluminium in seawater; another is that of the structural materials used in gas turbines.

Stress corrosion cracking

In the section of Chapter 2 dealing with stress corrosion cracking (§2.15) we discussed several mechanisms that can give rise to this. It is a mode of damage that can carry very heavy penalties.

A current example of great practical importance is that of the steam-generating tubes in pressurised-water nuclear power stations; for which the alloy 600, containing *15%* chromium, has been used; this alloy has proved sensitive to stress corrosion cracking in these particular conditions of use, with serious economic consequences. It is thought that an alloy with higher chromium content, around *30%*, would solve the problem.

Another example, which illustrates the compromise between mechanical strength and resistance to stress corrosion cracking, is that of high-strength aluminium alloys: in these, which are hardened by precipitates, the corrosion resistance tends to be minimal since mechanical strength is more important (Fig. 3.44).

Studies of propagation rates of stress corrosion cracks are usually made with test pieces of the types already described, immersed in a medium whose electro-chemical properties can be controlled. The results are usually expressed as crack-propagation laws, each valid for the particular association of metal and surrounding medium for which it was observed; these are of the form shown in Fig. 3.44.

Fig. 3.44 shows again the existence of a threshold below which there is no propagation; written K_{ISCC}. The more-or-less well-defined plateaux are explained qualitatively by the competition between passivation and depassivation mentioned previously in Chapter 2, §2.15, and also by the difficulty of the corrosive liquid gaining access to the tip of the crack.

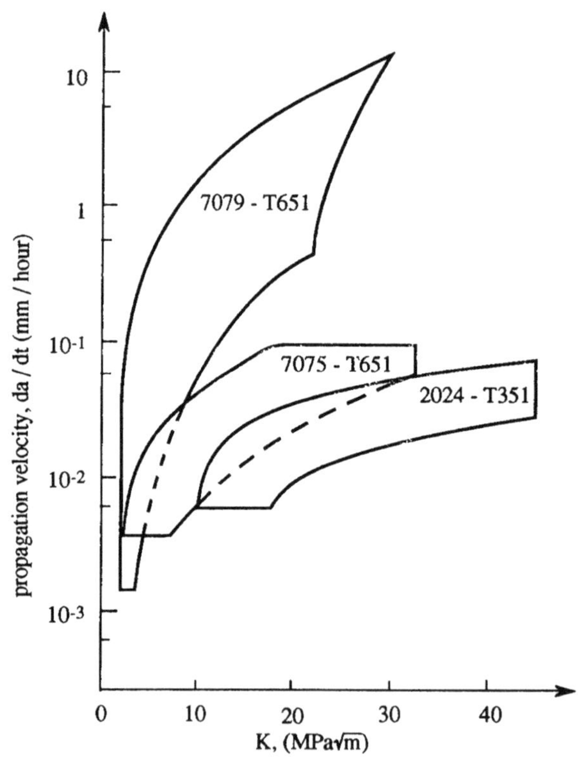

Fig. 3.44 Rate of propagation of stress corrosion cracks in high-strength aluminium alloys, along the short transverse direction of rolled plates tested in salt water. Notice that the highest rates are found in the 7000 alloys (Al-Cu-Zn) in T6 condition. (Brown (1977)).

Corrosion fatigue

The depassivation produced by plastic deformation at the crack tip can proceed faster when the loading is cyclic than when it is steady; this is expressed by the term *corrosion-fatigue*, and is closely related to stress-corrosion cracking. In general, three possible cases can be distinguished (Fig. 3.45), according to the value of K_{max} reached in the load cycle in relation to K_{ISCC}:

(1) if $K_{max} < K_{ISCC}$ (Fig. 3.45a) what is observed is pure corrosion-fatigue, corresponding to the synergetic-effects described in §2.15.2.

(2) if $K_{max} > K_{ISCC}$ (Fig. 3.45b) the damage results from a combination of fatigue and stress-corrosion cracking, the latter acting during that part of the cycle in which $K > K_{ISCC}$.

(3) Fig. 3.45c illustrates the case in which pure corrosion-fatigue and stress-corrosion cracking occur simultaneously during the cycle.

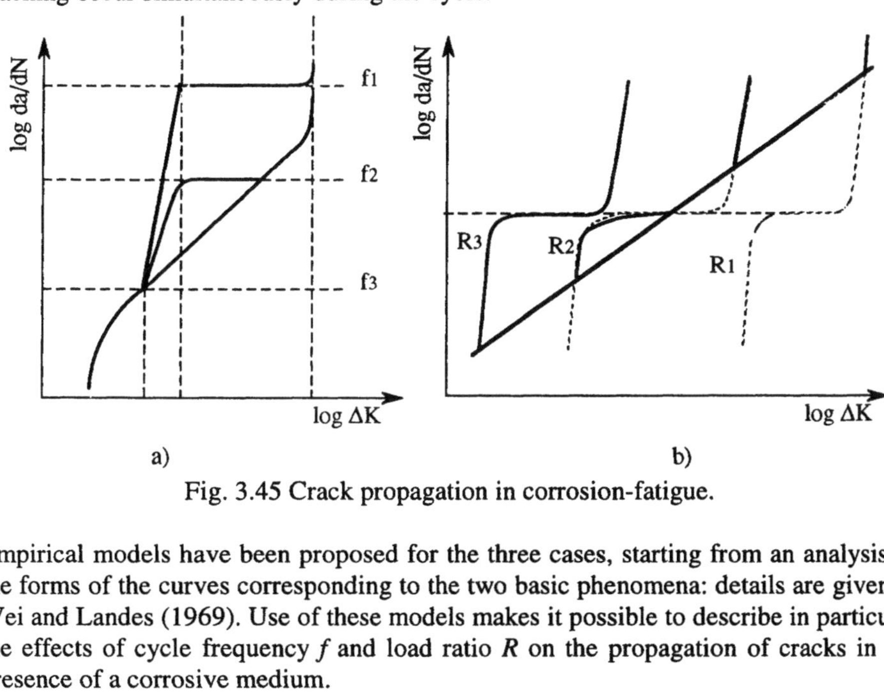

Fig. 3.45 Crack propagation in corrosion-fatigue.

Empirical models have been proposed for the three cases, starting from an analysis of the forms of the curves corresponding to the two basic phenomena: details are given in Wei and Landes (1969). Use of these models makes it possible to describe in particular the effects of cycle frequency f and load ratio R on the propagation of cracks in the presence of a corrosive medium.

Fig. 3.46 summarises these effects.

These models are still very incomplete. They need to be improved and refined, in particular by better measurements of the variation of the mechanical and electro-chemical parameters at the crack tip.

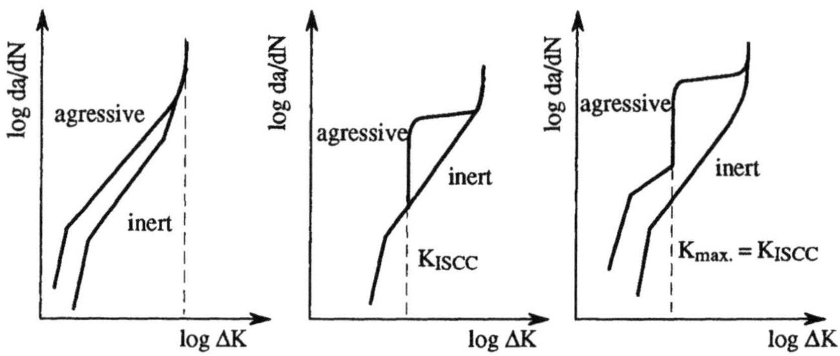

a) true corrosion fatigue (TCF) b) stress corrosion fatigue (SCF) c) SCF → TCF

Fig. 3.46 Diagrammatic representation of the effects of (a) cycle frequency *f* (b) load ratio *R* on the propagation of corrosion-fatigue cracks.Notice that at very low frequencies the depassivation is insufficient and there is no pure corrosion-fatigue effect.

Table 3.1 Stress intensity factors for various cases

Sketch	Description	K value
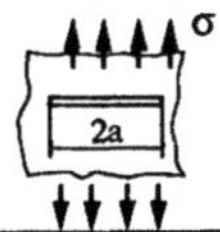	crack of length 2a in an infinite plate under uniaxial homogeneous stress	$K_I = \sigma\sqrt{\pi a}$
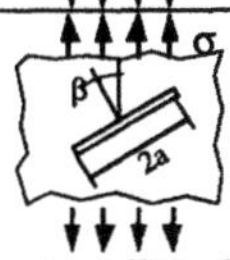	inclined crack of length 2a in an infinite plate under uniaxial homogeneous stress	$K_I = \sigma\sqrt{\pi a}\cos^2\beta$ $K_{II} = \sigma\sqrt{\pi a}\cos\beta\sin\beta$
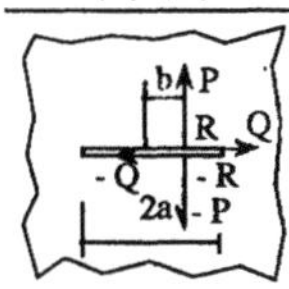	cracks of length 2a in an infinite plate under concentrated loads at a distance b from the axis	$K_I = \frac{P}{\sqrt{\pi a}}\left(\frac{a+b}{a-b}\right)^{1/2}$ $K_{II} = \frac{Q}{\sqrt{\pi a}}\left(\frac{a+b}{a-b}\right)^{1/2}$ $K_{III} = \frac{R}{\sqrt{\pi a}}\left(\frac{a+b}{a-b}\right)^{1/2}$
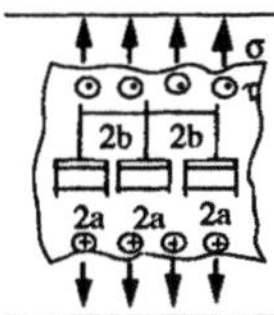	crack of length 2a, 2b apart, in an infinite plate under uniaxial homogeneous stress	$K_I = \sigma\sqrt{\pi a}\left(\frac{2b}{\pi a}\,tg\,\frac{\pi a}{2b}\right)^{1/2}$ $K_{III} = \tau\sqrt{\pi a}\left(\frac{2b}{\pi a}\,tg\,\frac{\pi a}{2b}\right)^{1/2}$
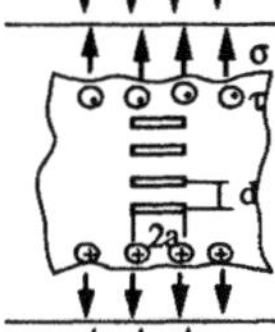	parallel cracks of length 2a, d apart, in an infinite plate under uniaxial homogeneous stress	$\frac{a}{d} < 0.5$ $K_I = \sigma\sqrt{\pi a}\left[1 - \frac{1}{2}\left(\frac{\pi a}{d}\right)^2 + \frac{3}{8}\left(\frac{\pi a}{d}\right)^4\right]$ $K_{III} = \tau\sqrt{\pi a}\left(\frac{d}{\pi a}\,tg\,\frac{\pi a}{d}\right)^{1/2}$
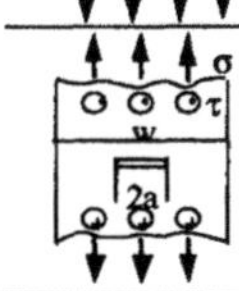	crack of length 2a in the center of plate of width w under uniaxial homogeneous stress	$K_I = \sigma\sqrt{\pi a}\left(\cos\frac{\pi a}{w}\right)^{1/2}$ $K_{III} = \tau\sqrt{\pi a}\left(\frac{w}{\pi a}\,tg\,\frac{\pi a}{w}\right)^{1/2}$
	edge crack of depth a in a semi-infinite plate under uniaxial homogeneous stress	$K_I = 1.122\sqrt{\pi a}$ $\approx 2\sigma\sqrt{a}$
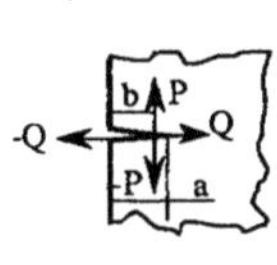	edge crack of depth a in a semi-infinite plate under concentrated loads at a distance b from the surface	$\frac{K_I}{P} = \frac{K_{II}}{Q} = \frac{2}{\sqrt{\pi a}}\left[1 - (b/a)^2\right]^{-1/2}$ $\left[0.2945 - 0.3912\left(\frac{b}{a}\right)^2 + 0.7685\left(\frac{b}{a}\right)^4 - 0.9942\left(\frac{b}{a}\right)^6 + 0.5094\left(\frac{b}{a}\right)^8\right]$
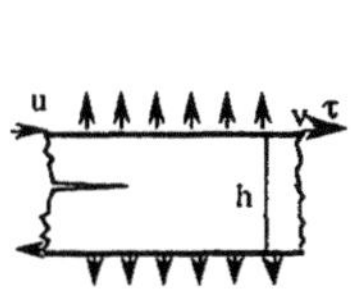	**plane stress** semi-infinite crack in a strip of height h under prescribed displacement **plane strain**	$K_I = \sqrt{\frac{2}{h}}\,Ev\ (\tau = 0)$ $K_I = \sqrt{\frac{2}{h}}\frac{Ev}{\sqrt{1-\nu^2}}\ (u = 0)$ $K_I = \sqrt{\frac{2}{h}}\frac{Ev}{1-\nu^2}\ (\tau = 0)$ $K_I = \sqrt{\frac{2}{h}}\frac{Ev}{(1+\nu)\sqrt{1-\nu^2}}\ (u = 0)$

Sketch	Description	K value
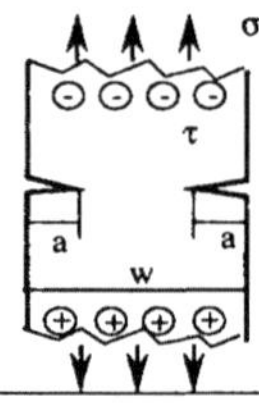	two edge cracks of depth a in a plate of width w under uniaxial homogeneous stress	$K_I = \sigma\sqrt{\pi a}\,\left[1 + 0.122\cos^2\left(\frac{\pi a}{w}\right)\right]\left(\frac{w}{\pi a}\,tg\frac{\pi a}{w}\right)^{1/2}$ $K_{III} = \tau\sqrt{\pi a}\left(\frac{w}{\pi a}\,tg\frac{\pi a}{w}\right)^{1/2}$
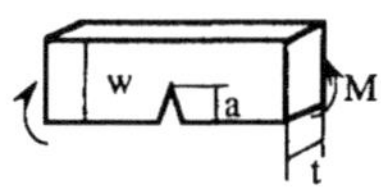	an edge crack of depth a in a beam of height w under constant bending moment	$K_I = \frac{6M}{t\,w^2}\sqrt{\pi a}\left[1.122 - 1.4\frac{a}{w} + 7.33\left(\frac{a}{w}\right)^2 - 13.08\left(\frac{a}{w}\right)^3 + 14.0\left(\frac{a}{w}\right)^4\right]$
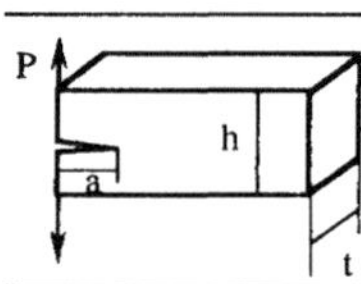	double cantilever beam (DCB)	$K_I = \frac{P}{t\sqrt{h}}\,4\sqrt{6}\left(0.32 + \frac{a}{h}\right)$
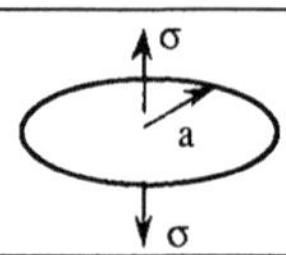	penny-shaped crack of radius a in an infinite medium under uniform uniaxial stress	$K_I = \sigma\sqrt{\pi a}\,\frac{2}{\pi}$
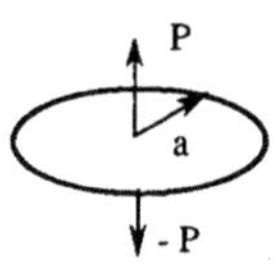	penny-shaped crack of radius a in an infinite medium under concentrated load at its center	$K_I = \frac{P}{(\pi a)^{3/2}}$
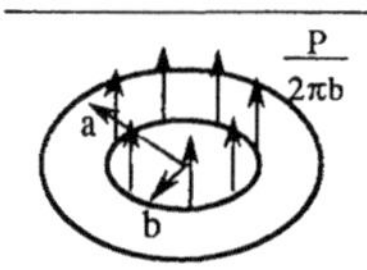	penny-shaped crack of radius a in an infinite medium under concentrated loads on a circle of radius b	$K_I = \frac{P}{(\pi a)^{3/2}}\left[1 - \left(\frac{b}{a}\right)^2\right]^{-1/2}$
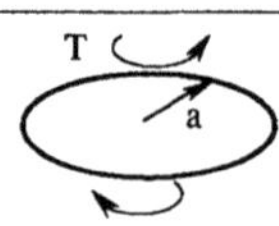	penny-shaped crack of radius a in an infinite medium under a torsion moment	$K_{III} = \frac{8\,T}{3\,\pi^{3/2}a^{5/2}} = \tau\sqrt{\pi a}\,\frac{4}{3\,\pi}$
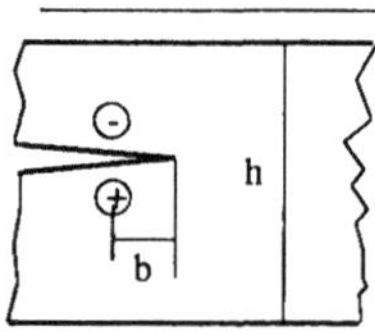	semi-infinite crack in a strip of height h under concentrated loads (mode III) at a distance b from the tip	$K_{III} = S\frac{2}{\sqrt{h}}\left[1 - \exp\left(-\frac{2\pi b}{h}\right)\right]^{1/2}$ $h \to \infty$ $K_{III} \to S\sqrt{\frac{2}{\pi b}}$
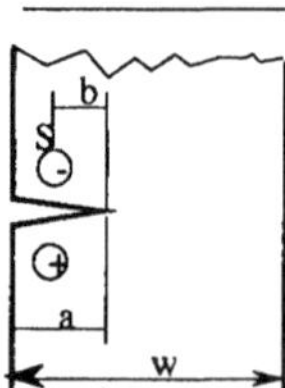	edge crack of depth a in a plate of width w under concentrated loads at a distance b from the tip (mode III)	$K_{III} = S\sqrt{\frac{2}{w}}\,\frac{\cos\left(\frac{\pi}{2}\frac{b}{w}\right)\sqrt{tg\left(\frac{\pi a}{2w}\right)}}{\left[\cos^2\left(\frac{\pi}{2}\frac{b}{w}\right) - \cos^2\left(\frac{\pi a}{2w}\right)\right]^{1/2}}$

ANNEX

THE PLASTIFIED ZONE AT THE CRACK TIP IN MODE III
The case of small-scale yielding

The problem is described by Fig. 3.47. Solutions in the elastic zone were given in §3.2.3; in the plastic zone the stresses must satisfy the plasticity criterion, that is

$$\sigma_{13}^2 + \sigma_{23}^2 = k^2 \qquad (3.A1)$$

where $k = R_p/2$ for a material to which the Tresca criterion applies.

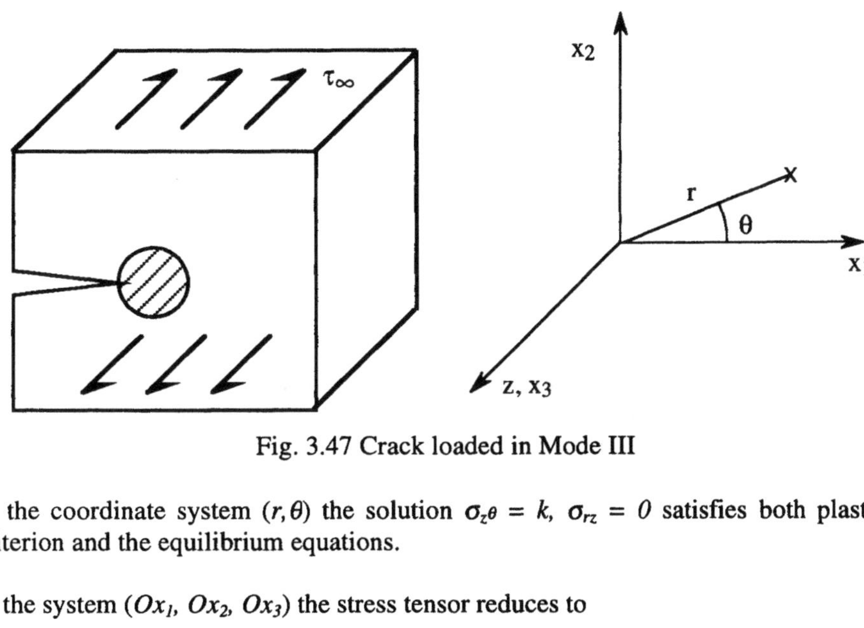

Fig. 3.47 Crack loaded in Mode III

In the coordinate system (r, θ) the solution $\sigma_{z\theta} = k$, $\sigma_{rz} = 0$ satisfies both plasticity criterion and the equilibrium equations.

In the system (Ox_1, Ox_2, Ox_3) the stress tensor reduces to

$$\sigma_{13} = -k \sin\theta, \ \sigma_{23} = k \cos\theta$$

In fact, we have

$$\begin{pmatrix} \sigma_{11} & \sigma_{12} & \sigma_{13} \\ \sigma_{12} & \sigma_{22} & \sigma_{23} \\ \sigma_{13} & \sigma_{23} & \sigma_{33} \end{pmatrix} = P^{-1} \begin{pmatrix} \sigma_{rr} & \sigma_{r\theta} & \sigma_{rz} \\ \sigma_{\theta r} & \sigma_{\theta\theta} & \sigma_{\theta z} \\ \sigma_{rz} & \sigma_{\theta z} & \sigma_{zz} \end{pmatrix} P$$

where

$$P = \begin{pmatrix} \cos\theta & \sin\theta & 0 \\ -\sin\theta & \cos\theta & 0 \\ 0 & 0 & 1 \end{pmatrix}$$

From the constitutive equations in plasticity we can write $d\varepsilon_{rz}^{P} = \Lambda\sigma_{rz} = 0$ where Λ is the plastic multiplier.

If we neglect elastic deformations we have $\varepsilon_{rz} = 0$, from which we deduce that the only non-zero component u_z is a function of θ only, that is

$$u_z = f(\theta) \tag{3.A2}$$

In cylindrical coordinates

$$\varepsilon_{rz} = \frac{1}{2}\left(\frac{\partial u_r}{\partial z} + \frac{\partial u_z}{\partial r}\right), \quad \varepsilon_{z\theta} = \frac{1}{2}\left(\frac{\partial u_\theta}{\partial z} + \frac{1}{r}\frac{\partial u_z}{\partial \theta}\right) = \frac{1}{2r}\frac{du_z}{d\theta}$$

Let $R(\theta)$ be the plastic boundary; by continuity we must have there

$$\varepsilon_{z\theta} = \frac{k}{2\mu} = \frac{1}{2r}\frac{du_z}{d\theta} = \frac{1}{2R(\theta)}\frac{du_z}{d\theta}$$

from which we deduce $\quad u_z = \dfrac{k}{\mu}\displaystyle\int_0^\theta R(\beta)d\beta$

Thus there is a discontinuity in the displacement at the crack tip; this represents the *Crack Tip Opening Displacement* (CTOD) δ of §3.3.3.1 and is given by

$$\delta = u_z(\theta = \pi/2) - u_z(\theta = -\pi/2) = \frac{k}{\mu}\int_{-\pi/2}^{\pi/2} R(\beta)d\beta$$

We can assume that if we move a little away from the crack tip the elastic singularity continues to exist; thus (see §3.2.3) if ζ is large we have a solution of the form

$$\sigma_{23} + i\sigma_{13} = \tau = \frac{K_{III}}{(2\pi\varsigma)^{1/2}} \quad \text{where} \quad \varsigma = x_1 + ix_2$$

As in Irwin's model, we postulate that a simple translation will give a solution to the problem, so

$$\tau = \sigma_{23} + i\sigma_{13} = \frac{K_{III}}{\left[2\pi(\varsigma - \lambda)\right]^{1/2}} \quad \text{where } \lambda \text{ is real}$$

and the condition for continuity of stress along the plastic boundary is

$$\tau = k\cos\theta - ik\sin\theta = ke^{-i\theta} = \frac{K_{III}}{\left[2\pi(\varsigma - \lambda)\right]^{1/2}}$$

It can be shown (see Exercise 13) that this is satisfied by

$$\lambda = \frac{K_{III}^2}{2\pi k^2}, \quad R(\theta) = \frac{K_{III}^2}{\pi k^2}\cos\theta$$

a solution suggested by the Irwin problem; it follows that the plastic zone is a circle of diameter $(1/\pi)(K_{III}/k)^2$.

EXERCISES

1. Show that the elastic solution given by equations 3.2 and 3.3 satisfies the equilibrium equations and the boundary conditions.

2. A transmission shaft , for example the rotor of a turbo-alternator, has a crack in its external surface (Fig. 3.48). It is acted on by a steady driving torque and a cyclic bending moment. Find the modes of opening along the crack front, particularly at A and B.

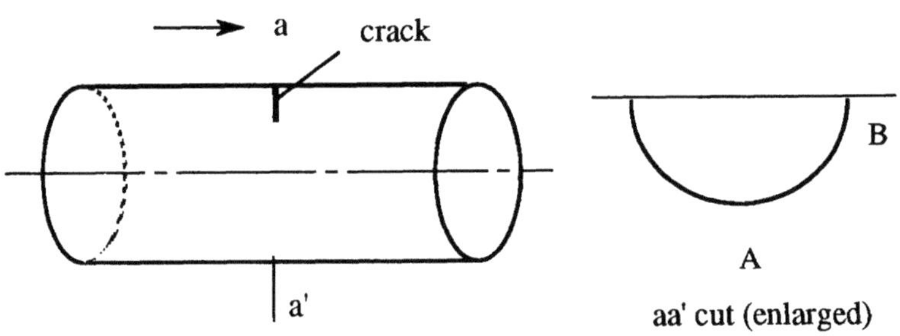

Fig. 3.48 Crack in the cross-section plane of a transmission shaft.

3. This is a plane problem in mixed-mode loading. A plate under tension has a central crack of length $2a$, inclined at an angle β to the direction of the tension (Fig. 3.49).

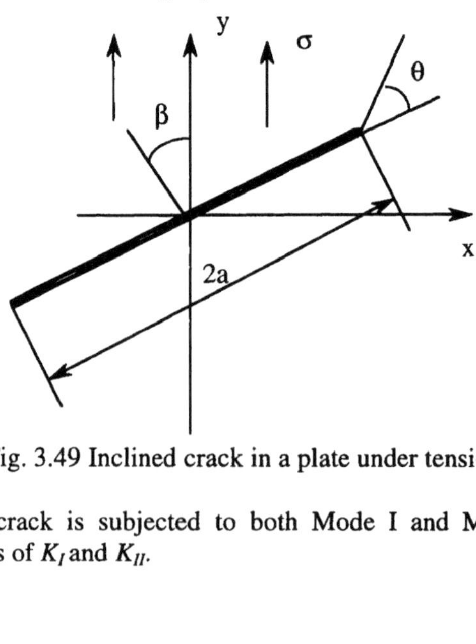

Fig. 3.49 Inclined crack in a plate under tension.

3.1 Show that the crack is subjected to both Mode I and Mode II; calculate the corresponding values of K_I and K_{II}.

3.2 The crack can be assumed to propagate radially in the direction of maximum tangential stress and to become unstable when the component $\sigma_{\theta\theta}$ reaches a critical value at a characteristic distance X_c, when $K_{Ic} = \sqrt{2\pi X_c}\sigma_{\theta\theta_{max}}$. Find the bifurcation angle θ for these conditions. Assuming pure Mode II, find the corresponding value of θ and of K_{IIc}.

3.3 Suggest experimental equipment that will enable tests to be made in pure Mode II.

3.4 Suggest experiments using a simple sheet of paper that will give the answers to Question 3.2.

4. A large plate (Fig. 3.50) acted on by a uniform tensile stress σ has a hole at the edge of which a crack is initiating.

4.1 Using Table 3.I suggest simple approximations for the stress intensity factor K.

4.2 Compare these with the forms due to Tada, Paris and Irwin (1985) given in Fig. 3.51, restricting your treatment to the case $\tau_l = 0$. How do you explain the effect observed when the stress $\lambda\sigma$ is applied parallel to the plane of the crack?

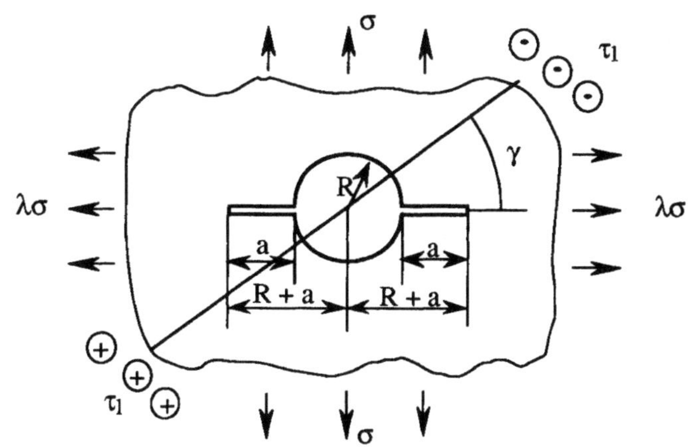

Fig. 3.50 Crack starting at the edge of a hole in a plate subjected to various stresses.

5. A tube in which a hole is drilled is acted on by an alternating torque $\pm C$. Assuming that cracks will develop in the directions of maximum tangential stress, find these directions. Find the corresponding value of K_I for the cases of (a) small cracks (b) cracks much larger than the radius of the hole. State clearly any assumptions made.

6. You are required to obtain some measurements of toughness. Two geometries are available, a plate with a central crack and a 3-point bend-test piece: knowing that the powers of the test machines are necessarily limited, which do you consider the more suitable? Use the relations given in Table 3.I in making your decision.

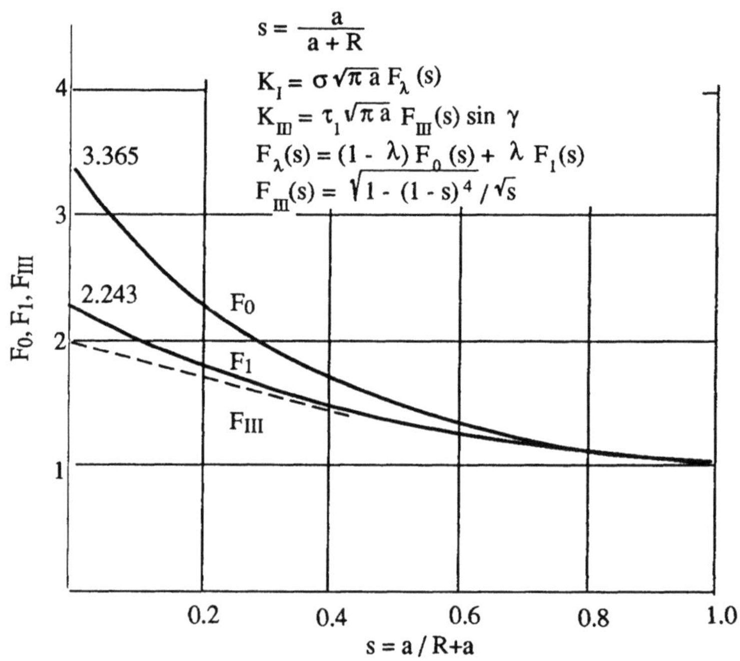

Fig. 3.51 Expressions for calculating values of stress intensity factors.

7. Fig. 3.52 represents the geometry of a Double Cantilever Beam (DCB) test piece, which can be opened by an amount 2λ by means of some device such as a screw; assuming that the two parts are subjected to pure bending, it follows from beam theory that $\lambda = 4Pa^3/Eh^3$.

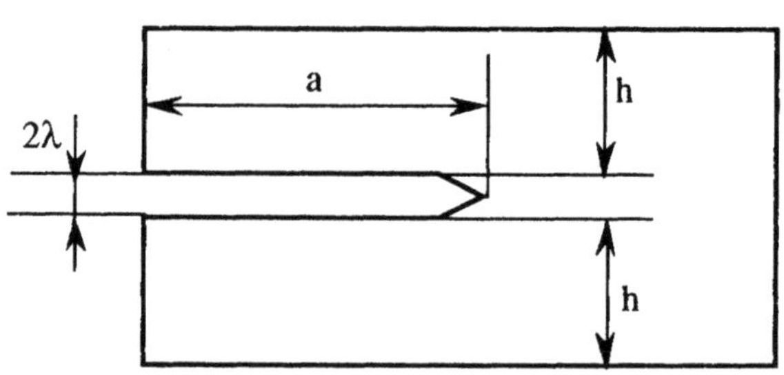

Fig. 3.52 Double Cantilever Beam (DCB) test piece.

7.1 If the displacement λ is prescribed, find an expression from which K_I can be calculated as a function of a.

7.2 How do you value such a test piece?

7.3 How would you change the shape so that K stays constant as the crack progresses?

8. Calculate the compliance C and G for the plate with a central crack represented in Fig. 3.53. Tada, Paris and Irwin (1985) give the following relation, valid for $0 < a/b < 0.20$:

$$K = \sigma\sqrt{\pi a}\ F(a/b) \text{ where } F(a/b) = (1 - a/2b)/\sqrt{1 - a/b}$$

8.1 Find the compliance of the test piece, giving the numerical value for $a/b = 0.20$ and $L = b$, taking $E = 200{,}000\ MPa$ and the thickness $B = 10\ mm$.

8.2 A spring of compliance C_R is inserted at the point where the load F is applied. Find the general expression for G as a function of the total displacement of the spring.

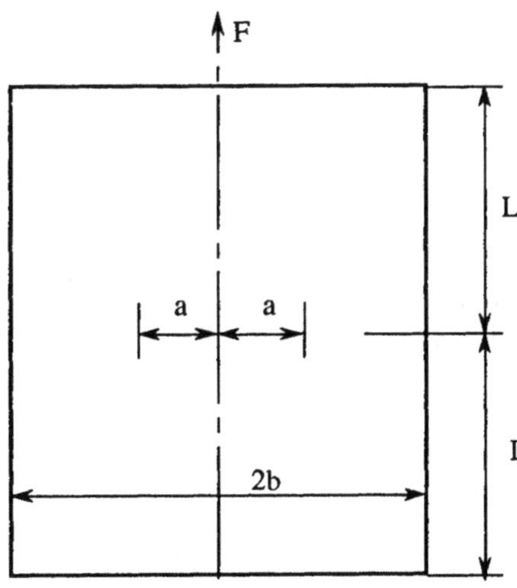

Fig. 3.53 Plate with a central crack.

9. Using the principle of superposition and the weight function method for a central crack in an infinite plate (Equation 3.24), show that the stress intensity factor is indeed of the form $K_I = \sigma\sqrt{\pi a}$

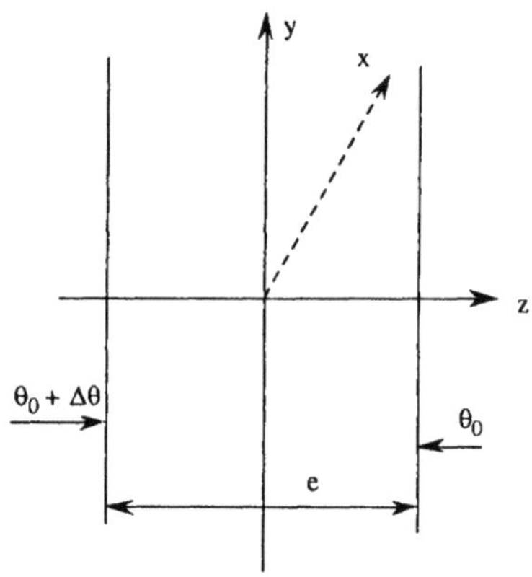

Fig. 3.54 Plate with temperature gradient.

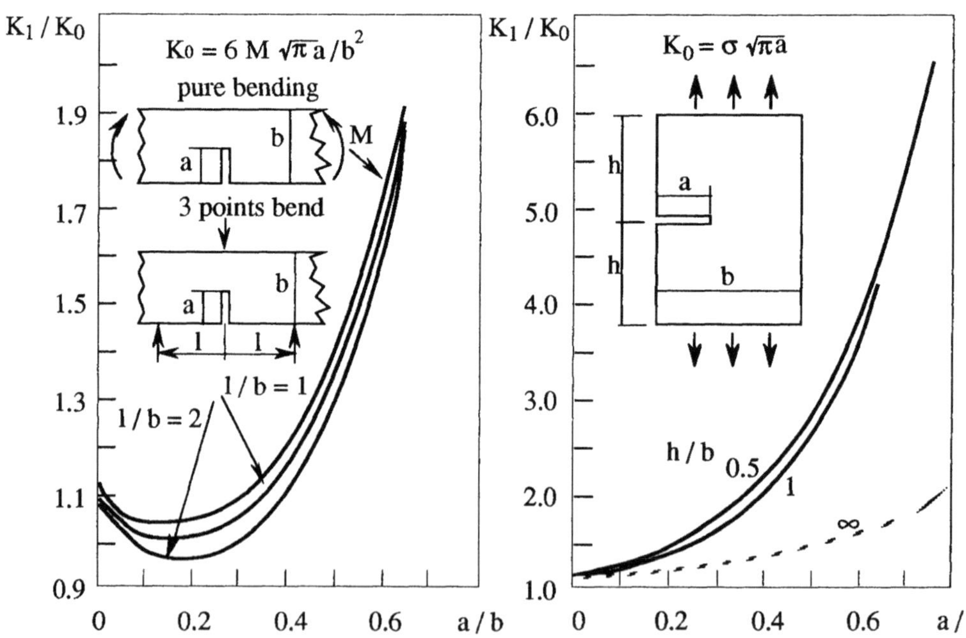

Fig. 3.55 Stress intensity factors for (a) bending (b) tension.

10. This problem concerns approximating the value of K associated with a thermal field (Fig. 3.54). A plate of thickness e in which there is a linear temperature gradient $\Delta\theta$ across the thickness is constrained so that it cannot bend.

10.1 Show that the temperature distribution generates a stress field such that

$$\sigma_{xx} = \sigma_{yy} = [E\alpha(\Delta\theta)/(1 - \nu)](z/e)$$

where E is Young's modulus, α is the coefficient of thermal expansion and ν is Poisson's ratio. It can be assumed that the plate is in plane stress and that $\sigma_{xx} = \sigma_{yy}$.

10.2 A part-through crack of length $a = 0.2e$ forms on the external surface; to which mode of opening, tension or compression, is it subjected?

10.3 To a first approximation it can be assumed that the stress distribution is not perturbed by the crack, and thus that the crack-free part is subject to the linear stress field found in 10.1, confined to this part. Show that this distribution is equivalent to the sum of a bending moment M per unit of width and a compression F per unit of length such that

$$M = 0.043E\alpha(\Delta\theta)e^2/(1 - \nu) \text{ and } F = -0.08E\alpha\Delta\theta e/(1 - \nu)$$

What is your opinion of the model used?

10.4 This reduces the problem to two problems with known solutions, given by the curves of Fig. 3.55. From these find estimates for K_I for a range of values of the ratio a/b.

11. Show that the two complex functions Φ, Ψ of Equation (3.25) give the solution of the problem of that section.

12. Starting from the solutions for Mode I opening in conditions of elasticity for a material to which the von Mises criterion applies, find an approximation for the shape of the plastic zone at the tip of the crack, in plane stress and plane strain respectively. Compare your results with those given in Fig. 3.20.

13. Show that the solution τ given in Annex 1 does indeed satisfy the conditions for the Mode III problem. Using the expression given there for the CTOD δ, show that in Mode III this can be put in the form $\delta = 4(1 + \nu)K_{III}^2/\pi\mu k$, where k is the yield strength in shear.

14. Prove that the relation (3.22) gives the extent of the plastic zone in the Dugdale-Barenblatt model. For this, use the fact that the stress intensity factor at A and A' in Fig. 3.22 for the loading resulting from the superposition of the two fields is zero. Deduce the shape of the crack shown in Fig. 3.23.

15. As an application of the extremal theorems to determine the limit load for a structure containing a crack, consider the plate with a central crack shown in Fig. 3.25. Show that the field of slip lines is statically and kinematicaly admissible and deduce the expression (3.35) for the limit load.

16. Prove that the slip lines at the crack tip shown in Fig. 3.28 enable the Hencky relations to be satisfied: that is, that $p + 2k\varphi$ is constant along an α line and $p - 2k\varphi$ is constant along a β line.

17. As an example of the calculation of J from the area under the loading curve, derive the relation (3.39) by integration by parts.

18. This problem concerns the "Leak Before Break" situation.
It is required to build a cylindrical gas-storage tank, for which the dimensions of height *6* meters and radius *0.90* meters are fixed and the storage pressure is to be *10 MPa* (*100 bar*). Three materials are available, with the following properties:

	R_p (MPa)	K_{Ic} (MPa $\sqrt{m}$)
A	1000	180
B	700	100
C	1250	100

The problem is to find the optimum thickness, knowing that:

(a) The safety regulations allow stresses equal to $0.75\ R_p$ to be prescribed,

(b) There can be cracks in the material, for which it can be assumed that $K = 2\sigma\sqrt{a}$ (see Fig. 3.56).

Because of the purpose of this vessel, safety requires that the critical length of a crack should be equal at least to the thickness of the wall, for otherwise there would be leakage before fracture, a very dangerous situation in view of the amount of energy stored in the vessel when filled with gas under pressure.

Decide which material(s) is (are) to be preferred, giving your reasons.

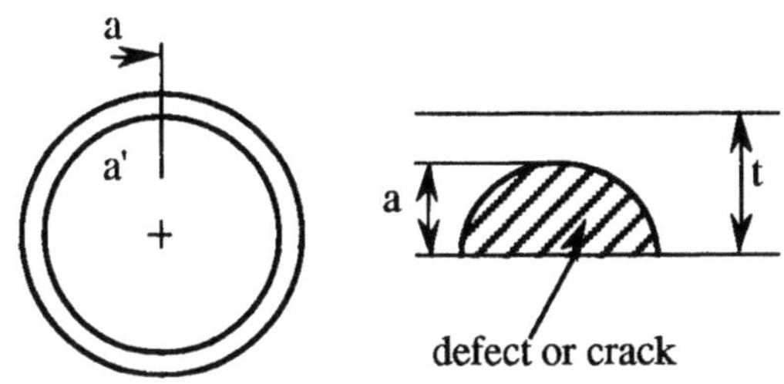

Fig. 3.56 Crack in the cross-section of a tube.

19. The toughness of a high-strength (Maraging) steel decreases as the yield strength increases according to the relation $K_{Ic} = 360 - 0.15\, R_p$, where K_{Ic} and R_p are measured in $MPa\sqrt{m}$ and MPa respectively. A sheet $2.5cm$ thick is to be used in an application in which no crack propagation whatever can be allowed; a factor of safety of 0.75 is laid down, meaning that the working stress σ_w can be $0.75\, R_p$, and the methods of non-destructive testing available enable surface cracks of $5 \pm 2.5\ mm$ to be detected. Two types, I and II (see Fig. 3.57), are considered possible.

It is required to find which material will have the greatest mechanical strength and then to decide the most appropriate heat treatment; for types I and II the relations $K_I = 2\sigma\sqrt{a}$ and $\sigma\sqrt{\pi a}$ can be assumed, respectively.

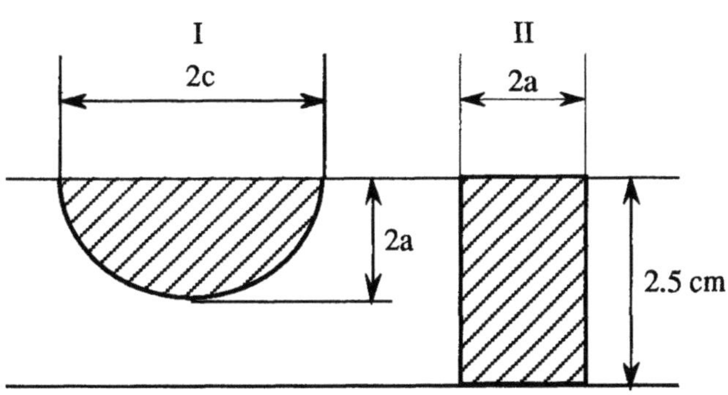

Fig. 3.57 Two possible types of crack in a thin plate.

20. The problem concerns propagation of fatigue cracks.
It is intended to use a medium-strength steel with the following properties for a welded structure:

$$R_p = 690\ MPa,\ K_{Ic} = 165\ MPa\sqrt{a}$$

$$da/dN = 1.7\ 10^{-10}\ (\Delta K_I)^{2.25}\ m/cycle$$

The structure is to be subjected to an oscillating load such that

$$\sigma_{max} = 310\ Mpa,\ \sigma_{min} = 172\ MPa.$$

20.1 Calculate the critical crack size.

20.2 Suppose there is already a crack of length a_0 when the loading starts; find the number N of cycles that the structure can survive, giving the numerical value for $a_0 = 7.6\ mm$.

20.3 With the conditions of 20.2, which of the following would you choose so as to ensure that the structure survived at least 10^5 cycles?

(a) Use instead a steel having roughly the same propagation law but higher toughness.

(b) Reduce the length a_0 of the initial crack by instituting better control of the manufacturing process.

NB: You may take $K_I = 1.12\ \sigma\sqrt{\pi a}$

21. This problem concerns prove testing.
A structure is built from a high-strength steel; before putting into service it is subjected to a proof test in which a stress $\sigma_T = \alpha\sigma_W$ is applied, where σ_W is the stress to which it will be subjected in service and α is usually *1.20-1.50.* Find the stress that should be applied so as to ensure that the structure will not fracture suddenly in use. In service, cracks of two types may propagate: fatigue, for which $da/dN = C(\Delta K)^m$ and fatigue-corrosion, for which $da/dt = Bk^n$.

CHAPTER 4 CONTACT MECHANICS: FRICTION AND WEAR

4.1 INTRODUCTION

This final chapter is devoted to the study of a number of mechanical interactions between the surfaces of two solids in contact, interactions which lead to important phenomena on the macroscopic scale such as friction and wear. This field, that of *tribology*, is very wide and can be studied on several scales:

- the atomic scale, where the interest is in the interactions between the atoms,

- the medium or micro-metric scale, where what is studied is the inevitable roughness of the surfaces in contact,

- the macroscopic scale, where the interest is in the stress and strain fields in the two solids, resulting from the contact.

We shall limit our study mainly to the third of these, although, as will appear in the treatment of friction and wear, we shall touch on the second also. Further, we shall be concerned only with *non-conforming contact*, that is, contact in which the two surfaces have different profiles and therefore do not interlock perfectly.

4.1.1 Relative movement of the surfaces

Fig. 4.1 represents the contact between the surfaces. Oxy is the common tangent plane and Oz the common normal; the shapes of the solid before deformation are given by

$$z_1 = f_1(x, y),\ z_2 = f_2(x, y) \tag{4.1}$$

and the distance separating the surfaces by

$$h = z_1 + z_2 = f(x, y) \tag{4.2}$$

We consider only two types of relative movement, *sliding* and *rolling*; in sliding the displacement vector has two components

$$\Delta v_x = v_{x1} - v_{x2},\ \Delta v_y = v_{y1} - v_{y2} \tag{4.3}$$

whilst in rolling there are two angular components

$$\Delta\omega_x = \omega_{x1} - \omega_{x2},\ \Delta\omega_y = \omega_{y1} - \omega_{y2} \tag{4.4}$$

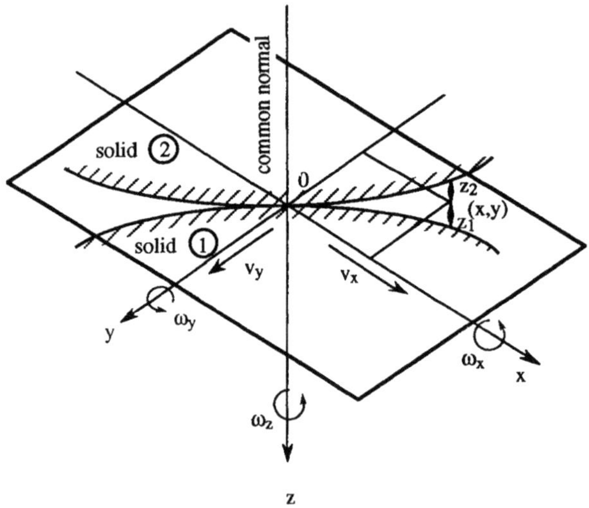

Fig. 4.1 Solids 1 and 2 in contact at *O*.

There can also be a relative "spin" in which either or both rotate about the axis *Oz*.

4.1.2 Forces transmitted at the point of contact

The resultant force transmitted from one surface to the other has two components: *normal*, a compression *P* acting in the direction of the common normal, and *tangential*, a force *Q* acting in the common tangent plane and itself resolvable into components Q_x, Q_y (Fig. 4.2)

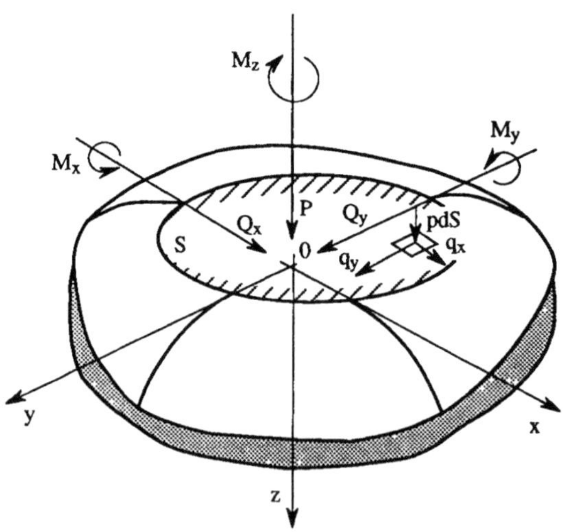

Fig. 4.2 Forces and moments acting on the contact area *S*.

If there is no sliding Q must be less than μP where μ is the coefficient of friction; whilst if the contact is one of pure sliding Q must reach this limiting value in the direction opposite to that of the relative velocity and we must have

$$Q_x = -(\Delta v_x/|\Delta v|)\mu P,\ Q_y = -(\Delta v_y/|\Delta v|)\mu P \tag{4.5}$$

The contact force will deform the solids in such a way as to produce a *contact area*; thus the contact will transmit not only a force but also moments, M_x and M_y of rolling and M_z of spin.

The forces and the moments are transmitted to the solids by surface distributions p and q of normal and tangential stress respectively, such that

$$P = \int_S p\, dS,\ Q_x = \int_S q_x\, dS,\ Q_y = \int_S q_y\, dS \tag{4.6}$$

and

$$M_x = \int_S py\, dS,\ M_y = -\int_S px\, dS\ M_z = \int_S (q_y x - q_x y)\, dS \tag{4.7}$$

4.2 CONTACT MECHANICS

4.2.1 Elastic analysis

4.2.1.1 The problem

We find that we have to study problems which differ in their 2-dimensional or 3-dimensional aspects; further, we have to consider four different types of boundary conditions, according to the type of problem; with u^*_x, u^*_z representing prescribed values for displacements, these are:

- *type I* $p(x)$ and $q(x)$ are given,

- *type II* $u^*_z(x)$ for $z = 0$ and $q(x)$ are given, or $u^*_x(x)$ and $p(x)$ are given; frictionless contact ($q(x) = 0$) belongs to the first case,

- *type III* u^*_z and $u^*_x(x)$ for $z = 0$ are given, when the corresponding $p(x)$ and $q(x)$ are to be found,

- *type IV* $u^*_z(x)$ is given and there is a relation $q(x) = \pm\mu p(x)$; this is the case for sliding with friction.

Most of the solutions to these problems have been obtained since the end of the last century and can be found in, for example Timoshenko and Goodier (1951); a very good recent source is Johnson (1985), from which we have taken much material. We shall give only the solutions, without going into the details of proof: these can be found in Gladwell (1980).

4.2.1.2 Line load on a semi-infinite solid

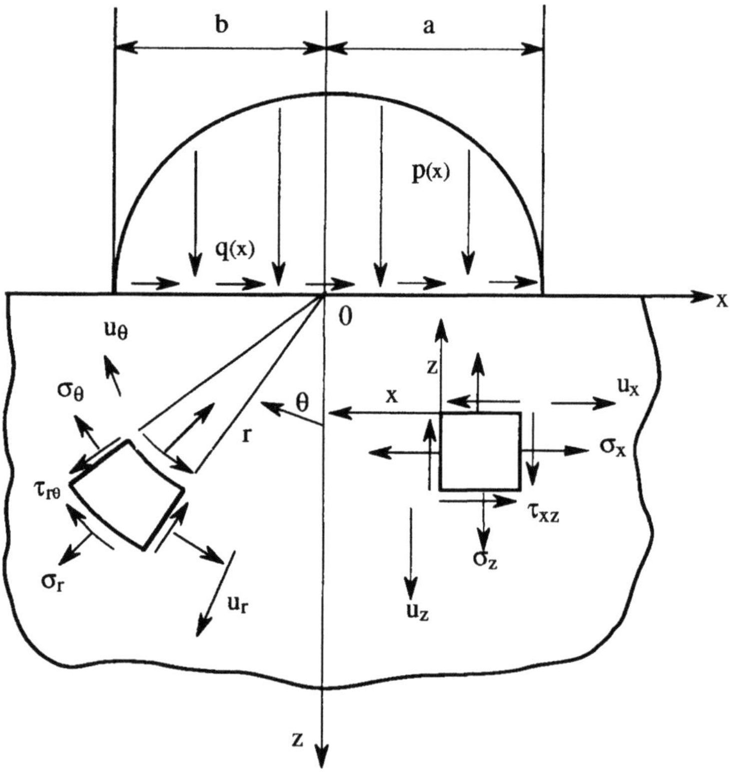

Fig. 4.3 Line loads $p(x)$ normal and $q(x)$ tangential.

The load is applied along the direction of a single straight line, which we take as along Oy; it is distributed over a band of width $a + b$ in the direction of Ox (Fig. 4.3) and we assume conditions of plane strain, $\varepsilon_{yy} = 0$. This gives a problem in plane elasticity which can be solved in terms of Airy functions.

We have to find a function Φ that satisfies the biharmonic equation with the boundary conditions at $z = 0$ (the asterisk, as before, denoting a prescribed value):

$$\sigma^*_{zz} = \tau^*_{xz} = 0, \text{ for } x < -b \text{ and } x > a$$
$$\sigma^*_{zz} = -p(x),\ \tau^*_{xz} = -q(x) \text{ for } -b < x < a \tag{4.8}$$

The stresses are obtained from this function. In cartesian coordinates

$$\sigma_{xx} = \partial^2\Phi/\partial z^2\ \sigma_{zz} = \partial^2\Phi/\partial x^2,\ \tau_{xz} = -\ \partial^2\Phi/\partial x\partial z,\ \sigma_{yy} = \nu(\sigma_{xx} + \sigma_{zz}) \tag{4.9}$$

and in cylindricals

$$\sigma_{rr} = \partial\Phi/r\partial r + \partial^2\Phi/r^2\partial\theta^2,\ \sigma_{\theta\theta} = \partial^2\Phi/\partial r^2,\ \tau_{r\theta} = -\ \partial/\partial r(\partial\Phi/r\partial\theta) \tag{4.10}$$

(a) Concentrated normal line force

Let P be the force per unit length; it can be shown (Exercise 1) that the solution is of the form $\Phi = Ar\theta \sin\theta$, where A is a constant to be determined; from this

$$\sigma_{rr} = (2A/r)\cos\theta,\ \sigma_{\theta\theta} = \tau_{r\theta} = 0 \tag{4.11}$$

The load is purely radial and the stress has a singularity like $1/r$. The value of A is found from the equilibrium condition:

$$-P = \int_{-\pi/2}^{\pi/2} \sigma_{rr}\, r\cos\theta\, d\theta = A\pi \tag{4.12}$$

whence

$$\sigma_{rr} = -\frac{2P}{\pi}\frac{\cos\theta}{r} \tag{4.13}$$

Thus the radial stress has a constant value $-2P/\pi d$ on a circle of diameter d passing through the origin O (Fig. 4.4). The maximum shear stress τ_l at (r, θ) has the value $\sigma_{rr}/2$ and acts on planes inclined at $45°$ to the radial direction.

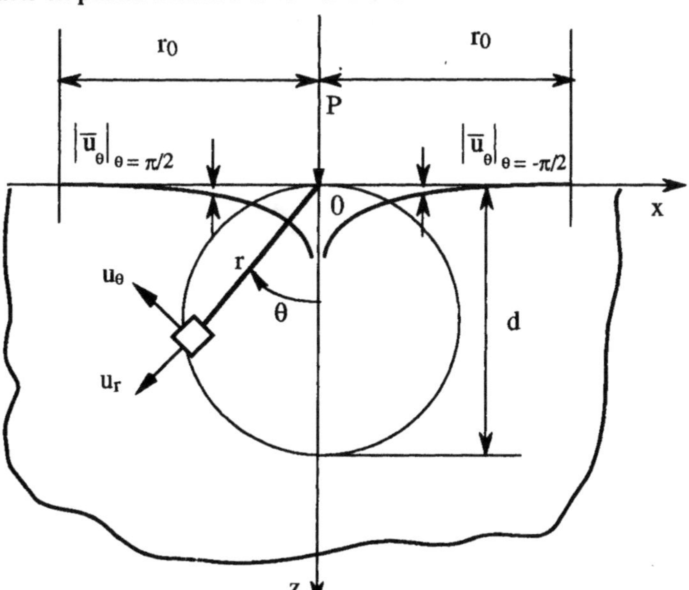

Fig. 4.4 Concentrated normal line load corresponding to applied force P per unit length.

The deformation of the surface is shown diagrammatically in Fig. 4.4; this also has a singularity of the order of log r. In cartesians, the stresses are:

$$\sigma_{xx} = \sigma_{rr}\sin^2\theta = -\frac{2P}{\pi}\frac{x^2 z}{(x^2+z^2)^2}$$

$$\sigma_{zz} = \sigma_{rr}\cos^2\theta = -\frac{2P}{\pi}\frac{z^3}{(x^2+z^2)^2} \tag{4.14}$$

$$\tau_{xz} = \sigma_{rr}\sin\theta\cos\theta = -\frac{2P}{\pi}\frac{xz^2}{(x^2+z^2)^2}$$

(b) Concentrated tangential line force

A tangential force Q per unit length generates a stress field that is similar to that for a normal force but is turned through $90°$; if we measure θ clockwise from Ox we have

$$\sigma_{rr} = -\frac{2Q}{\pi}\frac{\cos\theta}{r}, \quad \sigma_{\theta\theta} = \tau_{r\theta} = 0 \tag{4.15}$$

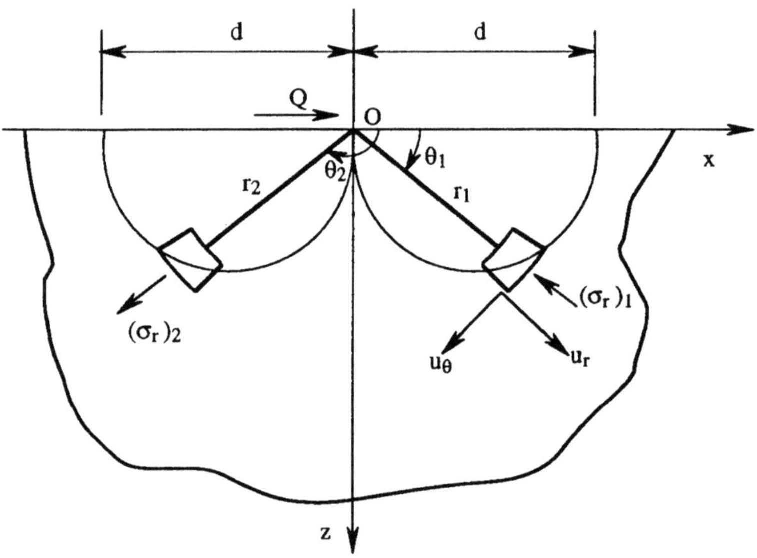

Fig. 4.5 Concentrated tangential line load, force Q per unit length.

The stress contours, shown in Fig. 4.5, are circles. In cartesians

$$\sigma_{xx} = -\frac{2Q}{\pi}\frac{x^3}{(x^2+z^2)^2}$$

$$\sigma_{zz} = -\frac{2Q}{\pi}\frac{xz^2}{(x^2+z^2)^2} \tag{4.16}$$

$$\tau_{xz} = -\frac{2Q}{\pi}\frac{x^2 z}{(x^2+z^2)^2}$$

(c) Distributed normal and tangential line forces

Here we assume that the loads $p(s)$ and $q(s)$ are distributed, as shown in Fig. 4.6; the resulting stresses are obtained from the previous equations by integrating the elementary efforts $p(s)ds$ along the *x-axis*. Thus if the distributions $p(x)$ and $q(x)$ are known then the solution can in principle be found by integration; the result, however, will not always be expressible in closed form. Here we shall give the results only for uniform distributions.

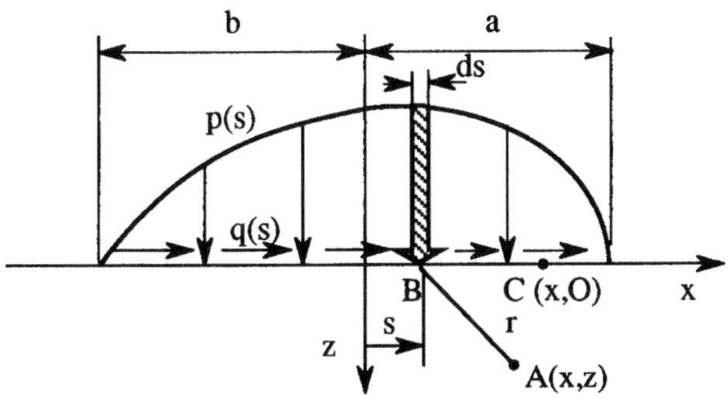

Fig. 4.6 Distributed line loads: $p(s)$ normal, $q(s)$ tangential.

Constant normal pressure (Fig. 4.7)

$$\begin{aligned}
\sigma_{xx} &= -\frac{p}{2\pi}\left\{2(\theta_1-\theta_2)-(\sin 2\theta_1-\sin 2\theta_2)\right\}\\
\sigma_{zz} &= -\frac{p}{2\pi}\left\{2(\theta_1-\theta_2)+(\sin 2\theta_1-\sin 2\theta_2)\right\}\\
\tau_{xz} &= -\frac{p}{2\pi}\left(\cos 2\theta_1-\cos 2\theta_2\right)\\
\tan\theta_{1,2} &= z/(x\pm a)
\end{aligned} \tag{4.17}$$

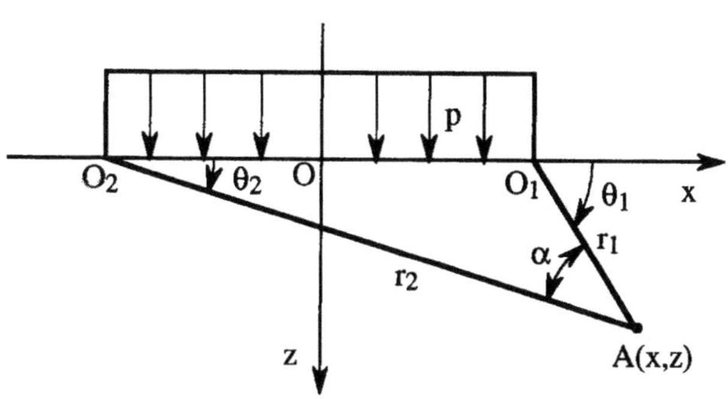

Fig. 4.7 Constant normal pressure.

It can be shown that if $\alpha = \theta_1 - \theta_2$ the principal stresses on the Mohr circle (Fig. 4.8) can be expressed simply as

$$\sigma_{1,2} = -(p/\pi)(\alpha \pm \sin\alpha),\ \tau_1 = (p/\pi)\sin\alpha \tag{4.18}$$

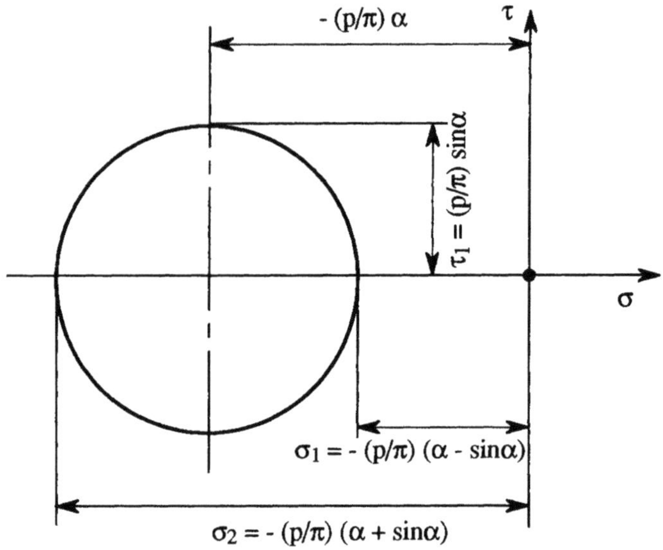

Fig. 4.8 Mohr circle for constant pressure *p*.

Thus the contours for maximum $\sigma_{1,2}$ and τ_1 are circles through O_1 and O_2 and that for maximum shear, corresponding to $\alpha = \pi/2$, is a circle with O_1O_2 as diameter (Fig. 4.9). Beneath the loaded surface, that is along O_1O_2, for which $\alpha = \pi$, the shear vanishes and the stress state is one of equal compression along the two axes, $\sigma_{xx} = \sigma_{zz} = -p$. This provides a qualitative explanation for the observation that if there is plasticity this is found not immediately under the line of application of the load but at some distance below the surface.

Fig. 4.10 shows the deformation of the surface; this is given by

$$u^*_z = -\frac{(1-\nu^2)p}{\pi E}\left\{(x+a)\log\left(\frac{x+a}{a}\right)^2 - (x-a)\log\left(\frac{x-a}{a}\right)^2\right\} + C \tag{4.19}$$

where C is a constant of integration, to be chosen; the curve of Fig. 4.10 corresponds to the conditions $u^*_z = 0$ for $x = \pm c$.

It is important to realise what these imply. The stresses are finite throughout the body of the solid except at O_1 and O_2 where there is a discontinuity in σ_{xx}, which is zero everywhere except under the contact line, where it is $-p$. Similarly τ_{xz} has a discontinuity, vanishing everywhere except under the contact line where it is p/π. The surface deformations are finite everywhere but the slope is theoretically infinite at O_1 and O_2.

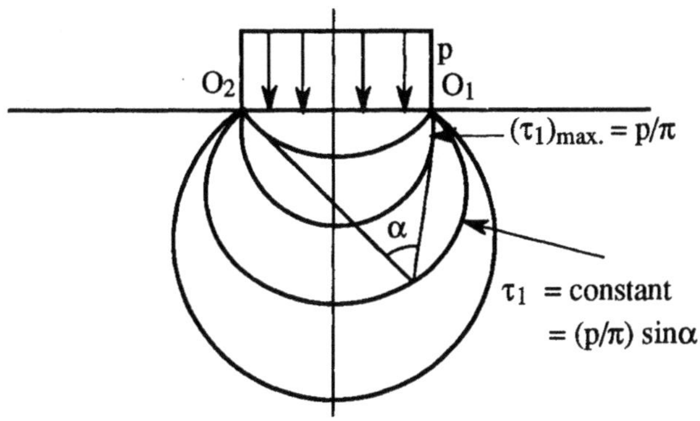

Fig. 4.9 Contours of maximum shear stress for constant normal pressure.

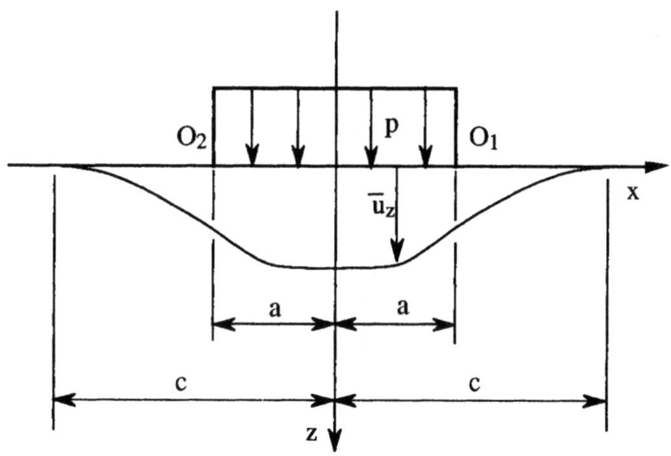

Fig. 4.10 Deformation of the surface produced by constant normal pressure.

Constant tangential tension The solution is

$$\sigma_{xx} = \frac{q}{\pi}\left\{\pi \log\left(\frac{r_1}{r_2}\right) - (\cos 2\theta_1 - \cos 2\theta_2)\right\}$$

$$\sigma_{zz} = \frac{q}{2\pi}(\cos 2\theta_1 - \cos 2\theta_2) \tag{4.20}$$

$$\tau_{xz} = -\frac{q}{2\pi}\{2(\theta_1 - 2\theta_2) + (\sin 2\theta_1 - \sin 2\theta_2)\}$$

$$\text{where} \quad r_{1,2} = \left\{(x \pm a)^2 + z^2\right\}$$

The distribution of the associated surface stresses τ^*_{xz} and σ^*_{xx} is shown in fig. 4.11. The shear is now distributed uniformly ($\tau^*_{xz} = -q/\pi$) but the stress σ^*_{xx} has a singularity at O_1 and at O_2. These stress concentrations certainly play an important role in the *fretting fatigue*- that develops when a pair of surfaces in contact oscillate relative to each other (see §4.2.1.5).

4.2.1.3 Point loading of a semi-infinite solid

(a) General methods of solution

The load now is *2*-dimensional; the resulting stress distribution will be *3*-dimensional and in general all *6* components of the stress tensor will be non-zero. Problems of this type - the Boussinesq-Cerruti problem - can be solved with the aid of potential theory.

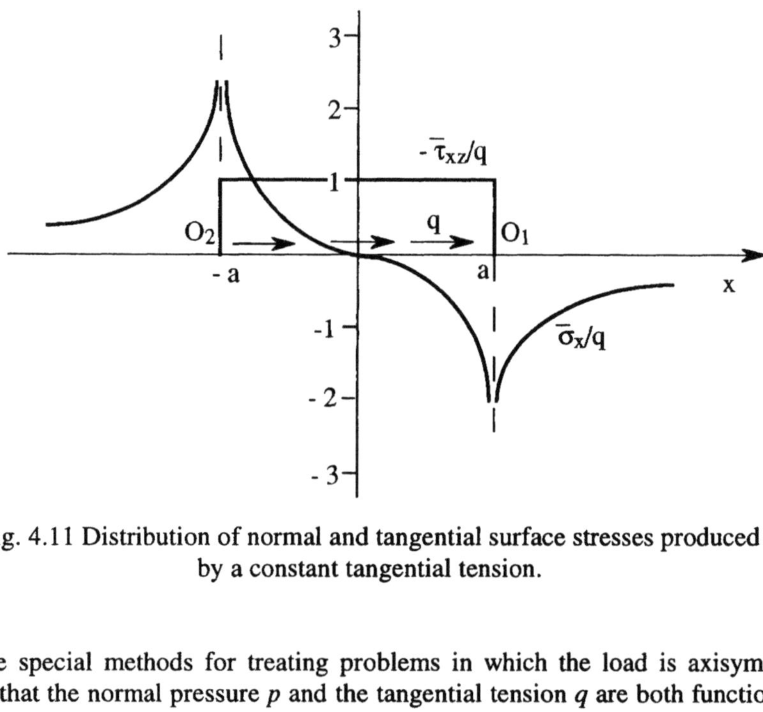

Fig. 4.11 Distribution of normal and tangential surface stresses produced by a constant tangential tension.

There are special methods for treating problems in which the load is axisymmetric, meaning that the normal pressure p and the tangential tension q are both functions of r only, independent of θ. The solutions are then simpler, since the shear components $\tau_{r\theta}$ $\tau_{\theta z}$ are both zero and the other components of the tensor are independent of θ. We shall give this case only.

(b) Concentrated normal force P

The stresses are as follows.

- in cartesians ($r^2 = x^2+y^2$, $\rho^2 = x^2+ y^2+ z^2$):

$$\sigma_{xx} = \frac{P}{2\pi}\left[\frac{(1-2v)}{r^2}\left\{\left(1-\frac{z}{\rho}\right)\frac{x^2-y^2}{r^2}+\frac{zy^2}{\rho^2}\right\}-\frac{3zx^2}{\rho^5}\right]$$

$$\sigma_{yy} = \frac{P}{2\pi}\left[\frac{(1-2v)}{r^2}\left\{\left(1-\frac{z}{\rho}\right)\frac{y^2-x^2}{r^2}+\frac{zx^2}{r^2}\right\}-\frac{3zy^2}{\rho^5}\right]$$

$$\sigma_{zz} = -\frac{3P}{2\pi}\frac{z^3}{\rho^5}$$

$$\tau_{xy} = \frac{P}{2\pi}\left[\frac{(1-2v)}{r^2}\left\{\left(1-\frac{z}{\rho}\right)\frac{xy}{r^2}-\frac{xyz}{\rho^3}\right\}-\frac{3xyz}{\rho^5}\right]$$

$$\tau_{xz} = -\frac{3P}{2\pi}\frac{xz^2}{\rho^5} \tag{4.21}$$

$$\tau_{yz} = -\frac{3P}{2\pi}\frac{yz^2}{\rho^5}$$

- in cylindricals

$$\sigma_{rr} = \frac{P}{2\pi}\left\{(1-2v)\left(\frac{1}{r^2}-\frac{z}{\rho r^2}\right)-\frac{3zr^2}{\rho^5}\right\}$$

$$\sigma_{\theta\theta} = -\frac{P}{2\pi}(1-2v)\left(\frac{1}{r^2}-\frac{z}{\rho r^2}-\frac{z}{\rho^3}\right) \tag{4.22}$$

$$\sigma_{zz} = -\frac{3P}{2\pi}\frac{z^3}{\rho^5}$$

$$\tau_{rz} = -\frac{3P}{2\pi}\frac{rz^2}{\rho^5}$$

The stresses σ_{zz} and τ_{rz}, which act on planes parallel to the surface, are independent of Poisson's ratio; the resultant stress Σ on these planes is

$$\Sigma = \left(\sigma_{zz}^2+\tau_{rz}^2\right)^{1/2} = \frac{3P}{2\pi}\frac{z^2}{\rho^4} = \frac{3P}{2\pi d^2} = \text{constant} \tag{4.23}$$

This stress acts in a radial direction passing through the origin, so that this case has a certain analogy with that of a concentrated line load - compare Figs. 4.4 and 4.12. In general, however, Σ is not a principal stress, the principal stresses are not radially directed and the surfaces of maximum constant shear stress are not spherical.

The displacements are

$$u_r = \frac{P}{4\pi\mu}\left\{\frac{rz}{\rho^3} - (1-2\nu)\frac{\rho - z}{\rho r}\right\}$$
$$u_z = \frac{P}{4\pi\mu}\left\{\frac{z^2}{\rho^3} + \frac{2(1-\nu)}{\rho}\right\} \tag{4.24}$$

and the displacement at the free surface ($z = 0$) given by the values of u^*_r and u^*_z, is

$$u_r{}^* = -\frac{(1-2\nu)}{4\pi\mu}\frac{P}{r}, \qquad u_z{}^* = \frac{(1-\nu)}{2\pi\mu}\frac{P}{r} \tag{4.25}$$

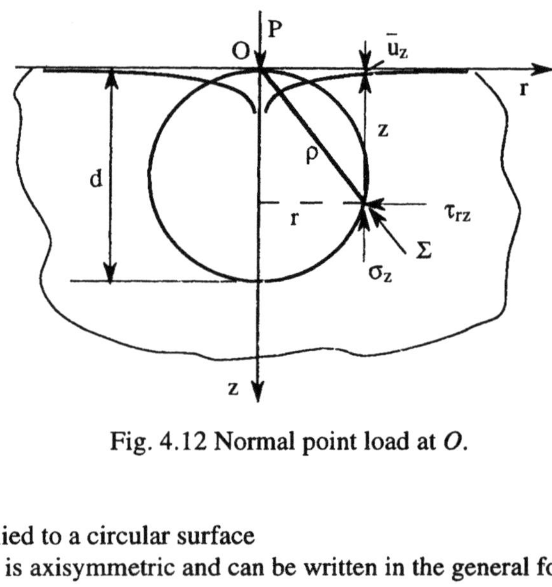

Fig. 4.12 Normal point load at O.

(c) Pressure applied to a circular surface
This distribution is axisymmetric and can be written in the general form

$$p = p_0(1 - r^2/a^2)^n \tag{4.26}$$

Solutions have been obtained for various values of n; $n = 0$ corresponds to a uniform pressure, $n = -1/2$ to a uniform displacement. In the next section (§4.2.1.4) we discuss the theory of contact due to Hertz, to which the value $n = 1/2$ applies; consequently we consider only this value here. With this distribution the total load over a circle of radius a is $P = 2\pi p_0 a^2/3$.

Given the solution in (b) above we can, by superposition, calculate the stress field and the corresponding displacement for the present case. The stresses along Oz can be calculated without difficulty by considering a ring of force concentrated on the circumference of a circle of radius r; these are

$$\frac{\sigma_{rr}}{p_0} = \frac{\sigma_{\theta\theta}}{p_0} = -(1+\nu)\left\{1 - \frac{z}{a}\tan^{-1}\frac{a}{z}\right\} + \frac{1}{2}\left(1 + \frac{z^2}{a^2}\right)^{-1}$$
$$\frac{\sigma_{zz}}{p_0} = -\left(1 + \frac{z^2}{a^2}\right)^{-1} \tag{4.27}$$

The principal stresses along Oz are σ_{rr}, $\sigma_{\theta\theta}$ and σ_{zz}; the shear stress $\tau_1 = (1/2)|\sigma_{zz} - \sigma_{rr}|$ is maximum at a distance $z = 0.48a$ below the surface at the value (Exercise 3)

$$\tau_{1max} = 0.31p_0 = 0.47P/\pi a^2 \tag{4.28}$$

The stresses on the surface of contact are as follows:

- in the interior of the circle

$$\sigma_{rr} / p_0 = \frac{1-2\nu}{3}(a^2 / r^2)\left\{1 - \left(1 - r^2 / a^2\right)^{3/2}\right\} - \left(1 - r^2 / a^2\right)^{1/2}$$
$$\sigma_{\theta\theta} / p_0 = -\frac{1-2\nu}{3}(a^2 / r^2)\left\{1 - \left(1 - r^2 / a^2\right)^{3/2}\right\} - 2\nu\left(1 - r^2 / a^2\right)^{1/2} \tag{4.29}$$
$$\sigma_{zz} / p_0 = -\left(1 - r^2 / a^2\right)^{1/2}$$

- outside the circle

$$\sigma_{zz} = 0, \quad \sigma_{rr} / p_0 = -\sigma_{\theta\theta} / p_0 = (1-2\nu)a^2 / 3r^2 \tag{4.30}$$

The radial stress here is thus tensile; it is maximum on the circumference (r = a) where it has the value

$$(\sigma^*_{rr})_{max} = (1-2\nu)p_0/3 \approx 0.13p_0 \tag{4.31}$$

This radial stress plays an important role in the fracture of brittle solids crushed one against the other; in ductile materials, in contrast, the shear stress produces a plastic sub-layer (see §4.2.2)

The deformation of the surface is given by -

- for $r < a$

$$u^*_r = -\frac{(1-2\nu)(1+\nu)}{3E}\frac{a^2}{r}p_0\left\{1-\left(1-r^2/a^2\right)^{3/2}\right\}$$

$$u^*_z = \frac{1-\nu^2}{E}\frac{\pi p_0}{4a}(2a^2-r^2) \qquad (4.32)$$

for $r > a$

$$u^*_r = -\frac{(1-2\nu)(1+\nu)}{3E}p_0\frac{a^2}{r^2} \qquad (4.33)$$

4.2.1.4 Hertz's theory of normal contact

This theory, developed in 1881, enables us to determine the dimensions of the contact area and the distributions of stress and deformation, and to relate these to the contact pressure and to the geometry and elastic moduli of the solids concerned. We give here a simplified account.

(a) *Geometry of the problem.* We assume that the surfaces of the solids in contact are regular and that their profiles close to the origin can be adequately represented by a quadratic form. With an appropriate choice of axes the initial profiles can then be expressed in terms of the principal radii of curvature of the solids, R_1' and R_1'' for the one and R_2' and R_2'' for the other, thus:

$$z_1 = x_1^2/2R_1'^2 + y_1^2/2R_1''^2,\ z_2 = -(x_2^2/2R_2'^2 + y_2^2/2R_2''^2) \qquad (4.34)$$

The separation of the surfaces is $h = z_1 - z_2$ and the coordinate system (x, y) can be chosen so that this can be put in the form

$$h = Ax^2 + By^2 = x^2/2R' + y^2/2R'' \qquad (4.35)$$

where A, B are positive constants and R', R'' are the *relative principal radii of curvature*. It can be shown that if the axes (x_1, y_1), (x_2, y_2) are inclined at an angle α

$$A + B = \tfrac{1}{2}(1/R' + 1/R'') = \tfrac{1}{2}(1/R_1' + 1/R_1'' + 1/R_2' + 1/R_2'')$$
$$B - A = \tfrac{1}{2}\{(1/R_1' - 1/R_1'')^2 + (1/R_2' - 1/R_2'')^2 + 2(1/R_1' - 1/R_1'')(1/R_2' - 1/R_2'')\cos 2\alpha\}^{1/2} \qquad (4.36)$$

An *equivalent radius* R_e is often introduced, defined by

$$R_e = (R'R'')^{1/2} = \tfrac{1}{2}(AB)^{1/2} \qquad (4.37)$$

These radii are taken as positive for convex surfaces and negative for concave.

For convenience, we consider two convex surfaces in contact under the effect of a normal load P (Fig. 4.13). Before compression the distance between a pair of points $\delta_1(x, y, z_1)$, $\delta_2(x, y, z_2)$ is given by (4.35). During the compression two points T_1, T_2 approach one another by amounts δ_1, δ_2, and if the solids were not deformed they would overlap each other as shown by the heavy lines in the diagram. A contact pressure is created, under the influence of which each point on the surface of the solid is displaced parallel to Oz by a quantity u^*_{z1} or u^*_{z2}; if after the deformation the points S_1, S_2 within the contact area coincide then

$$u^*_{z1} + u^*_{z2} + h = \delta_1 + \delta_2 = \delta \tag{4.38}$$

Thus the elastic displacements at the surface can be expressed

$$u^*_{z1} + u^*_{z2} = \delta - Ax^2 - By^2 \tag{4.39}$$

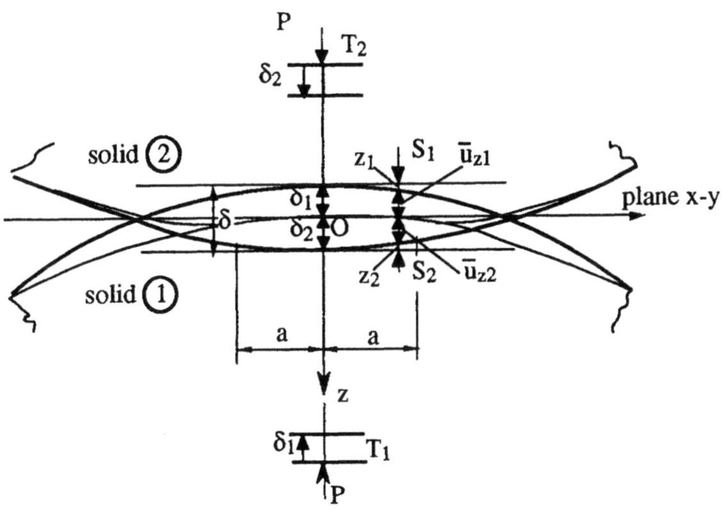

Fig. 4.13 Hertz contact.

If S_1, S_2 are located outside the contact area they will not come into coincidence and for these

$$u^*_{z1} + u^*_{z2} < \delta - Ax^2 - By^2 \tag{4.40}$$

To solve the problem we have to find the distribution of the pressure transmitted between the solids over the contact area, such that the elastic displacements normal to this surface satisfy these boundary conditions.

(b) *Approximate solution.* To simplify the problem we consider only solids of revolution for which the contact area is a circle of radius a, or cylinders whose axes are parallel, for which this area is a band of width $2a$.

For the first case we have from Fig. 4.13 $\delta_i = u^*_{z1}(0)$, $\delta_2 = u^*_{z2}(0)$; from (4.38)

$$(\delta_1 - u^*_{z1})/a + (\delta_2 - u^*_{z2})/a = h/a = \tfrac{1}{2}(1/R_1 + 1/R_2)(x^2/a) \qquad (4.41)$$

and therefore

$$[u^*_{z1}(0) - u^*_{z1}]/a + [u^*_{z2}(0) - u^*_{z2}]/a = \tfrac{1}{2}(1/R_1 + 1/R_2)(x^2/a) \qquad (4.42)$$

Putting $x = a$, the left side of (4.42) gives an approximation for the deformation of the pair of solids: if $d_i = u^*_{zi}(0) - u^*_{zi}$ we have

$$(d_1 + d_2)/a = (a/2)(1/R_1 + 1/R_2) \qquad (4.43)$$

from which the mean contact pressure p_m is given by

$$p_m/E_1 + p_m/E_2 \approx a(1/R_1 + 1/R_2) \qquad (4.44)$$

that is

$$p_m \approx a(1/R_1 + 1/R_2)/(1/E_1 + 1/E_2)$$

For the second case (contact of cylinders) $P = 2ap_m$, whence

$$a \propto \{P(1/E_1 + 1/E_2)/(1/R_1 + 1/R_2)\}^{1/2}$$
$$p_m \propto \{P(1/R_1 + 1/R_2)/(1/E_1 + 1/E_2)\}^{1/2} \qquad (4.45)$$

showing that both the pressure and the contact area increase like $\sqrt{P}$.

For contact between two spheres $P = \pi a^2 p_m$, giving

$$a \approx \{P(1/E_1 + 1/E_2)/(1/R_1 + 1/R_2)\}^{1/3}$$
$$p_m \approx \{P(1/R_1 + 1/R_2)/(1/E_1 + 1/E_2)\}^{1/3} \qquad (4.46)$$

with a and p_m now varying like $P^{1/3}$

(c) *Elastic contact.* The following assumptions are made:

(i) the surfaces are continuous and non-conforming,
(ii) the strains are small,
(iii) each solid can be regarded as a semi-infinite medium: $a << R_{1,2}$,
(iv) there is no friction: $q_x = q_y = 0$.

We confine our treatment to solids of revolution ($R_1' = R_2' = R_1$, $R_1'' = R_2'' = R_2$) and cylinders in contact along a common generator, and give only results.

Solids of revolution
The contact area is a circle of radius a and $A + B = \tfrac{1}{2}(1/R_1 + 1/R_2)$. Defining the *relative curvature* R by $1/R = 1/R_1 + 1/R_2$, the boundary conditions are

$$u^*_{z1} + u^*_{z2} = \delta - (1/2R)r^2 \qquad (4.47)$$

These can be satisfied by a parabolic pressure distribution of the form given in §4.2.1.3, since the corresponding deformation has the desired form. If we define an *effective modulus* E^* by $1/E^* = 1/E_1 + 1/E_2$ we have

$$(\pi p_0/4aE^*)(2a^2 - r^2) = \delta - r^2/2R \tag{4.48}$$

from which we find

$$a = \pi p_0 R/2E^*,\ \delta = \pi a p_0/2E^* \tag{4.49}$$

The total applied load is $P = \int_0^a p(r).2\pi r\, dr = 2\pi p_0 a^2/3$ (4.50)

This leads to the following expressions for the contact area a, the distance of approach δ of the two solids and the pressure p_0, in terms of the load P:

$$a = (3PR/4E^*)^{1/3},\ \delta = (a^2/R)(9P^2/16RE^{*2})^{1/3},\ p_0 = 3P/2\pi a^2 = (6PE^{*2}/\pi^3R^2)^{1/3} \tag{4.51}$$

where $1/E^* = 1/E_1 + 1/E_2$ and $1/R = 1/R_1 + 1/R_2$

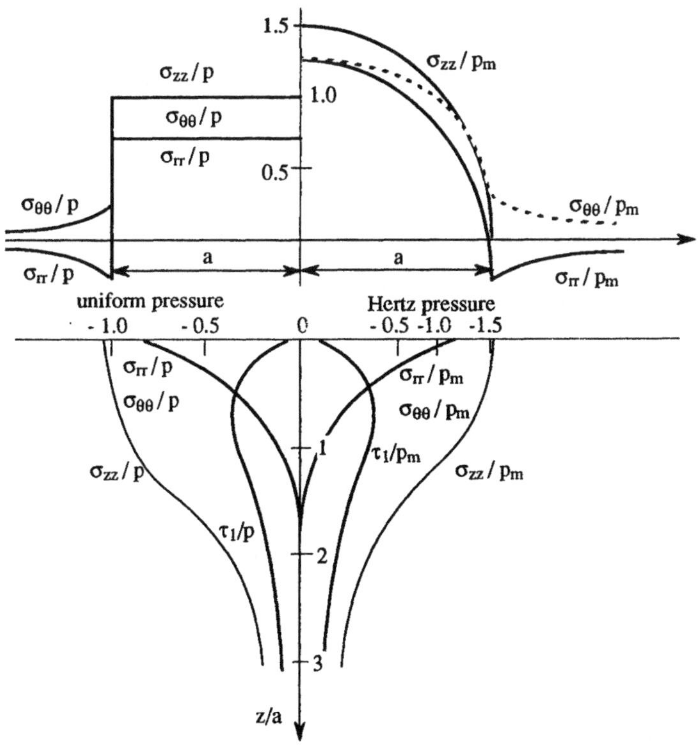

Fig. 4.14 Stress distribution on the surface and along the axis of symmetry due to (on the left) a uniform pressure and (on the right) a Hertz parabolic distribution.

We have already given the stress distribution within the solids for this case (§4.2.1.3); it is shown in Fig. 4.14, with the distribution due to a uniform pressure for comparison, the stresses being given in relation to the mean pressure $(2/3)p_0$. The following important relations should be noted:

(i) The stresses over the contact area are all compressive except on the boundary circle where σ_{rr} is a tensile . The maximum value of this stress is $(1-2\nu)p_0/3$, at $r = a$, $z = 0$.

(ii) The stress state at the center ($r = 0$, $z = 0$) is

$$\sigma_{rr} = \sigma_{\theta\theta} = (1 + 2\nu)p_0/2, \ \sigma_{zz} = p_0 \tag{4.52}$$

For an incompressible solid ($\nu = 0.5$) the stress state is pure hydrostatic.

(iii) For $\nu = 0.30$ the shear stress $\tau_1 = ½|\sigma_{rr} - \sigma_{\theta\theta}|$ on the axis of symmetry $r = 0$ is maximum at $z = 0.48a$, where the value is approximately $0.31p_0$; at the origin ($z = 0$) it is only $0.10p_0$ and on the boundary of the contact area ($r = a$, $z = 0$) it is $0.13p_0$. We shall see later (§4.2.2) that this accounts for the development of a plastic sub-layer.

2-dimensional contact of a pair of cylinders. For this we can use the solution given in §4.2.1.2 to the problem of the line loading of a semi-infinite solid: we have

$$h = z_1 + z_2 = Ax^2 = ½(1/R_1 + 1/R_2)x^2 = x^2/2R \tag{4.53}$$

For points below the contact area

$$u^*_{z1} + u^*_{z2} = \delta - Ax^2 = \delta - x^2/2R \tag{4.54}$$

whilst for those outside this region

$$u^*_{z1} + u^*_{z2} > \delta - x^2/2R \tag{4.55}$$

We now assume, with Hertz, that the displacements u^* can be determined by considering each cylinder as a semi-infinite medium. However, the presence of the logarithmic singularity gives rise to a difficulty in the calculation of the displacements, requiring the relative displacements to be measured from a point at infinity; this means that the local contact stresses alone no longer determine uniquely the distance δ of approach, but that the complete stress distribution in the bulk of the solids, and thus their geometry, must be taken into account. Solutions have been found for a number of specific problems, for example, a small cylinder compressed between two larger cylinders along diametrically opposite generators.

It can be shown that the pressure distribution remains parabolic, with

$$p(x) = (2P/\pi a)(1 - x^2/a^2) \tag{4.56}$$

As in the case of solids of revolution, we can expect that the stress distribution does not differ greatly from that produced by a uniform pressure distribution; with this assumption the solution is

$$a = 2(PR/\pi E^*)^{1/2},\ p_0 = 2P/\pi a = (4/\pi)p_m = (PE^*/\pi R)^{1/2} \qquad (4.57)$$

Along Oz the stresses are

$$\sigma_{xx} = -(p_0/a)\{(a^2 + 2z^2)(a^2 + z^2)^{-1/2} - 2z\}$$

$$\sigma_{zz} = -p_0 a(a^2 + z^2)^{-1/2} \qquad (4.58)$$

$$\tau_1 = \tfrac{1}{2}|\sigma_{zz} - \sigma_x| = p_0 a\{z - z^2(a^2 - z^2)^{/1/2}\}$$

The shear stress τ_1 reaches its maximum value of $0.30p_0$ at $z = 0.87a$; the stress profile is qualitatively close to that obtained previously (Fig. 4.14). Fig. 4.15 shows how the principal stresses σ_{xx}, σ_{zz} vary along Oz and gives contours of equal values of the shear stress τ_1. It can be checked that on the contact surface

$$\sigma^*_{xx} = \sigma^*_{zz} = -p \qquad (4.59)$$

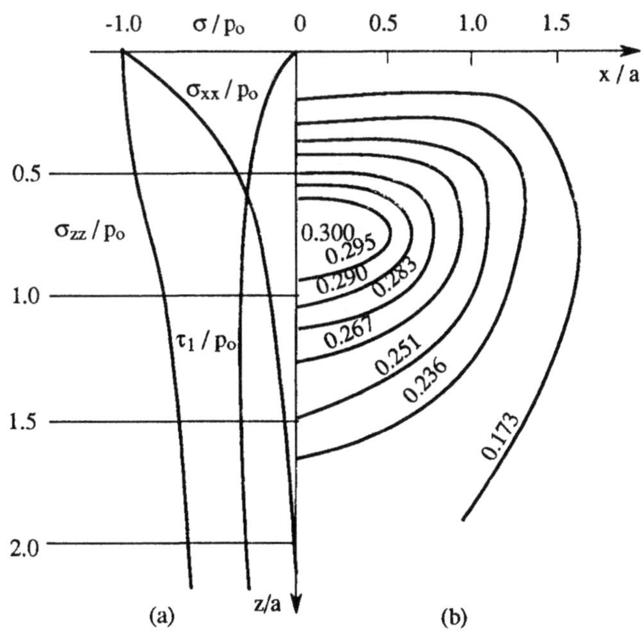

Fig. 4.15 Contact of two cylinders. (a) Stresses in a sub-layer along the axis of symmetry (b) contours of shear stress τ_1.

NOTES

1. Hertz treated also the more general case in which the contact surface is an ellipse. If a, b are the major and minor axes respectively and $c = \sqrt{ab}$ is an "equivalent" contact radius, the contact area, maximum pressure and relative displacement can be given in the same form as for contact over a circle, with multipliers F_1, F_2 which are functions of R'/R''; these have been tabulated and for moderate values of this ratio does not differ greatly from 1. Thus:

$$\begin{aligned} c &= (ab)^{1/2} = [3PR_e/4E^*]^{1/3}\,[F_1(R'/R'')]^{1/3} \\ p_0 &= 3P/2\pi ab = [6PE^{*2}/(\pi^3 R_e^2)]^{1/3}\,[F_1(R'/R'')]^{-2/3} \\ \delta &= [9P^2/16E^{*2}R_e]^{1/3}\,[F_2(R'/R'')] \end{aligned} \qquad (4.60)$$

2. We have not taken into account any friction that may be generated by the contact; this will occur only when the solids have different elastic constants, since only then can the pressure generate different tangential displacements in the surfaces in contact. This problem has been treated to some extent; in the next section we discuss the effects of sliding.

4.2.1.5. Tangential loading and sliding

(a) *Relative sliding of two solids: case of two cylinders.* In the introduction to this chapter (§4.1.1) we distinguished between *sliding* and *rolling*: sliding is characterised by a relative linear velocity at the periphery of the contact area and rolling by a relative angular velocity. The two can occur simultaneously.

We shall consider only linear sliding (Fig. 4.16). The initial point of contact is taken as the origin and remains fixed; solid 1, below the contact surface, moves with linear velocity V from left to right with respect to solid 2 above, which remains fixed. In the absence of friction the normal force P would produce the contact area given by Hertz's theory and the solution would be that given above.

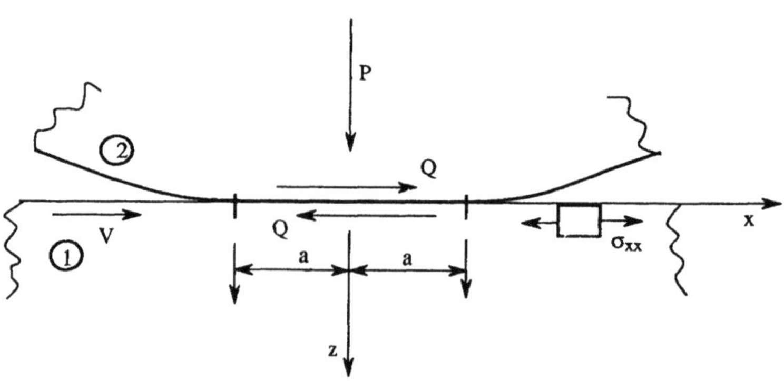

Fig. 4.16 Sliding contact.

The friction force Q, acting tangentially, is opposed to this relative movement and will alter the contact stresses. We assume at the outset that the two solids are in steady linear relative motion, expressed by saying that we have *kinetic friction*; later we consider the onset of sliding, referred to as *static friction*, corresponding to $Q < \mu P$ where μ is the coefficient of friction.

Our problem is to determine the contact area, in both shape and extent, and the pressure distribution resulting from the combination of P and Q. If the elastic constants are the same in the two solids the distribution of normal pressure is not changed from what we found previously. If they differ there is an interaction between P and Q, which can be neglected so long as it is valid to superpose the effects of normal pressure and tangential force associated with sliding, taken separately.

It is usual to take Coulomb's law for the friction, that is, that p and q are in a constant ratio at all points of the contact area:

$$|q(x, y)|/p(x, y) = |Q|/P = \mu \tag{4.61}$$

The only relative movement we shall consider is that of one cylinder in a direction perpendicular to its axis - this is a 2-dimensional problem.

For the two cylinders in contact the Hertz solution for the distribution of normal pressure is (§4.2.1.4)

$$p(x) = (2P/\pi a)(1 - x^2/a^2)^{1/2} \tag{4.62}$$

from which it follows that the tangential stress distribution is

$$q(x) = \pm(2\mu P/\pi a)(1 - x^2/a^2)^{1/2} \tag{4.63}$$

where the - sign is associated with positive values of V.

The conditions that obtain over the contact area, a band of width $2a$, are given by superposition of these distributions. As in the case of normal or tangential line loadings, or uniformly distributed loadings (§4.2.1.2), there is a certain analogy between the stress distributions due to P on the one hand and to Q on the other. This is expressed by two relations:

$$(\sigma_{zz})_q/q_0 = (\tau_{xz})_p/p_0\,,\ (\tau_{xz})_q/q_0 = (\sigma_{xx})_p/p_0 \text{ with } q_0 = \mu p_0 \tag{4.64}$$

the suffixes p, q denoting the stresses resulting from $p(x)$ and $q(x)$ respectively.

Thus from the solution given by Hertz we can obtain the components $(\sigma_{zz})_q$ and $(\tau_{xz})_q$; σ_{xx}, however, has to be determined separately. We have seen (§4.2.1.2) that for a uniform distribution $q(x) = q$ over the contact surface, σ^*_{xx} has a singularity, whilst for a parabolic distribution such that $q(x) = 0$ for $x = \pm a$ the situation is very different and this stress remains bounded. It can be shown that the value is

$$(\sigma^*_{xx})_q = -2q_0x/a \qquad \text{for } |x| \leq a$$

$$= -2q_0\left\{x/a \pm (x^2/a^2 - 1)^{-1/2}\right\} \qquad \text{for } x > a \tag{4.65}$$

Thus, the distribution within the moving solid, is shown in Fig. 4.17.

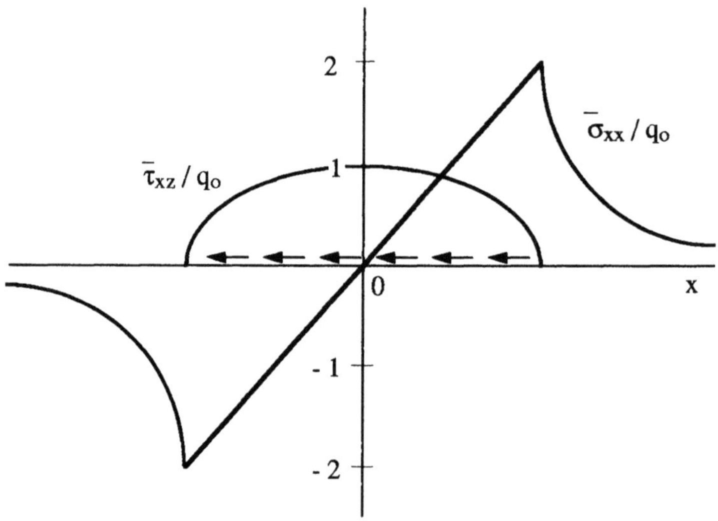

Fig. 4.17 Surface stresses generated by friction.

Over the contact surface the stresses due to the joint action of the load P and the friction force Q are obtained by superposing those due to the separate loads; the result is:

$$\sigma^*_{xx} = -p_0\left\{(1 - x^2/a^2)^{1/2} + 2\mu x/a\right\}$$

$$\sigma^*_{zz} = -p_0(1 - x^2/a^2)^{1/2}$$

$$\sigma^*_{yy} = \nu(\sigma^*_{xx} + \sigma^*_{zz})$$

$$\tau^*_{xz} = -\mu p_0(1 - x^2/a^2)^{1/2} \tag{4.66}$$

The tensile stress σ^*_{xx} is maximum at the right-hand edge of the sliding surface, where it has the value

$$(\sigma^*_{xx})_{max} = -2\mu p_0 \tag{4.67}$$

In the plane of contact the maximum shear stress τ_l is

$$\tau_l = \tfrac{1}{2}\{(\sigma^*_{xx} - \sigma^*_{zz})^2 + 4\tau_{xz}^{\ 2}\}^{1/2} = \mu p_0 \tag{4.68}$$

As is to be expected, the effect of sliding is to generate shear stresses on the contact surface; later we shall see the influence of these on the plasticity (§4.3.4).

The effect of sliding on the distribution of τ_l below the contact surface is to move the position of its maximum to a point closer to the surface and displaced in a direction parallel to that of slip; Fig. 4.18 shows this for the case of $\mu = 0.20$; in this diagram the stress contours are given for normalised values τ_l/p_0. The maximum is only very slightly above the value 0.30 given by the Hertz theory (cf. Fig. 4.15).

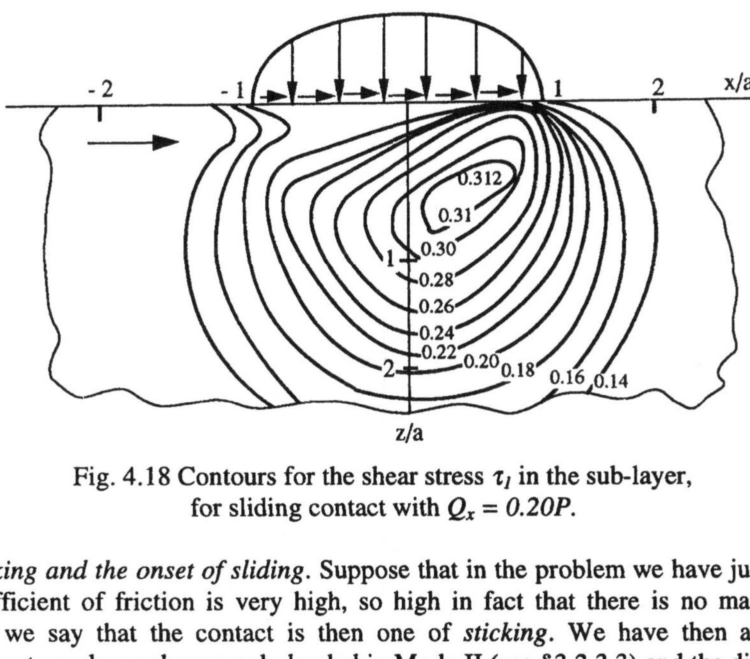

Fig. 4.18 Contours for the shear stress τ_l in the sub-layer, for sliding contact with $Q_x = 0.20P$.

(b) *Sticking and the onset of sliding*. Suppose that in the problem we have just studied the coefficient of friction is very high, so high in fact that there is no macroscopic sliding; we say that the contact is then one of *sticking*. We have then a problem analogous to a planar shear crack, loaded in Mode II (see §3.2.3.3) and the distribution $q(x)$ of tangential stress will have a singularity at the edges of the contact band. The difference between the surface displacements of the two solids must be constant, that is

$$u^*_{x1} - u^*_{x2} = \delta = constant \text{ for } -a < x < a \tag{4.69}$$

The corresponding tangential stress distribution $q(x)$ is

$$q(x) = Q/\pi(a^2 - x^2)^{1/2} = q'' \tag{4.70}$$

This would result in a distribution of infinite stresses at the edge of the contact band, which is physically impossible. To overcome this difficulty we can relax the condition of no sliding by assuming that there is *microsliding* over two bands bordering the edges, of width $(a - c)$, as in Fig. 4.19, in which zone the distribution will be such that $q' = q - q'' = \mu p$, whilst in the central zone the contact remains sticking and $q(x)$ is essentially the q'' above. The extent of the microslip zone is found from the equilibrium conditions,
thus:

$$Q = \int_{-a}^{a} q(x)\, dx = \int_{-a}^{a} q'(x) dx + \int_{-c}^{c} q''(x) dx$$

$$= \mu P - (c^2 / a^2)\mu P \tag{4.71}$$

whence $$c / a = \sqrt{1 - Q / \mu P} \tag{4.72}$$

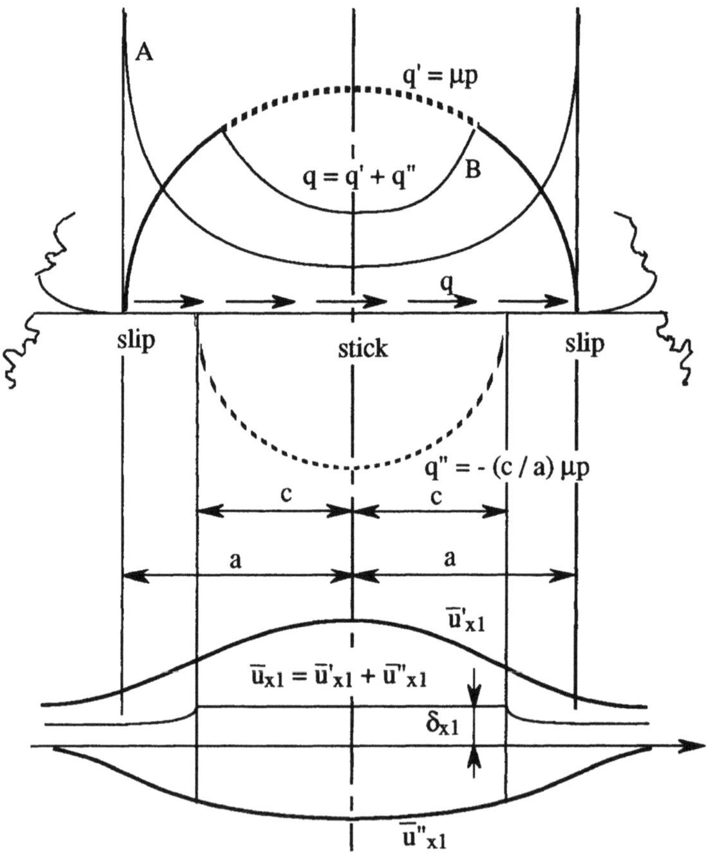

Fig. 4.19 Two cylinders in contact along a common generator (a) without slip (b) with partial slip.

NOTE: For contact between a sphere and a plane, the zone in which there is sticking is a circle of radius c and that in which there is microsliding is a ring of radii c and a (Fig. 4.20) such that

$$c/a = \left(1 - Q/\mu P\right)^{1/3} \tag{4.73}$$

The corresponding stress distribution has been calculated; Fig. 4.21 gives it for the values $\mu = 0.25$ and 0.30, for the case in which fixed solid is uppermost.

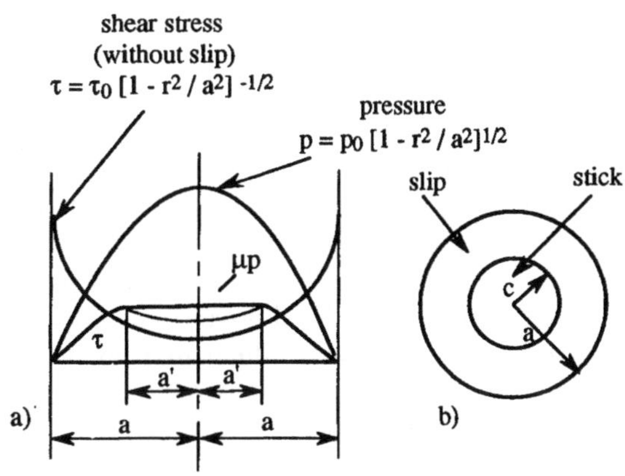

Fig. 4.20 Contact between a sphere and a plane: (a) section (b) plan view.

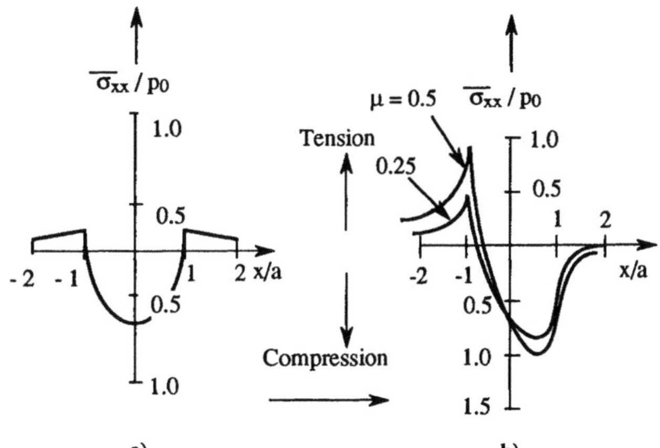

Fig. 4.21 Contact between a sphere and a plane: distribution of the stress $\overline{\sigma}_{xx}$ in the contact plane under conditions of (a) no sliding (b) microsliding.

Qualitatively, it is easily seen that such a distribution follows from the previous results. In the microslip zone the stress σ^*_{xx} ($z = 0$) is a tension acting on the leading edge of the zone in relative motion, with the maximum value $2\mu p_0$ previously obtained; in the central zone, in particular at $x = 0$, it is slightly less than p_0.

Thus the surface tensile stress σ^*_{xx} is much greater when there is microsliding than in the case of simple normal contact. It is thought that this is one of the sources of *fretting fatigue*, a form of damage which results from two surfaces rubbing against one another and which is often encountered in certain mechanical assemblies. We discuss this mechanical aspect of fatigue in more detail in the next section

(c) *Fretting fatigue*: see Waterhouse (1981) or Hills and Nowell (1994) for a detailed study. This can occur in cables, for example. In general, fatigue in the usual sense of the term results from a solid being subjected to an oscillatory tensile stress σ. In the situation we are considering it is in contact with a second solid and is acted on over the contact area by a constant normal pressure P and an inclined force F which also is oscillatory. The tangential component of F, although not great enough to produce macrosliding, may give rise to microsliding, which may in turn generate wear by abrasion (§4.3.4) and corrosion. The local friction conditions will depend strongly on the nature of the oxidation products formed, on the solids in contact and on the environment. This is a physico-chemical and tribological aspect of the problem of fretting fatigue; Fig. 4.22 shows the various factors which contribute to this form of damage.

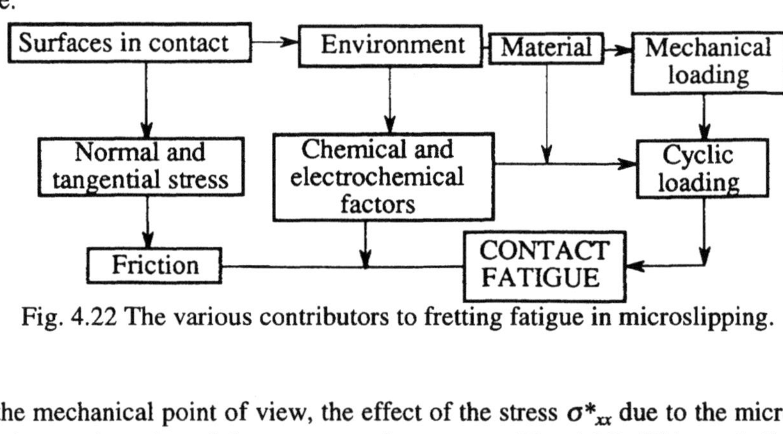

Fig. 4.22 The various contributors to fretting fatigue in microslipping.

From the mechanical point of view, the effect of the stress σ^*_{xx} due to the microsliding is superimposed on that of the macroscopic stress σ_T (Fig. 4.23). This, as indicated by Fig. 4.24, can have a strong effect on the endurance under fatigue, in some cases reducing the lifetime by a factor of *2* or *3*.

The overall effect is a function of many parameters, in particular the amount of microsliding, which will increase as the angle α between P and F increases (see Fig. 4.23). The zone over which microsliding occurs is shown clearly by the presence of oxide formed there and, often, a large amount of debris.

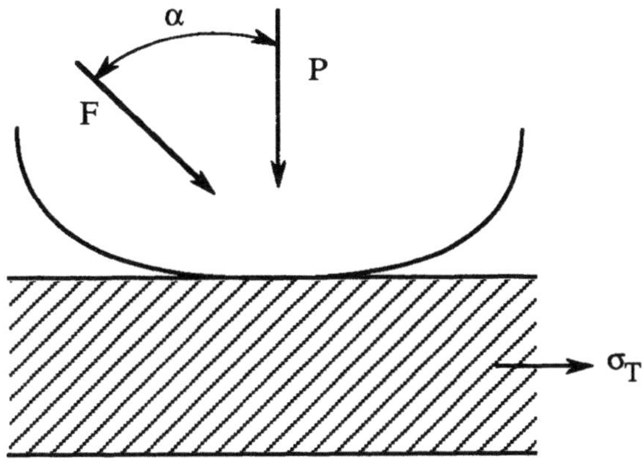

Fig. 4.23 Solid subjected to fretting fatigue.

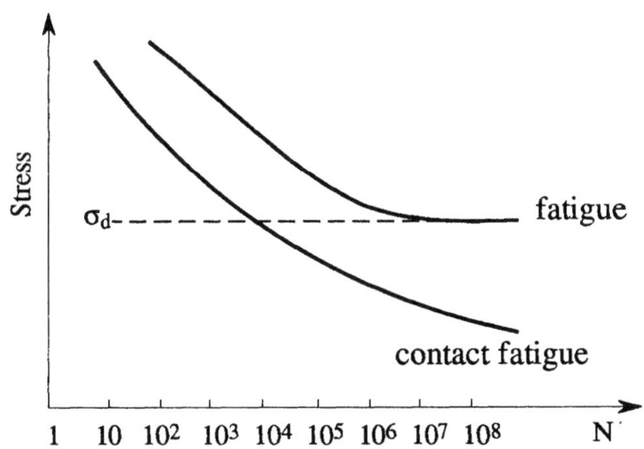

Fig. 4.24 Reduction in resistance to fatigue brought about by fretting fatigue.

This particular mode of damage has been the subject of a number of studies. More recently interest has concentrated on the formation of the cracks which develop preferentially in the microslip zone (Fig. 4.25), a form of damage which is important for the integrity of a number of structures and which is still not fully understood. As Fig. 4.22 suggests, it is complex; its practical importance warrants more study.

If D is the amplitude of the microsliding then for any given material three regions can be distinguished in the P - D plane, as indicated in Fig. 4.26a: if D is small and P large there is sticking; for large D there is complete slipping; and between the two is a zone in which cracking occurs. The results of recent observations made with an aluminum alloy are shown in Fig. 4.26b (Shaw et al (1992).). Such diagrams, when established, are very valuable for optimising the use of materials which will be subjected to fretting corrosion.

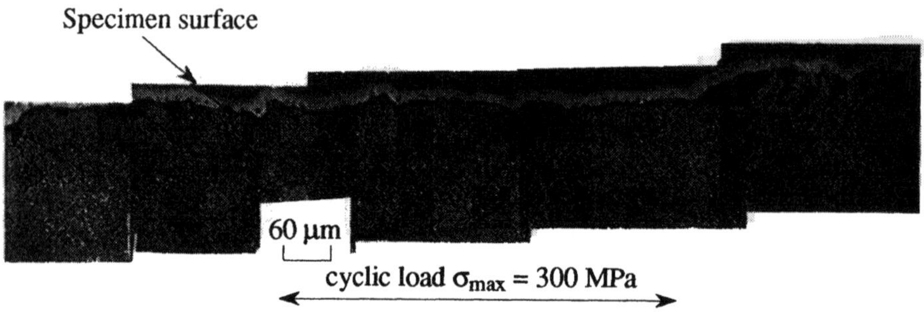

Fig. 4.25 3.5 Ni-Cr-Mo-V steel with cracks initiated by fretting fatigue. A protective layer of nickel has been deposited on the rubbing surface to make observation of the microcracks easier. Notice that the cracks make an angle of about *45°* with the direction of the applied load.

a)

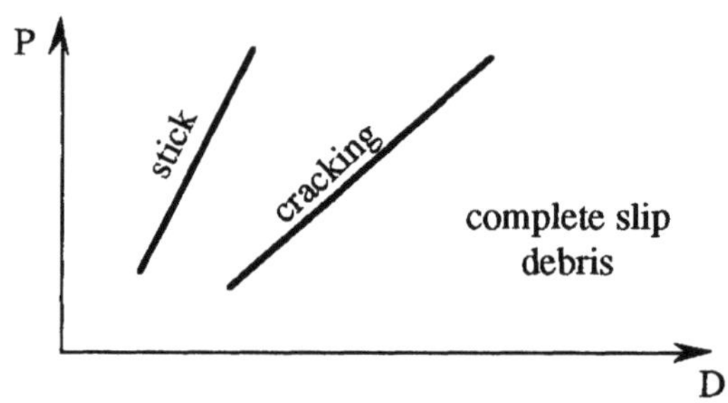

b)

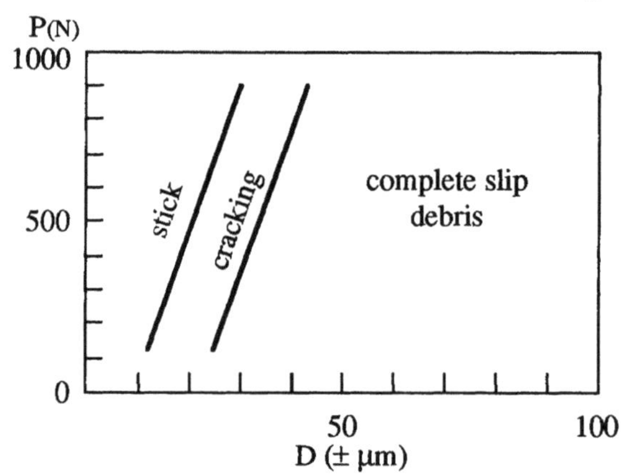

Fig. 4.26 Fretting corrosion. (a) General - specification of the three regions (b) Results for a high-strength aluminum alloy.

4.2.2. Conditions of plasticity

The analysis of the problem in the presence of plasticity will clearly be much more complex than what we have just given for the case of elasticity. There is non-linearity from two sources: the geometry of the Hertz contact, and the constitutive equations. Further, as in all problems involving plasticity, the choice of the constitutive equations is a matter of great importance; consequently we shall just make a few simple comments and limit our treatment to the case of *perfectly plastic solids*. We shall study only the onset of the plastification; the question of the limit load in the case of normal contact was treated in the section of Volume I dealing with the method of bounds in plasticity (I - §3.4.3.3).

4.2.2.1 Onset of plastification

A simple approach to the problem starts from the solutions already obtained for elasticity. We assume Tresca's conditions for plasticity: $\tau_1 = k = R_p/2$, where k and R_p are the yield strengths for shear and tension respectively.

For *normal contact of two cylinders* we showed (Fig. 4.15 and Eq. 4.58) that τ_1 has a maximum value of $0.30p_0$ at a point $0.78a$ below the contact surface; it follows that the limiting Hertz pressure corresponding to the onset of plasticity is

$$(p_0)_L = 3.33k = 1.67R_p \tag{4.74}$$

The corresponding normal limit load P is then found from the Hertz relation between p_0 and P (Eq. 4.57):

$$P_L = \pi R p_0^2/E^* = 2.79R_p^2(\pi R/E^*) \tag{4.75}$$

This shows that if, for a given geometry, what is required is a load which is as high as possible but below the plastic threshold in the sub-surface material, it is best to use a material having a high yield strength and a low elastic modulus. The second requirement - low modulus - follows from the relation between the modulus and the area of the contact surface.

This effect is still more marked in the case of two spherical solids. We know that here τ_1 has its maximum value $0.31p_0$ at $z = 0.48a$ (Exercise 3), from which we find

$$(p_0)_L = 3.23k = 1.61R_p \tag{4.76}$$

and

$$P_L = \pi^3R^2p_0^3/6E^{*2} = 4.17R_p^3(\pi^3R^2/6E^{*2}) \tag{4.77}$$

Returning to the case of the pair of cylinders, for *contact with sliding* we have seen that not only is the distribution of $(\tau_1)_{max}$ changed in the layers below the contact surface but that on the surface this stress can reach very high values if the coefficient of friction is high. In the contact plane we have simply $\tau_l = \mu p_0$, so that plastic conditions will begin to appear on the surface as soon as $p_0 = (p_0)_L = k/\mu = R_p/2\mu$. Below the surface the condition for plasticity will again be a function of μ (Fig. 4.18). To a first approximation, keeping the value $\tau_l = 0.30p_0$ obtained for normal contact, the condition for sub-surface plasticity is

$$p_0 = (p_0)_L = 1.67R_p = 3.33k \tag{4.78}$$

which is (4.74).

Thus taking together these limiting cases we have the curve in the $[(p_0)_L/k, \mu]$ plane shown in Fig. 4.27.

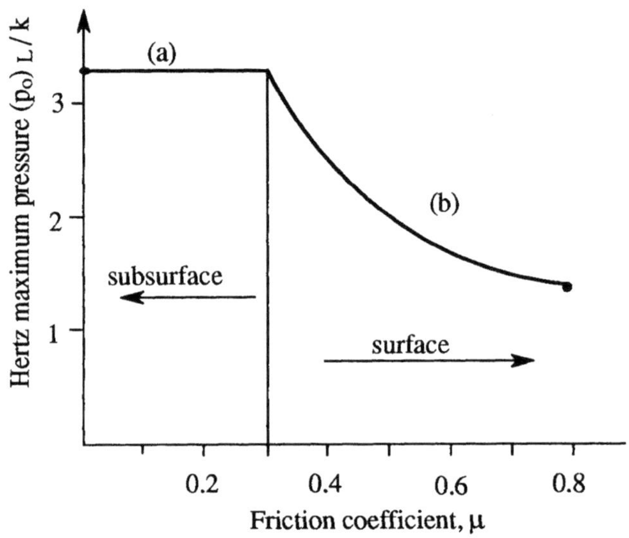

This simple argument enables us to give the orders of magnitude.

Fig. 4.28 gives the results of a fuller treatment, which shows that the effect of the increase in τ_l below the contact surface, resulting from slip, is to give a curvature to the part (a) of the curve of Fig. 4.27. For large values of μ the curve is essentially that of Fig. 4.27.

If the normal and tangential forces are applied repeatedly it is possible that cyclic plastic deformations will result, which will be sources of energy dissipation, possibly leading to damage. If this does not occur the structure is said to have *accommodated*: this is shown by the heavy lines in Fig. 4.28. More details are given by Johnson (1985).

4.2.2.2 Implications for damage due to contact

In certain mechanical assemblies it is possible that parts will be brought into contact repeatedly: for example, rolling mechanisms which have to support radial loads, for which the term *rolling fatigue* is sometimes used. Whilst, as in the case of fretting fatigue, not all damage reduces solely to considerations of a mechanical nature, the phenomenon of sub-surface damage in materials that are sensitive to plastic deformation is well known - for example, rolling fatigue in rails, under certain conditions.

Fig. 4.29 is an example of cracking in the head of a rail, in which it is seen that the crack starts below the surface. Whilst not everything can be explained by contact mechanics, and in this particular case residual stresses and the possible presence of inclusions may be important, the example shows the sub-surface effect of the load resulting from the contact.

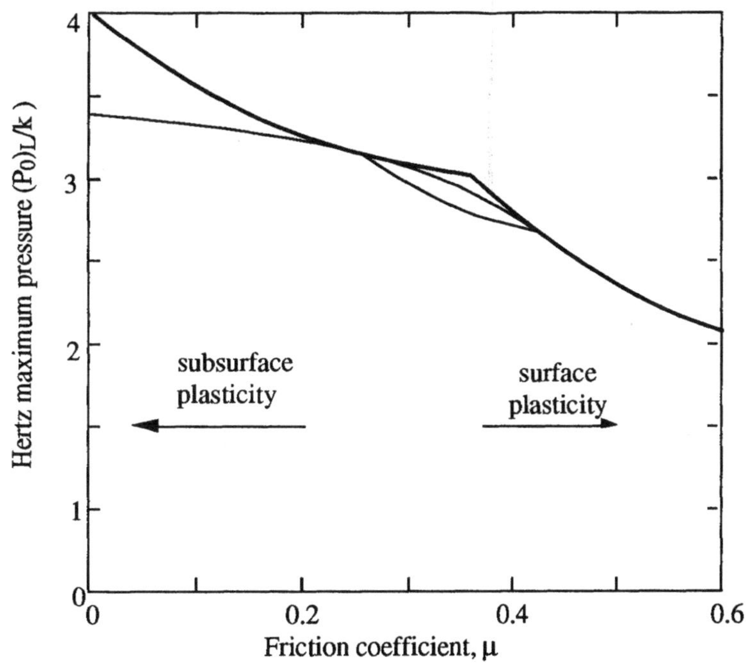

Fig. 4.28 Sliding with friction: maximum contact pressure corresponding to the onset of plasticity

Similarly, as we have said previously, in brittle materials such as glasses, ceramics and polymers below the vitreous transition temperature T_g fracture is due essentially to the maximum principal stress. Thus in normal contact we often see a fracture cone, whose base is the circumference of the circle of contact; the tensile stress $\sigma^*rr = -0.10p_0$ is the origin of this form of cracking. This result is used to measure the fracture toughness, cf. Lawn and Wilshaw (1975).

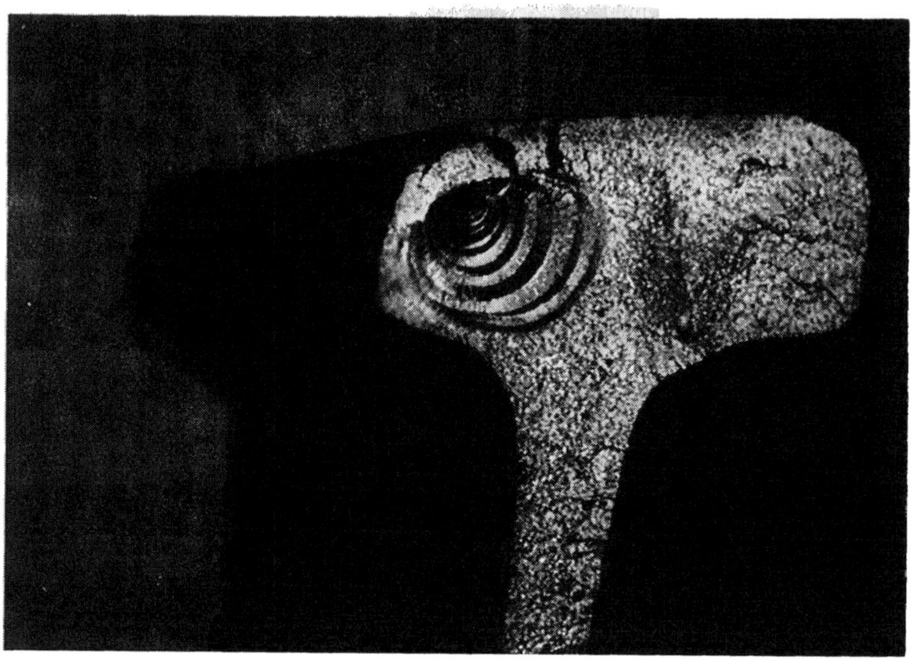

Fig. 4.29 Fatigue cracking in a rail, leading to brittle fracture.

4.3 FRICTION

4.3.1 Introduction

The practical importance of friction is obvious; it is the origin, for example, of the phenomena of wear which we shall be discussing later in this section. So far we have treated friction as a macroscopic phenomenon, in terms of the coefficient μ in the Coulomb law; we now study surface interactions on a smaller scale - nanometer to micrometer - so as to take into account the large variations in the coefficient of friction that are encountered. In this we shall deal only with dry friction. The presence of a lubricant between a pair of solids in relative sliding or rolling motion changes the coefficient, and especially the wear; study of these effects (of hydrodynamical lubrication in particular) makes use of the theory of fluid mechanics and is outside the scope of this chapter.

Friction and wear are still only incompletely understood; many aspects have to be taken into account - the topography and the modes of deformation and degradation of the solids on the scale of the surface roughness (micrometer), and the adhesion of the surfaces in contact. We shall look at these in turn before coming to the expression of the relevant laws.

4.3.2. The true area of contact: influence of surface topography and the nature of the contact

Real surfaces are never perfectly smooth; all have irregularities and asperities, of which Fig. 4.30 is typical, showing, diagrammatically the kind of surface that results from a careful machining operation such as turning with small passes. Such asperities remain even if the surface is then polished carefully.

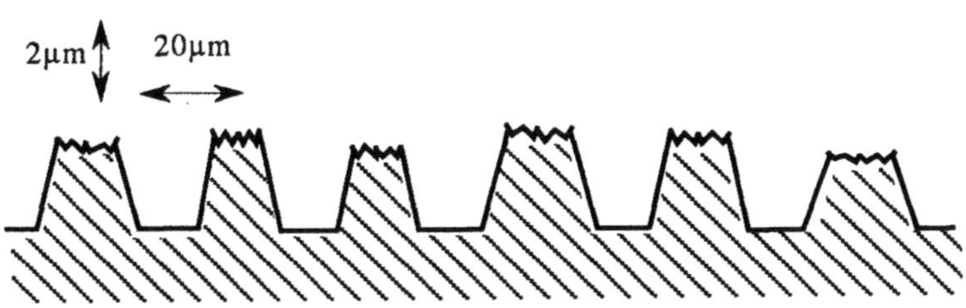

Fig. 4.30 Diagram illustrating roughness of a machined surface.

Because of these asperities, when a normal load P is applied the surfaces are in contact over a set of discrete zones, as indicated in Fig. 4.31; and the real area of contact A_r is some fraction of the apparent area A_a.

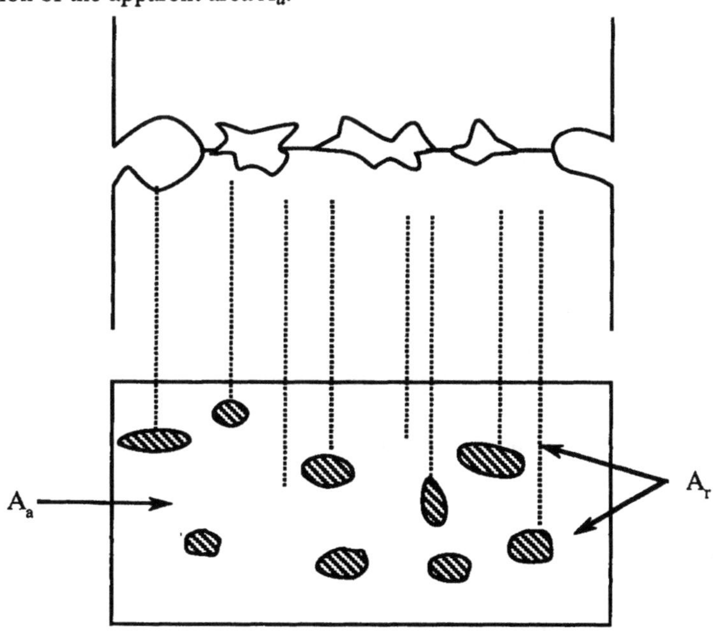

Fig. 4.31 Real (A_r) and apparent (A_a) areas of contact of two bodies.

The real area A_r is related to the microgeometry of the asperities. If we can assume that this is on the micron scale we can apply the Hertz theory, giving A_r as an increasing function of P; for contact between spheres it will be proportional to $P^{2/3}$ and between cylinders to P. General laws of this type have been proposed, that is

$$A_r = const. \times (P/E)^i, \text{ with } i \approx 1 \tag{4.79}$$

Purely elastic contact cannot be maintained for long on this scale, for the asperities will be deformed plastically. As we showed previously (§4.2.2.1), for a pair of spheres of radius R, if R_p is the yield strength the limit normal load is

$$P_L = const. \times (R_p^{\,3}/E^2)R^2 \tag{4.80}$$

showing that this limit load is reached the more easily the smaller the radius of curvature of the asperities.

Another criterion for plastification that is sometimes used involves the microgeometry of the surface irregularities If R_i is the mean radius of the asperities and h_i the standard deviation of their heights the dimensionless parameter Ψ is defined by

$$\Psi = (E/R_p)(h_i/R_i)^{1/2} \tag{4.81}$$

If $\Psi < 0.6$ the deformation of the asperities remains purely elastic and the real area of contact is proportional to the value given by the Hertz theory:

$$A_r \neq \pi a^2 = \pi\left(\tfrac{3}{4}PR_i/E\right)^{2/3} P^{2/3} \tag{4.82}$$

For $\Psi > 1$, which is the case in most situations, the asperities are deformed plastically and the real area of contact is

$$A_r = P/R_p \tag{4.83}$$

Since there is some punching R_p, strictly, can be greater here than the simple elastic limit - see, for example, the treatment in Volume I, §3.4.3.5 (Fig. 3.112) of the problem of a flat punch driven into a semi-infinite body.

For $0.6 < \Psi < 1$ the state is one of transition between purely elastic and purely plastic microcontacts.

The occurrence of plastic deformation, resulting from activation of dislocation movements, implies dissipation of energy. The mechanical energy provided by the load in forming the real contact area is not transferred as a whole across the interface, as it is in the idealised Hertz case of perfect elasticity; other mechanisms can intervene to bring about energy dissipation. During the two transient contact states, those of loading (formation of the contact) and unloading (subsequent separation), changes occur in the distribution of pressure which can entail microsliding and possibly (in the case of rapid loading, or of impact) propagation of elastic waves . As we saw previously, microsliding can occur in Hertz contact if the two bodies have significantly different elastic properties.

The repeated relative movements of two bodies rubbing against each other will result in wear. The materials are strongly work-hardened over a thin layer, and debris will be detached if the movement is great enough (Fig. 4.26); contact is then maintained through the intermediary of a bed of debris, which forms the "third body" often introduced in tribological analysis.

4.3.3. Adhesion

As well as mechanical, there are interactions of molecular type; the relative approach of the atoms of two solids, helped by plastic deformation at the microcontacts, enables the forces of molecular attraction to act across the interface.

The origin of these forces is the sudden interruption of the periodic structure of the lattice at a free surface, releasing bonds. This results in interactions of the types that contribute to the cohesion of solids - *metallic* (mobility of free electrons), *ionic* (electrostatic attraction) and *covalent* (sharing of valency electrons). These are all *short-range forces* (< *1 nm*) and depend on the nature of the material. There are also the *Van der Waals forces* (electromagnetic attraction, due to fluctuations in the electron distribution), which although long-range and independent of the type of bond are relatively very weak except in the case of polymers, where they are the only forces acting. The short-range forces, which can be expressed in terms of surface energy, are sufficient to constitute permanent bonds which are strong enough to create adhesive junctions.

The only way to investigate experimentally these adhesive interactions is to break the bonds and measure the force F_a needed to separate the two surfaces, having first removed the load P which brought them into contact. The result is expressed in terms of a coefficient of adhesion μ_a defined by

$$\mu_a = F_a/P \tag{4.84}$$

For two perfectly smooth surfaces of the same material, F_a is theoretically equal to the force needed to fracture the material in bulk. However, the adhesion is strongly affected by the reactions of the surfaces with the environment, both physical (absorption) and chemical (chemisorption - formation of an oxide film), so much so that these adhesive interactions are best seen in especially clean metallic single crystals under vacuum.

The adhesion coefficients thus measured are strongly anisotropic, since they are influenced by the structure as well as by the crystallographic orientation of the material. This aspect is related to the number of slip systems that can be activated by the normal load, that is, to the possibilities for plastic deformation which can increase significantly the size of the junctions where adhesive forces can develop.

Although adhesion results essentially from molecular forces limited to the interface, the coefficient is affected by the mechanical properties of the material. Thus it is found to decrease with increasing hardness and elastic moduli; high hardness, in fact, results in a reduced area of contact, and a high elastic modulus will delay the onset of plastic deformation and limit the area of contact over which the adhesive forces can act. On the other hand, a reduction in surface energy seems to imply a lower surface reactivity, which is unfavourable to the establishment of multiple bonds across the interface.

Finally, the mechanical behaviour of the material, not only determines, through the mode of elastic or plastic deformation, the extent of the junctions at the moment of establishing the contact; it also perturbs most of these and breaks them, at the moment of separation, where the elastic stresses play a role.

4.3.4 Models and laws of friction

Resistance to sliding results from interactions between the asperities, where phenomena associated with energy dissipation are observed during the various stages of the formation and separation of the microcontacts.

Models of friction are based on the principle that the macroscopic friction force is the sum of all the microscopic forces which act on the individual microcontacts; several approximations have to be made in view of the insufficiency of the available data on the surface properties of solids.

On the scale of the microgeometry of the contacts the sliding can be divided into three main stages (Fig. 4.32). In the first the microcontacts are deformed, first elastically and then plastically, with interpenetration of the asperities; this leads into the second stage, in which adhesive bonds are developed. In the third stage the junctions are sheared, followed by elastic unloading of the asperity.

These various contributions to the friction result from mechanisms of two main types, of deformation (elastic and plastic) and of adhesion, respectively. Although in general they are not independent, in certain cases one of them can dominate and as a first approximation the others can be neglected. In the next paragraph we consider the simple case of geometrical interaction of two asperities, and in §4.3.4.2 the adhesion between asperities.

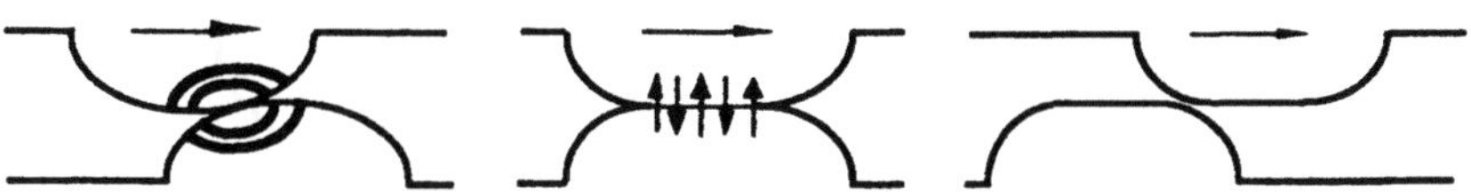

STAGE 1	STAGE 2	STAGE 3
(formation of contact) elastic strain plastic strain interpenetration	(development of adhesive bonds)	(separation of contact boundary shear elastic release)

Fig. 4.32 Elementary slip of an asperity.

4.3.4.1 Geometrical and mechanical interactions of asperities

The geometrical interaction of surface asperities during a displacement has been suggested as the source of the resistance which determines the work needed to overcome these obstacles. This implies the existence of some process strongly dependent on the roughness: thus an asperity with mean inclination θ would contribute an amount proportional to tan θ. Doubt is cast on the validity of this explanation by the fact that positive and negative angles of contact are equally likely, thus cancelling the respective contributions.

There is another type of interaction when the surfaces sliding with respect to each other have very different hardnesses: the softer is scratched by the asperities of the harder penetrating its surface, as in Fig 4.33, where A is the softer and B the harder. This is the cause of *wear by abrasion.*

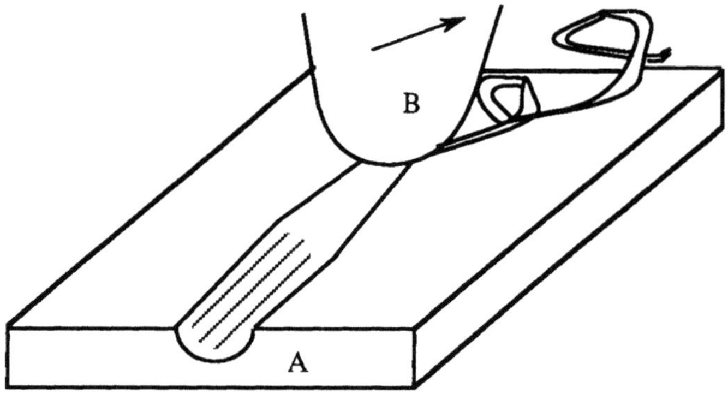

Fig. 4.33 Wear by abrasion.

Fig. 4.34 Section *AB* of Fig. 4.33, and projection *Ap* of the contact area A_F for a conical asperity

From a consideration of a particular geometry, such as that of Fig. 4.34 for a conical asperity, we can convince ourselves that the coefficient of friction is related only to the topographical details of the surface. The equilibrium conditions for the normal and tangential forces are

$$P = A_p R_p,\ Q = A_F R_p \tag{4.85}$$

from which $\mu = A_F/A_p$. For Fig. 4.34 this gives $\mu = 2/\pi \tan\theta$.

4.3.4.2 Physico-chemical interactions; adhesion of asperities

The resistance to sliding corresponds to the forces needed to shear the adhesive boundaries, especially in the case of a pair of metallic surfaces of only moderate hardness. We can write

$$F = A_r \, \tau_A \tag{4.86}$$

where τ_A is the mean value of the shear stresses over the real area of contact, which will not differ greatly from the resistance to shear τ_Y in the bulk of the softer material. The coefficient of friction is then

$$\mu = Q/P = A_r \tau_Y / A_r Y, \text{ that is } \mu = \tau_Y / Y \tag{4.87}$$

The stresses τ_Y and Y are the resistances to plastic deformation in shear ($\tau_Y = k \approx R_p/2$) and to compression respectively (Y is approximately the yield strength, at most $3R_p$). This gives values for μ of the order of *0.15* to *0.5 independent of the applied load and of the roughness of the surfaces in contact.* The validity of these conclusions - the range and the independence - has been checked in many cases, and they are often advanced as support for the theory of friction by adhesion.

However, as Table 4.1 shows, for very clean metallic surfaces μ can reach values of over *4*, whilst a lubricant can reduce it to values below *0.04*. Further, this simple model does not take into account contamination of the surfaces, whilst the environment can affect the formation of the junctions.

Material	μ
Pure metals *in vacuo*	*> 5*
in dry air	*0.8 - 2*
in moist air	*0.1 - 0.5*
Steel on other metal (lead, bronze)	*0.1 - 0.5*
Steel on ceramics (diamond, glass)	*0.1 - 0.5*
Ceramics on ceramics (e.g. carbides)	*0.05 - 0.5*
Polymers on polymers	*0.05 - 1*
Metals and ceramics on polymers (PE, PTFE, PVC)	*0.04 - 0.5*
Lubricants (MoS_2, graphite)	*0.05 - 0.2*
Hydrodynamic lubrication	*0.001 - 0.005*

Table 4.1 Coefficients of friction.

The mechanisms which cause deterioration of the surfaces of a pair of solids in contact are often very complex, and the terms used to describe them are correspondingly imprecise or ambiguous. Two types of classification have been suggested: one in terms of the four possible types of relative movement of the solids - rolling, sliding, impact and fretting (oscillation) - the other in terms of the mechanisms of wear - abrasion, adhesion, tribochemical wear and fretting fatigue. The first relates to type of external loading, in the second the interest focusses on the degradation mechanisms on the microscopic scale. The two are complementary since in practice each type of relative motion is associated with a particular one or group of the wear mechanisms, according to the loading conditions or the environment.

We shall consider only the case of sliding of metallic surfaces; here the wear is due mainly to *abrasion* and *adhesion*, for which we describe the elementary mechanisms and their modeling. We make only brief reference to the other types of wear; fretting fatigue has already been discussed, in §4.2.1.5.

4.4.1 Wear due to abrasion

This is an essentially mechanical mode, resulting from one surface penetrating the other.

Two situations can lead to abrasive wear: one material being significantly harder than the other, or the presence of hard, angular particles between the surfaces (the "third body"), either introduced deliberately - e.g. for polishing - or the result of chemical reactions in the wear debris (tribochemical origin). Material is removed and the deterioration of the surface is shown by the polishing effect or the creation of scratches or furrows, accompanied or not by production of fine debris, shavings or flakes.

We can form an estimate of the volume of material removed by taking the asperities of the harder material to be conical; with the notation of Fig. 4.35 this gives $P \approx \frac{1}{4}\pi d^2 R_p$ and the volume dV removed in a displacement dL is

$$dV = \tfrac{1}{2}d^2 \tan\theta \, dL = (2P/\pi R_p) \tan\theta \, dL \qquad (4.88\text{-}89)$$

From this we can find the total amount removed by all the asperities over a distance L. The result shows that the *wear coefficient* $K = (R_p/P)\, dV/dL$ depends only on the shape of the asperities: for the case we are considering it is $(2/\pi) \tan\theta$. Typically, the order of the numerical value is between 10^{-2} and 10^{-1}. The result shows also that the volume removed varies linearly with the applied load and the distance traversed, and inversely with the hardness of the soft material, which is the only mechanical property involved.

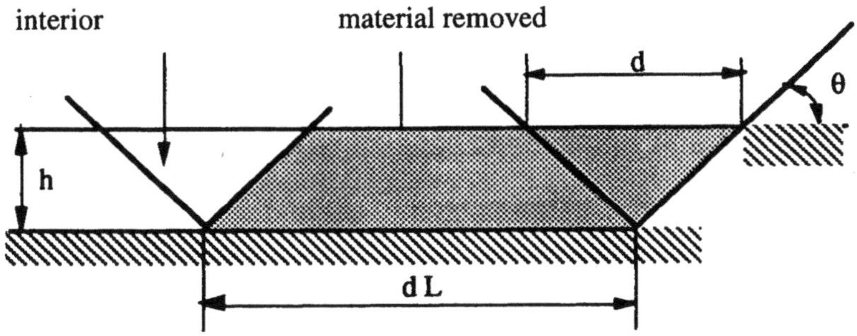

Fig. 4.35 Abrasive wear: estimation of volume of material removed by a conical asperity

However, the true dependence of the volume removed on the distance traversed seems to be more complex than this. Such a linear relation is observed when the abrasive material is always meeting a fresh surface ("open circuit") of which the properties do not change; but in repeated passages ("closed circuit") debris tends to accumulate between the asperities and so to insulate the surface from the abrader. This effect is usually expressed by a relation of the form

$$V = V_i(1 - e^{-\frac{L}{\delta}}) \qquad (4.90)$$

where V_i is the total volume that would be removed if the sliding were maintained indefinitely, and δ is a constant.

4.4.2 Wear by adhesion

This occurs when strong bonds are formed at the junctions, as predicted by the theory of adhesion; its manifestation is the transfer of material to the surface of the attacking body.

Junctions are successively created and broken during the sliding;. Usually a junction is weaker than the material itself and is broken at the interface; the wear is then very slight and the particles removed are of sub-micron size. However, it can happen that a junction is more resistant, for example when it is not parallel to the direction of the sliding; shear is then generated at a short distance from the interface (Fig. 4.36) and the particles removed are of microscopic size.

With metallic materials this can result in heavy wear, and possibly in seizure of the moving parts. The particles are taken preferentially from the material of lower resistance to shear and transferred to that of higher; however, small fragments of the harder material may be seen amongst the debris, suggesting local regions of lower resistance, as in Fig. 4.36 (b).

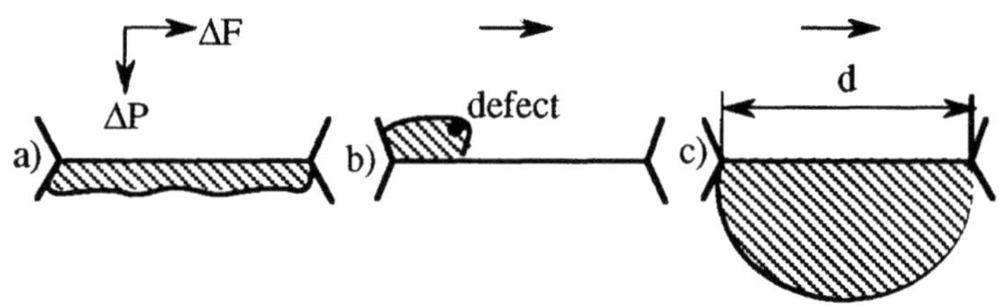

Fig. 4.36 (a) Breaking of a junction in the weaker material, close to the interface
(b) Breaking in the stronger material in the neighbourhood of a defect
(c) Modeling of the material removed.

An estimate for the volume of material removed for each junction formed can be derived from a simple model. The true area of contact at any given instant is $A_r = P/Y$; if the total number of junctions is j and we can assume them all to be circles of diameter d, we have

$$A_r = P/R_p = j\pi d^2/4, \text{ giving } j = 4P/\pi d^2 R_p \tag{4.91}$$

We next assume that each junction remains intact for a slip distance d, when it is broken and afterwards reformed, over a succession of locations where junctions are formed.

In a slip distance ΔL there will be many intervals of length d,. which is equivalent to saying that over a distance ΔL a junction will be formed $\Delta L/d$ times. Therefore the total number of junctions produced over a distance ΔL is $j\Delta L/d$.

If we assume that for each junction formed there is a constant probability k_{ad} that an adhesive particle is created, and that this particle is a hemisphere of diameter d (the same as that of the junction) (Fig. 4.36c) we find, on passing to the limit,

$$dV = k_{ad}(\pi d^3/12)(j/d)dL \tag{4.92}$$

and substituting for j from (4.91)

$$dV = k_{ad}(P/3R_p)dL \tag{4.93}$$

This has exactly the same form as that for abrasive wear, Equation 4.89; from which it follows that in any given case it will be difficult to decide, on the basis of these relations alone, which mechanism is operating.

Table 4.2 gives typical values for the wear coefficient, showing that this can vary considerably according to the materials concerned. Further, a lubricant can reduce the value by several orders of magnitude.

Conditions	$K\,(10^{-3})$
Zinc on zinc	*160.*
Copper on copper	*32.*
Stainless steel on stainless steel	*21.*
Copper on carbon steel	*1.5*
Bakelite on bakelite	*0.02*

Table 4.2 Wear coefficient $K = (\Delta V/\Delta L)/(P/3R_p)$

4.4.3 Other modes of wear

We discussed wear by *fretting fatigue* in §4.2.1.5; this is manifested by cracks which can result in large fragments of the worn material being broken off. As we saw there, it is difficult to separate this from environmental problems.

Physico-chemical aspects are equally relevant in the remaining mode, *tribo-chemical wear* (also referred to as *corrosion wear*). Here damage is dominated by the products of chemical reactions between the elements in contact and/or between these and the environment. Reactive films may form, and become detached when they reach a certain thickness; and oxidation of the surfaces and of the debris resulting from wear can alter the surface properties. It is difficult to model this type of damage quantitatively.

To summarise, in the behaviour of materials under wear the transfer and adhesion are determined on the one hand by their capacity for plastic deformation, with the consequential creation of a large real contact area, and on the other by their ability to form strong bonds across the interface. The friction force is more-or-less independent of the relative velocity, is proportional to the normal pressure and is almost independent of the apparent area and the roughness of the contact surface.

4.5 MATERIALS FOR USE IN CONDITIONS OF FRICTION AND WEAR

We are concerned here only with materials for bearings. Most of the time bearings are being lubricated with oil or grease, although some polymer-based materials or composites incorporating a solid lubricant are used dry. Ceramic bearings are used for some specific applications.

The choice of material for any particular application will involve a compromise between:

- "adaptability": if the alignment is poor the material should not be too hard; play in bearings is usually of the order of *0.025 mm*,

- resistance to fretting fatigue,

- resistance to erosion in the case of hydrodynamically-lubricated bearings.

4.5.1 Metallic materials

Very often the best results are obtained by depositing an *0.2 - 0.5 mm* layer of one of the following materials on a low-carbon steel substrate; this is the solution often adopted for engine bearings.

(a) *Alloys of lead or tin*

Alloys of Pb, Sb and Sn are often used, for example 10% Sb, 6% Sn; or of Sn, Sb and Cu, for example 8% Sb, 3% Cu. In the first of these the microstructure consists of Sb-Sn compounds and a pseudo-eutectic Pb-Sb-Sn; in the second, a hard compound Cu_6Sn_5 and a relatively soft Sn-Sb solid solution.

(b) *Cuprous alloys*

These are either Cu-Pb alloys or bronzes (Cu-Al) containing lead. In the first there are two practically distinct phases, one of copper and the other of lead, the solubility of either in the other being very low. As with lead- or tin-based alloys, the same principle applies, that of mixing a soft and a hard phase.

(c) *Aluminum alloys.*

These much-used alloys have the advantage of good resistance to corrosion. Several are available and two large families, Al-Sn and Al-Pb are particularly popular.

(d) *Porous bronzes.*

This much-used material is made by powder metallurgy. The process is such as to leave a residual porosity of *10* to *25%*, the pores being interconnected so that lubricating oil can circulate through the channels.

4.5.2 Polymers

These give a low coefficient of friction. They are often used in the form of composites consisting of asbestos fibers or cloths impregnated with a thermo-hardening polymer such as a polyester resin. A composite often used for dry bearings is porous bronze impregnated with a mixture of PTFE and lead.

4.5.3 Ceramics

An excellent surface can be obtained by using alumina or silicon nitride; moreover, these materials can be used in conditions of very high temperature or, as in nuclear reactors, irradiation. A great deal of research is now being directed at thermo-mechanical ceramics, with the aim of producing materials with a very high resistance to frictional wear.

EXERCISES

1. Show that the Airy function $\Phi = Ar\theta\sin\theta$ satisfies the biharmonic equation $\Delta^2 \Phi = 0$, where

$$\Delta^2\Phi = \left(\frac{\partial^2}{\partial r^2} + \frac{1}{r}\frac{\partial}{\partial r} + \frac{1}{r^2}\frac{\partial^2}{\partial \theta^2}\right)\left(\frac{\partial^2\Phi}{\partial r^2} + \frac{1}{r}\frac{\partial\Phi}{\partial r} + \frac{1}{r^2}\frac{\partial^2\Phi}{\partial \theta^2}\right)$$

2. Show that for a constant normal pressure distributed over a band, the principal stresses are of the form $\sigma_{1,2} = -(P/\pi)[\alpha \pm \sin\alpha]$, where α is defined in Fig. 4.7.

3. Show that for a parabolic pressure distribution $p = p_0(1 - x^2/a^2)^{1/2}$ applied over a circular surface the shear stress τ_1 has a maximum value $(\tau_1)_{max} = 0.31p_0$ at a distance $z = 0.48a$ below the contact surface.

4. Find the contact area corresponding to a locomotive wheel on a rail.

5. The typical rate of wear for piston rings in a car engine is $30\mu m$ per $10{,}000\ km$. Justifying any assumptions that you make, calculate the wear factor K. Is your value reasonable, in view of the values given in Table 4.2?

MECHANICAL BEHAVIOUR OF MATERIALS: Volume II, Tables

Table 1. Self-diffusion in metals (from Peterson (1968)).

Element	*Q* (*kcal/mole*)	D_0 (cm^2s^{-1})
Aluminium	*28.750*	*0.035*
Antimony ‖ c	*48,000*	*56.0*
⊥ c	*35,800*	*0.1*
Beryllium ‖ c	*39,400*	*0.62*
⊥ c	*37,600*	*0.52*
Cadmium ‖ c	*18,200*	*0.05*
⊥ c	*19,100*	*0.10*
Chromium	*73,700*	*0.20*
Cobalt (paramagnetic)	*67,700*	*0.83*
Copper	*49,560*	*0.62*
Germanium	*68,500*	*7.8*
Gold	*41,700*	*0.091*
Hafnium	*38,700*	*0.0012*
Indium ‖ c	*18,700*	*2.7*
⊥ c	*18,700*	*3.7*
Iron α (paramagnetic)	*57,300*	*2.0*
Iron γ	*67860*	*0.49*
	65,000	*0.18*
Iron δ	*57,000*	*1.9*
Lead	*26,060*	*1.37*
Lithium	*13,490*	*0.39*
Magnesium ‖ c	*32,200*	*1.0*
c	*32,500*	*1.5*
Molybdenum	*92,200*	*0.1*
Nickel	*68,000*	*1.9*
Niobium	*96,000*	*1.1*
Palladium	*63,600*	*0.21*
Platimum	*68,200*	*0.33*
Potassium	*9,750*	*0.31*
Silicon	*110,000*	*1800*
Silver	*44,270*	*0.44*
Sodium	*10,090*	*0.145*

Table 1 (continued)

Element	*Q(kcal/mole)*	$D_0(cm^2s^{-1})$
Tantalum	*98,700*	*0.124*
Thallium α ‖ c	*22,900*	*0.4*
⊥ c	*22,600*	*0.4*
Thallium β	*20,000*	*0.7*
Thorium α	*76,600*	*1.2*
Tin β ‖ c	*25,600*	*7.7*
⊥ c	*25,100*	*10.7*
Titanium β	*see below*	
Tungsten	*153,000*	*42.8*
Uranium γ	*26,500*	*0.00112*
Vanadium 880-1360°C	*73,650*	*0.36*
1360-1830°C	*94,140*	*214.0*
Zinc ‖ c	*21,900*	*0.13*
⊥ c	*23,000*	*0.18*
Zirconium	*see below*	

Titanium β: $D=3.58.10^{-4}exp(-31,200/RT)+1.09\exp(-60,000/RT)$

Zirconium β: $D=8.5.10^{-5}\exp(-27,700/RT)+1.34\exp(-65,200/RT)$

Table 2. Heterodiffusion in Nickel (from Adda&Philibert (1966)

Addition Element	Atomic compo-sition%	C_D atomic %	Tracer	Temperature °C	D cm^2s^{-1}	D_0 cm^2s^{-1}	Q kcal/ mole
Al		0.7		1099-1280		1.87	64
		0.08		800-920		1.1	59.5
Au	0	ε	Au^{198}	900-1100		2	65
		2		850-975		$1.8.10^{1}$	85.2
	0	ε	C^{14}	600-900		0.08	33
	0	ε	C^{14}			0.13	34.5
C	0.48	ε	C^{14}	600-1400		0.12	32.9
				700-1300		2.49	40
		0.17–0.12		700-1300		2.48	40
Co	0	ε	Co^{60}	1193	$1.71.10^{-10}$		66.3
				1095	$3.42.10^{-11}$		
	0	ε	Co^{60}	748-1192		0.75	64.7
	0	ε	Co^{60}	1154-1373		1.11	64.9
Co	0	ε	Co^{60}	900-1250		1.46	68.3
	5.2	ε	C^{14}	600-900		0.4	37
	0	ε	Cr^{51}	900	1.10^{-12}		
				1000	$9. 10^{-12}$		66
				1100	7.10^{-11}		
Cr	0	ε	Cr^{51}	1100-1268		1.1	65.1
	0	ε	Cr^{51}	600-900		0.03	40.8
	0	ε	Cr^{51}	350-600		$5.45. 10^{-9}$	13.7
	0.84	ε	C^{14}	600-900		0.15	34
	5.2	ε	C^{14}	600-900		0.3	37

Table 2. (continued)

Addition Element	Atomic compo-sition%	C_D *atomic %*	Tracer	Temperature *°C*	D cm^2s^{-1}	D_0 cm^2s^{-1}	Q *kcal/mole*
Cu	0	ε	Cu^{64}	1054-1359		0.57	61.7
	0	ε	Cu^{64}	800-1050		0.724	61
Fe	0	ε	Fe^{59}	950-1250		9.10^{-3}	50.7
	0	ε	Fe^{59}	950-1130		$8.4.10^{-3}$	51
	0	ε	$Fe^{55\text{-}59}$	900-1250		0.8	61
		0.1		1100-1300		0.34	67.5
Mo	0	ε	Mo^{90}	900-1200		$1.6.10^{-2}$	51
	3.14	ε	Co^{60}	1050-1250		0.078	66
	1.8Mo	ε	C^{14}	600-900		1	38
	0.07C	ε					
		1		1150-1400		3	68.9
W	0	ε	W^{185}	1100-1275		1.13	71
	0	ε	W^{185}	1100-1295		2	71.5
	1.7	ε	W^{185}	1100-1295		2.2	73.1
	5.3	ε	W^{185}	1100-1295		17	80.5
	9.2	ε	W^{185}	1100-1295		1.4	74.5

Table 3. Self-diffusion coefficients instrongly disoriented boundaries given by Gibbs&Harris (1969)

Element	Structure	$D_{0j}\ cm^2s^{-1}$	Q_j kcal/mole
Ag	FCC	0.03-0.12	20.2-21.5
		0.024	20.3
		0.063	18.9
Ni	FCC	0.018	26
		0.01	26.6
		0.07	27.4
			28.2
Pb	FCC	1.6	15.7
			4.7
Fe γ	FCC	$2.1.10^{-4}$	30.6
		3-5	39
		1.5	38.0
Fe α	BCC		45
		13	40
		0.014	33.7
		22.4	41.5
Cr	BCC		46
W	BCC	6.7	92
U γ	BCC	0.021	42.7
Zn	CPH	0.22-0.38	14.3-14.6
			12.3
Cd	CPH	1	13
Co	CPH	4	39
Te	hexagonal	0.22	20
Sn	quad.		9.4
		0.06	9.55
Sb	rhombohedral	3-6	22-23
U α	orthorhombic	320	44.3

GLOSSARY

Adiabatic shearing Localised shear strain, due to local softening as a result of overheating.

Blunting Increase in the CTOD (q.v.) due to plastic strain

Cavity A hole, initiated at an inhomogeneity or by vacancies condensing on grain boundary.

Cleavage Mode I fracture along crystallographic planes.

Corrosion fatigue Damage in which fatigue and corrosion are combined.

Crack hardening/softening. Increase/decrease of limit load due to cyclic loading.

Crack tip opening displacement (CTOD) Distance between the edges of a crack at its extremity.

Craze Microcracks in a polymer, bridged by the fibres.

Creep Deformation at high temperature under constant load.

CTOD Crack tip opening displacement (q.v.).

Dimples Depressions in a fracture surface, produced by the coalescence of cavities.

Dislocation climb Movement of a dislocation out of its slip plane.

Dynamic ageing (Portevin-Le Chatelier effect) Interaction between mobile dislocations and impurity atoms.

Embrittlement Lowering of resistance to fracture.

Endurance limit (for cyclic stressing). Amplitude of stress below which the material will sustain a very large number (10^6 to 10^8) of cycles without fatigue failure.

Energy release rate (for a crack). Change of elastic and potential energy per unit increase in the area of the crack.

Fatigue Damage under cyclic loading.

Fracture energy Energy absorbed in the breaking of the test sample in an impact test.

Fracture toughness Resistance to fracture.

Fretting fatigue Damage resulting from the relative movement of two solids in contact.

Glide Displacement parallel to a crystallographic plane.

Grain boundary. Boundary between two grains.

Intergranular cracking Fracture along a grain boundary.

Intergranular segregation Segregation of impurities on grain boundaries.

Limit load The load level at which plastic flow occurs in an unconsolidated material.

Low-cycle fatigue Fatigue in the plastic domain, leading to fracture after a small number of stress cycles.

Necking Local change in the cross section, with the deformation becoming unstable.

Nil ductility temperature (NDT) Temperature at which the ductility falls to zero.

Persistent slip band (PSB) Slip band formed by fatigue and persisting after polishing.

Propagation threshold Value of load below which cracks do not propagate.

Recovery Reduction in the internal work-hardening.

River Lines appearing on the surfaces exposed by cleavage fracture.

Stress corrosion cracking (SCC). Damage produced by the combination of mechanical stress and a harsh environment.

Stress intensity factor (SIF) Parameter expressing the strength of the singularity in the stresses and elastic strains at the tip of a crack.

Stress relaxation Fall in the stress in the course of a test in which the strain is prescribed.

Superplasticity Ability of a material to sustain a very large elongation.

Tongue A particular feature formed at the intersection of a cleavage plane and a mechanical twin.

Weight functions Functions which enable stress intensity factors to be calculated.

NOTATION

Scalar quantities

a	dimensions of crack or notch
	radius of dislocation loop
	radius of col
	Manson-Coffin exponent
a_c	critical radius of a crack
a_u	critical radius of a crack under stress σ_u
b	lattice spacing
	cavity size
	Basquin exponent
c	surface length of a crack
	surface concentration of particles
	concentration of solute atoms
c_0	thickness of cementite platelet
d	grain size
d_f	fibre diameter
f	volume fraction
	force
	plastic surface flow
f_G	volume fraction of granules
f_v	volume fraction of vacancies
g	potential of vacancies
h	thickness of a beam
	consolidation level
	distance between pile-ups
h_i	standard deviation of height of asperities
i	(mathematical) imaginary unit ($i^2 = -1$)
j	number of junctions
k	incompressibility modulus
	yield strength in shear
	Boltzmann's constant
	stiffness
k_D	incompressibility modulus for damaged material
k_y	Petch factor
k_{ad}	probability of creation of an adhesive particle
l	crack length
	length of a dislocation segment in a network of dislocations
l_c	critical length for fibre fracture
l_b	distance between cavities
$\bar{l}$	mean depth of cracks

m	exponent for dislocation velocity
	Weibull exponent
	exponent for secondary stage fatigue
	exponent for sensitivity to velocity
n	number of dislocations
	number of cracks per unit volume
	Norton-Hoff exponent
	work-hardening exponent
n'	cyclic work-hardening exponent
p	particle size
	normal surface stress
p_m	mean contact pressure
q	tangential surface stress
r	distance to the end of a pile-up or a crack
	radius of dislocation loop
	distance from the axis of a test piece
	cavity radius
	velocity of recovery
s_{ij}	component of deviatoric stress tensor
t	time
	line tension
t_R	time to fracture
u_D	energy of a dislocation loop
u_{AA}, u_{BB}, u_{AB}	bond energies betwen atoms A, B, AB
$\dot{u}$	velocity of displacement
$\dot{u}_r, \dot{u}_\theta, \dot{u}_z$	velocity components in cylindrical coordinates
v	dislocation velocity
	velocity of propagation of a crack
$\bar{v}$	mean dislocation velocity
v_a	activation volume
v_c	crack velocity
	activation velocity
v_t	activation volume
v_μ	velocity of a dislocation under stress μ
w	breadth of a beam
x	cartesian coordinate
y	distance to the neutral axis of a beam
z	(mathematical) complex variable
A	area of a crack
A_j	surface fraction of cavities on a grain boundary
A_{jc}	critical surface fraction of cavities on a grain boundary
A_{j0}	initial surface fraction of cavities on a grain boundary
A_r	area of true contact
A^*	activation area
B	breadth of a beam (cf. w)
C	compliance

	stiffness matrix
C_D	stiffness matrix for a damaged material
C_v	volume fraction of inclusions
C_{MG}	Monkman-Grant constant
C_0	equilibrium concentration of vacancies
C_S	concentration of vacancies
D	damage parameter
	extension at fracture
	self-diffusion coefficient
D_{ij}	component of damage tensor
D_t	parameter for damage in tension
D_c	parameter for damage in compression
D_j	inter-granular diffusion coefficient
D_s	surface diffusion coefficient
D_v	self-diffusion coefficient
D_L	vacancy diffusion coefficient
E	Young's modulus
	elastic energy
E_c	energy variation at the initiation of cleavage
E_D	Young's modulus for a damaged material
E_p	tangent modulus
$\dot{E}_r, \dot{E}_\theta, \dot{E}_z$	macroscopic deformation velocity components in cylindrical coordinates
$E(k)$	elliptic integral
E^*	effective modulus
G	rate of restitution of elastic energy
G_c	toughness, fracture energy
G_{IC}	toughness
H_A^{sub}, H_B^{sub}	sublimation enthalpies for A, B
H	height of a cylinder
I	moment of inertia
I_N	normalisation factor
J	flux of vacancies
	Rice-Cherepanov integral
K	stress intensity factor
K_I, K_{II}, K_{III}	stress intensity factors for modes I, II, III
K_T	stress concentration factor
K_σ	stress concentration factor
K_ε	strain concentration factor
K_{IC}	toughness
K_{max}	maximum stress concentration factor
K_{opn}	stress concentration factor at the start of opening of a crack

L mean depth of cracks
length of a beam
distance between cavities
L_0 mean distance between cavities
L_p distance between cavities
M moment
bending moment
M_L limiting moment
M_R fracture moment
M_x, M_y rolling moments
M_z spin moment
M_0 instant of start of plastification
N number (elements, test items, cycles, cavities, sources per unit volume)
N_R number of test items fractured
number of cycles to fracture
$\dot{N}$ rate of germination of cavities
N_m number of activation sites
P normal force
P_R probability of fracture
fracture load
P_0 probability of elementary fracture
P_{GY} plastification load
P_L limit of load without work-hardening
$\wp$ potential energy
Q tangential force
activation energy
Q_0 self-diffusion energy
Q_j inter-granular diffusion energy
R radius of a cavity or a notch
radius of curvature
dimension of plastic zone
R^* critical radius of a cavity
R', R'' principal radii of curvature
R_0 core radius, non-dimensional
initial size of cavity
R_c critical radius for growth of a cavity
R_{ci} characteristic ratio for inter-granular cleavage
R_D dimension of Dugdale plastic zone
Re (mathematical) real part (of complex quantity)
R_e equivalent radius of curvature
R_f fibre fracture stress
R_i mean radius of asperities
R_m load at fracture
R_p yield strength for flow
R_{pm} yield strength of matrix material

S	surface energy, non-dimensional
	area of a test piece
	length of a craze
	area of a cavity
S	area
S^*	damaged area
S_{Dijkl}	compliance matrix for a damaged material
S_j	area of section of a cavity cut by a grain boundary
T	absolute temperature (°K)
T_D	temperature of zero ductility
T_β	temperature of β-transformation
T_f	temperature (absolute) of fusion
T	(time) period
U	blunting energy, non-dimensional
V	volume, volume of a cavity
$\dot{V}$	rate of change of volume
V_0	volume element
V_c	velocity of crazing
V_{eff}	Weibull effective volume
W	elastic energy
	breadth/width of a plate
$\dot{W}$	plastic power dissipation
W_c	kinetic energy
Y	thermodynamic damage force
Y_c	yield strength in compression
Y_t	yield strength in tension
Y_t^0, Y_c^0	thermodynamic damage force thresholds for tension and compression
Y_m	work-hardening variables
z	(mathematical) complex variable
Z	necking
	coordination number

α	coefficient of proportionality
β	angular factor
	coefficient of interaction between cavities
β'	angular factor
δ	crack width
δ_j	thickness of diffusion layer along a grain boundary
δ_s	thickness of surface diffusion layer
ε	strain
$\dot{\varepsilon}$	velocity of strain
ε_{eq}	equivalent strain
$\dot{\varepsilon}_{eq}$	velocity of equivalent strain
ε_{ij}	component of strain tensor

$\dot{\varepsilon}_{ij}$	component of strain velocity tensor
ε_j	intergranular strain
ε^R	strain at fracture
ε_{rr}, $\varepsilon_{\theta\theta}$, ε_{zz}	strain components in cylindrical coordinates
ε^s	strain threshold for damage
ε^T	creep strain
ε^V	viscous strain
$\dot{\varepsilon}^V$	velocity of viscous strain
η	coefficient of viscosity for a grain boundary
$\overline{\eta}$	distribution density
γ	fracture energy
γ_A	antiphase energy
γ_{CL}	fracture energy for a liquid metal
γ_j	grain boundary energy
γ_j^0	boundary energy for a pure metal
γ_m	step energy
γ_p	plastic deformation energy
γ_R	fracture energy
$\dot{\gamma}_r$	slip velocitiy in system r
γ_S	surface energy
γ_S^0	fracture energy for a pure metal
γ_μ	fracture energy at constant chemical potential
γ_Γ	fracture energy at constant concentration
μ	shear modulus
	chemical potential
μ_a	coefficient of adhesion
μ_D	shear modulus for a damaged material
ν	Poisson's ratio
	activation frequency
ω	solution energy
	angular velocity
ϖ	volume density of complementary potential
π	the mathematical constant
κ	coefficient characterising a state of plane stress or strain
ρ	radius of curvature of a notch
	distance to the centre of a cavity
ρ_0	dislocation density
ρ_m	density of mobile dislocations
σ	stress
σ_0	Norton-Hoff stress
	flow stress
	initial stress
	Weibull stress threshold
	yield strength

$\overline{\sigma}$	equivalent stress
σ_1, σ_2, σ_3	principal stresses
σ_a	applied stress
σ_c	theoretical cleavage stress
$\sigma_{1R}^C, \sigma_{1R}^T$	axial fibre fracture stresses in compression and tension
$\sigma_{2R}^C, \sigma_{2R}^T$	normal fibre fracture stresses in compression and tension
σ_d	crack initiation stress
σ_e	effective stress for crack porpagation
σ_{eff}	effective stress
σ_f	propagation stress
σ_F	fracture stress
σ_m	hydrostatic stress
σ_{max}	maximum stress
σ_n	normal stress
σ_{nom}	nominal stress
σ_{rr}, $\sigma_{\theta\theta}$, σ_{zz}	stress components in cylindrical coordinates
σ_s	threshold creep stress
	internal stress
σ_u	Weibull mean stress
σ_v	viscous stress
σ_w	Weibull stress
τ	shear stress
	time of residence in a craze
τ_i	internal stress
τ_F	fracture stress in shear
τ_r	reduced shear for system r
τ_y	yield strength in shear
θ	angle
	polar angle
	angular opening of a notch
φ	angle
	plastic potential
Γ	integration contour
Γ_j	fractional number of solute atoms on a grain boundary
Γ_s	fractional area of solute atoms on a grain boundary
Δ	class interval
Δa	increment of crack length
ΔA	area swept by a dislocation
$\Delta \varepsilon$	amplitude of variation of strain
$\Delta \varepsilon^p$	amplitude of variation of plastic strain
ΔE_e	increment of elastic energy
ΔG	free enthalpy
ΔG^*	activation energy

ΔG_A activation energy
ΔG_f, ΔE_l energy of creation of a vacancy
ΔG_j enthalphy of segregation in a grain boundary
ΔG_s enthalpy of surface segregation
ΔH enthalpy of activation
ΔH_A enthalpy of activation
ΔK_s threshold for propagation of fatigue cracks
ΔS_A entropy of activation
ΔV volume element
$\Delta\sigma$ amplitude of variation of stress
$\Delta\tau_j$ relaxation time
$\Delta\tau_D$ relaxation time

Vectors

$\vec{b}$ Burgers vector
$\vec{g}$ vector normal to a crack
$\vec{n}$ (unit) normal vector
$\vec{T}$ stress vector
$\vec{u}$ displacement

Tensors

$\underline{\underline{D}}$ (components D_{ij}) damage
$\underline{\underline{s}}$ (components s_{ij}) deviatoric stress
S Eshelby tensor

$\underline{\underline{\varepsilon}}$ (components ε_{ij}) strain
$\underline{\underline{\varepsilon}}^p$ plastic strain
$\dot{\underline{\underline{\varepsilon}}}^p$ rate of plastic strain
$\underline{\underline{\sigma}}$ (components σ_{ij}) stress
$\underline{\underline{\sigma}}^+$, $\underline{\underline{\sigma}}^-$ positive and negative eigen-parts of $\underline{\underline{\sigma}}$

PHYSICAL CONSTANTS
and numerical factors

N	Avogadro's number $(6.02257 \pm 0.009)\ 10^{23}\ gmole)^{-1}$ (^{12}C = *12*)
n	Planck's constant $(6.62554 \pm 0.00015)\ 10^{-27}\ erg.sec$
$\hbar$	$h/2\pi = (1.05443 \pm 0.00004)\ 10^{-27}\ erg.sec$
m	mass of electron $(9.10904 \pm 0.00013)\ 10^{-28}\ g$
e	charge on electron $(4.80296 \pm 0.00006)\ 10^{-10}\ esu$
F	Faraday's constant = Ne $(2.89366 \pm 0.00003)\ (gmole)^{-1}$
λ_0	wavelength of X-rays corresponding to $1\ eV$ $(12398.04 \pm 0.12)\ 10^{-8}\ eVcm$
c	velocity of light $(299792.5 \pm 0.2)\ km\ sec^{-1}$
k	Boltzmann's constant $(1.38053 \pm 0.00006)\ 10^{-16}\ erg./°C = 8.62.10^{-5}\ eV/°C$
R	gas constant (perfect gases) $(8.31696 \pm 0.00034)\ J/mole\ °C$ $= 1.9869\ cal/mole\ °C$
V	standard volume of perfect gas $(22.4146 \pm 0.0006)\ 10^{3}\ cm^{3}/mole$
M	mass of atom of mass *1* atomic unit $(= 1/N)\ (1.66042 \pm 0.00003)\ 10^{-24}\ g$
	melting point of ice $273.16\ °K$

$1\ ev = (1.602095 \pm 0.000022)\ 10^{-19}\ J$

$1\ eV/mol = 23.05\ cal/mol$
$1\ kcal = 4.186.10^{10}\ erg$
$1\ kilobar = 100\ Mpa = 10^{9}\ dynes/cm^{2} = 986.92\ atm = 14504\ psi$
$1\ cal = 4.186\ joules,\ 1\ joule = 10^{7}\ ergs$
$1\ volt = 1/300\ esu$
$1\ inch = 2.54001\ cm$
$1\ Å = 10^{-10}\ m$
$1\ psi\ (pound/sq.inch) = 6897\ Newton/m^{2}$

BIBLIOGRAPHY

Y. Adda and J. Philibert. volume 2. Presses Universitaire de France, 1966.

A.S. Argon. Mechanisms and mechanics of fracture in creeping alloys. In B. Wilshire and D.R.J. Owen, editors, *Recent Advances in Creep and Fracture of Engineering Materials and Structures.* Pineridge Press, Swansea, 1982.

A.S Argon, I.W. Chen, and C.W. Lau. Creep - fatigue - environment interaction. In R.M. Pelloux and N.S. Stoloff, editors, *Intergranular Cavitation in Creep: Theory and Experiment*, pages 46–83. AIME, New York, 1980.

M.F. Ashby and D.R.H. Jones. *Engineering Materials - An Introduction to their Properties and Application.* Pergamon Press, Oxford, 1980.

M.F. Ashby and R.A. Verral. Diffusion-accommodated flow and superplasticity. *Acta. Metallurgica*, 21:149–163, 1973.

R.W. Bailey. Note on the softening of strain-hardened materials and its relation to creep. *Engineering*, 121:351–352, 1926.

S.B. Batdorf. In R.C. Bradt and D.P.H. Hasselman, editors, *Flaws and Testing*, pages 1–30. Plenum Press, New York, 1978.

S.B. Batdorf and H.L. Heinisch. Fracture statistics of brittle materials with surface cracks. *Eng. Fract. Mech.*, 11:831–891, 1978.

W. Beere. Cavities and cracks in creep and fatigue. In J. Gittus, editor, *Theoretical Treatment of Creep Cavity Growth and Nucleation*, pages 1–27. Applied Science Publishers, 1981.

W. Beere. Cavities and cracks in creep and fatigue. In J. Gittus, editor, *Mechanism Maps*, pages 29–57. Applied Science Publishers, 1981.

B.A. Bilby, A.H. Cottrell, and K.H. Swinden. The spread of plastic yield from a notch. *Proc. Roy. Soc. A.*, 272, 1963.

P.W. Bridgman. *Studies in Large Plastic Flow and Fracture.* McGraw Hill, New York, 1952.

D. Broek. *Elementary Engineering Fracture Mechanics.* Martinus Nijhoff, Dordrecht, 1986.

R.H. Brown. *ASTM STP*, 87(518), 1977.

C.B. Buchalet and W.H. Bamford. Stress intensity factor solutions for continuous surface flows in reactor pressure vessel. In J.R. Rice and P.C. Paris, editors, *Mechanics of Crack Growth*, volume 590 of *STP.* ASTM, Philadelphia, 1974.

B. Budiansky, J.W. Hutchinson, and S. Slutsky. In H.G. Hopkins and M.J. Sewell, editors, *Mechanics of Solids, The Rodney Hill 60th Anniversary Volume.* Pergamon Press, Oxford, 1982.

H.D. Bui. *Mécanique de la Rupture Fragile.* Masson, 1978.

J.D. Campbell and W.C. Ferguson. Temperature and strain-rate dependence of shear-strength of mild steel. *Phil. Mag.*, 21:63–82, 1970.

R. Cazaud, G. Pomey, P. Rabbe, and Ch. Janssen. *La Fatigue des Métaux.* Dunod, Paris, 1969.

I.W. Chen and A.S. Argon. Diffusive growth of grain-boundary cavities. *Acta Met.*, 29:1759–1768, 1981.

G.P. Cherepanov. *Mechanics of Brittle Fracture.* Nauka, Moscow, 1974. in Russian.

T.J. Chuang, K.I. Kagawa, J.R. Rice, and L.B. Sills. Non-equilibrium models for diffusive cavitation of grain interfaces. *Acta Met.*, 27:265–284, 1979.

R.L. Coble. A model for boundary diffusion controlled creep in polycrystalline materials. *J. Appl; Phys.*, 34:1679–1682, 1963.

I.G. Crossland, R.B. Jones, and G.W. Lewthwaite. The use of helically coiled springs in creep experiments with special reference to the case of Bingham flow. *J. Phys. D.: Appl. Phys.*, 6:1040–1046, 1973.

K. Dang Van. *Memorial de l'Artillerie Francaise*, 3, 1973.

N.N. Davidenkov. *Dinamicheskaya Ispytania Metallov.* Moscow, 1936.

N.N. Davidenkov and N.I. Spiridonova. The analysis of the state of stress in the neck of a tesion specimen. *Zavodsk, Labor.*, 11:583–593, 1945.

B.F. Dyson and D. McLean. *Metal Science Journal*, pages 37–45, 1977.

F. Erdogan and G.C. Sih. On the crack extension in plates under plane loading and transverse shear. *J. Basic Eng. Trans. ASME*, 85D:519–527, 1963.

R.W. Evans and B. Wilshire. *Creep of Metals and Alloys.* Institute of Metals, London, 1985.

D. Francois and L. Joly. *La Rupture des Métaux.* Masson, 1972.

J. Friedel. *Dislocations.* Pergamon, Oxford, 1964.

H.J. Frost and M.F. Ashby. *Deformation-Mechanism Maps: the Plasticity and Creep of Metals and Ceramics.* Pergamon Press, Oxford, 1982.

H.J. Frost, F. Spaepen, and M.F. Ashby. A second report on tilt boundaries in hard-sphere fcc crystals. *Scrip. Metal*, 16:1165–1170, 1982.

N.E. Frost. *Proc. I. Mech. Eng.*, 173:811–827, 1959.

F. Garofalo. *Fundamentals of Creep and Creep Rupture in Metals.* MacMillan, New York, 1965.

E.E. Gdoutos. *Fracture Mechanics Criteria and Applications.* Kluwer Academic Publishers, Dordrecht, 1990.

E.E. Gdoutos. *Fracture Mechanics, An Introduction.* Kluwer Academic Publishers, Dordrecht, 1993.

Gibbs and Harris. Melbourne, 1969. Interfaces Conference, Butterworth.

J.J. Gilman. Mechanism of ortho kink band formation in compressed zinc monocrystals. *J. Metals*, 6:621–629, 1954.

G.M.L. Gladwell. *Contact Problems in the Classical Theory of Elasticity*. Sijthoff and Noordhoff, Dordrecht, 1980.

J. Goodman. *Mechanics Applied to Engineering*. Longman Green, London, 1914.

A.A Griffith. The phenomenon of rupture and flow in solids. *Phil. Trans. Roy. Soc.*, A221:163–198, 1921.

J.R. Griffiths and D.R.J. Owen. Elastic-plastic stress analysis for a notched bar in plane strain bending. *J. Mech. Phys. Solids*, 19:419–431, 1971.

A.L. Gurson. Continuum theory of ductile rupture by nucleation and growth: Part I Yield criteria and flow rules for porous ductile media. *J. Eng. Materials and Technology*, 99:2–15, 1977.

D. Hall and D.E. Rimmer. The growth of grain-boundary voids under stress. *Phil. Mag.*, 4:673–687, 1959.

E.O. Hall. The deformation and ageing of mild steel II. Characteristics of the Lüders deformation. *Proc. Roy. Soc., London*, B64:742–747, 1951.

M.D. Hanna and G.W. Greenwood. Cavity growth and creep in copper at low stresses. *Acta. Metall*, 30:719–724, 1982.

D.R. Hayhurst, W.A. Trampczynski, and F.A. Leckie. On the role of cavity nucleation in creep deformation and fracture. *Acta. Metall.*, 31:1537–1542, 1983.

C. Herring. Diffusional viscosity of a polycrystalline solid. *J. Appl. Phys.*, 21:437–445, 1950.

D.A. Hills and D. Nowell. *Mechanics of Fretting Fatigue*. Kluwer Academic Publishers, Dordrecht, 1994.

P.D. Hilton and J.W. Hutchinson. Plastic intensity factors for cracked plates. *Eng. Fracture Mech.*, 3:435–451, 1971.

J.W. Hutchinson. Singular behavior at the end of a tensile crack in a hardening material. *J. Mech. Phys. Solids*, 16:13–31, 1968.

G.R. Irwin. Elasticity and plasticity. In S. Flügge, editor, *Encyclopaedia of Physics*, volume 6, pages 551–590. Springer Verlag, 1958.

Johnson. *Plasticity of Steels.* 19.

L.M. Kachanov. Rupture time under creep conditions. In *Problems in Continuum Mechanics, Muskhelishvili Memorial Volume*, Philadelphia, 1961.

A. Kelly. *Strong Solids.* Clarendon Press, Oxford, 1966.

J.F. Knott. *Fundamentals of Fracture Mechanics.* Butterworth, London, 1973.

J.F. Knott. Micromechanisms of fracture and the fracture toughness of engineering alloys. In D.M.R. Taplin, editor, *Fracture, 1977*, volume 1, pages 61–92. 1977.

R. Labbens. *Introduction á la Mécanique de la Rupture.* Pluralis, 1980.

R.C. Labbens, A. Pellisser-Tanon, and J. Heliot. Practical methods of calculating stress intensity factors through weight functions. In J.R. Rice and P.C. Paris, editors, *Mechanics of Crack Growth.* ASTM STP 590, Philadelphia, 1974.

R.F. Larson and J. Miller. A time-temperature relationship for rupture and creep stresses. *Trans. ASME*, 74:765–775, 1952.

B.R. Lawn. *Fracture of Brittle Solids.* Cambridge University Press, Cambridge, second edition, 1993.

F.A. Leckie and D.R. Hayhurst. Constitutive equations for creep-rupture. *Acta. Metall.*, 25:1059–1070, 1977.

J. Lemaitre and J.L. Chaboche. *Mechanics of Solid Materials.* Cambridge University Press, Cambridge, 1990.

N. Levy, P.V. Marcal, W.J. Ostergren, and J.R. Rice. Small scale yielding near a crack in plane strain: a finite-element analysis. *Int. J. Fracture Mech.*, 7:143–156, 1971.

C. Licht and P. Suquet. *C.R. Acad. Sciences, Paris*, II(306):327–330, 1988.

H. Liebowitz. *Fracture, an Advanced Treatise, 3 Volumes.* Academic Press, New York, 1971.

J.R. Low. *The Relation of Microstructure to Brittle Fracture*, pages 163–179. ASM, Novelty, Ohio, 1954.

S.S. Manson. *Pressure Vessels and Piping, Analysis and Design*, page 157. ASME, New York, 1972.

J. Mazars. *Thèsis de doctorat d'état, Université Paris.* PhD thesis, Université Paris, 1984.

F.A. McClintock. Plasticity aspects of fracture. In H. Liebowitz, editor, *Fracture, an Advanced Treatise*, volume 3, pages 47–225. Academic Press, New York, 1971.

F.A. McClintock and A.S. Argon. *Mechanical Behavior of Materials.* Addison Wesley, Reading, Mass., 1966.

D. McLean. *Mechanical Properties of Metals.* Wiley, New York, 1962.

P. Michel and P. Suquet. The constitutive law of nonlinear viscous and porous materials. *J. Mech. Phys. Solids*, 40:783–812, 1992.

A.K. Miller, editor. *Unified Constitutive Equations for Creep and Elasticity.* Elsevier, London, 1987.

F.C. Monkman and N.J. Grant. Empirical relation between rupture life and minimum creep rate in creep-rupture tests. *Proc. ASTM*, 56:955–967, 1956.

A.K. Mukherjee, J.E. Bird, and J.E. Dorn. Experimental correlations for high temperature creep. *Trans. Amer. Inst. Min.*, 62:155–179, 1969.

Y. Murakami, editor. *Stress Intensity Factors Handbook.* Pergamon Press, Oxford, 1987.

F.R.N. Nabarro. Dislocations of a simple cubic lattice. *Proc. Phys. Soc.*, 59:256–272, 1947.

F.R.N. Nabarro. The mathematical theory of stationary dislocations. *Advances in Physics*, 1:269–394, 1952.

E. Orowan and J.W. Scotland. Classification and nomenclature of internal stress. *J. Iron Steel Inst.*, 54:45–58, 1946.

P.C. Paris, M.P. Gomez, and W.E. Anderson. *A Rational Analytic Theory of Fatigue*, pages 9–14. University of Washington, 1961.

R. N. Parkins. *Br. Corros. J.*, 14, 1979.

R.N. Parkins and B.S. Greenwell. The interface between corrosive fatigue and stress-corrosion cracking. *Metals Science*, 11:405–413, 1977.

D.M. Parks. A stiffness derivative finite element technique for determination of crack tip intensity factors. *Int. J. Fracture*, 10:487–502, 1974.

E.V. Perevezentsev, Rybinm V.V., and V.N. Chuvil'deev. The theory of structural superplasticity. *Acta. Metall Mater*, 40:887–894; 895–905; 907–914; 915–924, 1992.

N.J. Petch. The cleavage strength of polycrystals. *J. Iron Steel Inst.*, 174:25–28, 1953.

N.L. Peterson. Diffusion in metals. *Solid State Physics*, 22:409–, 1968.

R.E. Peterson. *Stress Concentration Factors.* John Wiley and Sons, New York, 1974.

J.P. Poirier. *Creep of Crystals: High Temperature Deformation Processes in Metals.* Cambridge University Press, Cambridge, 1985.

R. Raj and M.F. Ashby. Grain-boundary sliding and diffusional creep. *Metallurg. Trans.*, 2:1113–, 1971.

J.R. Rice. A path independent integral and the approximate analysis of strain concentration by notches and cracks. *J. Appl. Mech.*, 35:379–386, 1968.

J.R. Rice. Mathematical analysis in the mechanics of fracture. In H. Liebowitz, editor, *Fracture, an Advanced Treatise*, volume 2. Academic Press, New York, 1971.

J.R. Rice and G.R. Rosengren. Plane strain deformation near a crack tip in a power-law hardening material. *J. Mech. Phys. Solids*, 16:1–12, 1968.

J.R. Rice and R. Thomson. Ductile versus brittle behavior of crystals. *Phil. Mag.*, 29:73–97, 1974.

J.R. Rice and D.M Tracey. On the ductile enlargement of voids in triaxial stress fields. *J. Mech. Phys. Solids*, 17:201–217, 1969.

R.O. Ritchie. Near-threshold crack-propagation in ultra high strength steel - influence of load-ratio and cyclic strength. *J. Eng. Materials*, 99:195–204, 1977.

D.P. Rooke and D.J. Cartwright. *Compendium of Stress Intensity Factors.* Hillingdon Press, Uxbridge, Middx, U.K., 1976.

G. Rousselier. Ductile fracture models and their potential in local approach of fracture. *Nuclear Engineering and Design*, 105:97–111, 1987.

M.P. Seah. Adsorption-induced interface decohesion. *Acta. Metall.*, 28:955–962, 1980.

T.T. Shih. *ASTM STP*, 3:590, 1976.

G.C. Sih. *Handbook of Stress Intensity Factors.* Institute of Fracture and Solid Mechanics, Lehigh University, Bethlehem, Pa., U.S.A., 1973.

E. Smith. The nucleation and growth of cleavage microcracks in mild steel. In *Proc. Conference on Physical Basis of Yield and Fracture*, pages 36–46, Oxford, 1966. Institute of Physics and Physical Society.

M.V. Speight and W. Beere. Vacancy potential and void growth on grain boundaries. *Met. Sci.*, 9:190–191, 1975.

P. Stanley. *Fracture Mechanics in Engineering Practice.* Applied Science Publishers, 1976.

A.N. Stroh. The formation of cracks as a result of plastic flow. *Proc. Roy. Soc., London*, A223:404–414, 1954.

H. Tada, P.C. Paris, and G.R. Irwin. *The Stress Analysis of Cracks Handbook.* Del Research Corporation, Hellerton, Pa., 1985.

A.S. Tetelman and A.J. McEvily. *Fracture of Structural Materials.* Wiley, New York, 1967.

S. Timoshenko and J.N. Goodier. *Theory of Elasticity.* McGraw Hill, 1951.

V. Tvergaard. Influence of voids on shear brand instabilities under plane strain conditions. *Int. J. Fracture*, 17:389–407, 1981.

V. Tvergaard. On localization in ductile materials containing spherical voids. *Int. J. Fracture*, 18:237–252, 1982.

R.B. Waterhouse. *Fretting Fatigue*. Applied Science Publishers, 1981.

J. Weertman. Comment on Garofalo, F., An empirical relation defining the stress dependence of minimun creep in metals. *Trans. Metall. Soc. AIME*, 227:1475–1476, 1963.

R.P. Wei and J.D. Landes. Correlation between sustained-load and fatigue crack growth in high-strength steels. *Mater. Res. Standards*, 9:25–, 1969.

W. Weibull. *J. Appl. Mech.*, 18(293), 19.

C. Zener. Micromechanism of fracture. In *Fracturing of Metals*, pages 3–31, Novelty, Ohio, 1949. ASM.

Z. R. Zhou, S. Fayeulle, and L. Vincent. Cracking behavior of various aluminum alloys during fretting wear. *Wear*, 15S:317–320, 1992.

NAME INDEX

Gives the first occurrence of the item

SUBJECT INDEX

Mechanics

SOLID MECHANICS AND ITS APPLICATIONS

Series Editor: G.M.L. Gladwell

Aims and Scope of the Series

The fundamental questions arising in mechanics are: *Why?*, *How?*, and *How much?* The aim of this series is to provide lucid accounts written by authoritative researchers giving vision and insight in answering these questions on the subject of mechanics as it relates to solids. The scope of the series covers the entire spectrum of solid mechanics. Thus it includes the foundation of mechanics; variational formulations; computational mechanics; statics, kinematics and dynamics of rigid and elastic bodies; vibrations of solids and structures; dynamical systems and chaos; the theories of elasticity, plasticity and viscoelasticity; composite materials; rods, beams, shells and membranes; structural control and stability; soils, rocks and geomechanics; fracture; tribology; experimental mechanics; biomechanics and machine design.

1. R.T. Haftka, Z. Gürdal and M.P. Kamat: *Elements of Structural Optimization.* 2nd rev.ed., 1990 ISBN 0-7923-0608-2
2. J.J. Kalker: *Three-Dimensional Elastic Bodies in Rolling Contact.* 1990 ISBN 0-7923-0712-7
3. P. Karasudhi: *Foundations of Solid Mechanics.* 1991 ISBN 0-7923-0772-0
4. *Not published*
5. *Not published.*
6. J.F. Doyle: *Static and Dynamic Analysis of Structures.* With an Emphasis on Mechanics and Computer Matrix Methods. 1991 ISBN 0-7923-1124-8; Pb 0-7923-1208-2
7. O.O. Ochoa and J.N. Reddy: *Finite Element Analysis of Composite Laminates.* ISBN 0-7923-1125-6
8. M.H. Aliabadi and D.P. Rooke: *Numerical Fracture Mechanics.* ISBN 0-7923-1175-2
9. J. Angeles and C.S. López-Cajún: *Optimization of Cam Mechanisms.* 1991 ISBN 0-7923-1355-0
10. D.E. Grierson, A. Franchi and P. Riva (eds.): *Progress in Structural Engineering.* 1991 ISBN 0-7923-1396-8
11. R.T. Haftka and Z. Gürdal: *Elements of Structural Optimization.* 3rd rev. and exp. ed. 1992 ISBN 0-7923-1504-9; Pb 0-7923-1505-7
12. J.R. Barber: *Elasticity.* 1992 ISBN 0-7923-1609-6; Pb 0-7923-1610-X
13. H.S. Tzou and G.L. Anderson (eds.): *Intelligent Structural Systems.* 1992 ISBN 0-7923-1920-6
14. E.E. Gdoutos: *Fracture Mechanics.* An Introduction. 1993 ISBN 0-7923-1932-X
15. J.P. Ward: *Solid Mechanics.* An Introduction. 1992 ISBN 0-7923-1949-4
16. M. Farshad: *Design and Analysis of Shell Structures.* 1992 ISBN 0-7923-1950-8
17. H.S. Tzou and T. Fukuda (eds.): *Precision Sensors, Actuators and Systems.* 1992 ISBN 0-7923-2015-8
18. J.R. Vinson: *The Behavior of Shells Composed of Isotropic and Composite Materials.* 1993 ISBN 0-7923-2113-8
19. H.S. Tzou: *Piezoelectric Shells.* Distributed Sensing and Control of Continua. 1993 ISBN 0-7923-2186-3

Kluwer Academic Publishers – Dordrecht / Boston / London

Mechanics

SOLID MECHANICS AND ITS APPLICATIONS

Series Editor: G.M.L. Gladwell

20. W. Schiehlen (ed.): *Advanced Multibody System Dynamics.* Simulation and Software Tools. 1993 ISBN 0-7923-2192-8
21. C.-W. Lee: *Vibration Analysis of Rotors.* 1993 ISBN 0-7923-2300-9
22. D.R. Smith: *An Introduction to Continuum Mechanics.* 1993 ISBN 0-7923-2454-4
23. G.M.L. Gladwell: *Inverse Problems in Scattering.* An Introduction. 1993 ISBN 0-7923-2478-1
24. G. Prathap: *The Finite Element Method in Structural Mechanics.* 1993 ISBN 0-7923-2492-7
25. J. Herskovits (ed.): *Advances in Structural Optimization.* 1995 ISBN 0-7923-2510-9
26. M.A. González-Palacios and J. Angeles: *Cam Synthesis.* 1993 ISBN 0-7923-2536-2
27. W.S. Hall: *The Boundary Element Method.* 1993 ISBN 0-7923-2580-X
28. J. Angeles, G. Hommel and P. Kovács (eds.): *Computational Kinematics.* 1993 ISBN 0-7923-2585-0
29. A. Curnier: *Computational Methods in Solid Mechanics.* 1994 ISBN 0-7923-2761-6
30. D.A. Hills and D. Nowell: *Mechanics of Fretting Fatigue.* 1994 ISBN 0-7923-2866-3
31. B. Tabarrok and F.P.J. Rimrott: *Variational Methods and Complementary Formulations in Dynamics.* 1994 ISBN 0-7923-2923-6
32. E.H. Dowell (ed.), E.F. Crawley, H.C. Curtiss Jr., D.A. Peters, R. H. Scanlan and F. Sisto: *A Modern Course in Aeroelasticity.* Third Revised and Enlarged Edition. 1995 ISBN 0-7923-2788-8; Pb: 0-7923-2789-6
33. A. Preumont: *Random Vibration and Spectral Analysis.* 1994 ISBN 0-7923-3036-6
34. J.N. Reddy (ed.): *Mechanics of Composite Materials.* Selected works of Nicholas J. Pagano. 1994 ISBN 0-7923-3041-2
35. A.P.S. Selvadurai (ed.): *Mechanics of Poroelastic Media.* 1996 ISBN 0-7923-3329-2
36. Z. Mróz, D. Weichert, S. Dorosz (eds.): *Inelastic Behaviour of Structures under Variable Loads.* 1995 ISBN 0-7923-3397-7
37. R. Pyrz (ed.): *IUTAM Symposium on Microstructure-Property Interactions in Composite Materials.* Proceedings of the IUTAM Symposium held in Aalborg, Denmark. 1995 ISBN 0-7923-3427-2
38. M.I. Friswell and J.E. Mottershead: *Finite Element Model Updating in Structural Dynamics.* 1995 ISBN 0-7923-3431-0
39. D.F. Parker and A.H. England (eds.): *IUTAM Symposium on Anisotropy, Inhomogeneity and Nonlinearity in Solid Mechanics.* Proceedings of the IUTAM Symposium held in Nottingham, U.K. 1995 ISBN 0-7923-3594-5
40. J.-P. Merlet and B. Ravani (eds.): *Computational Kinematics '95.* 1995 ISBN 0-7923-3673-9
41. L.P. Lebedev, I.I. Vorovich and G.M.L. Gladwell: *Functional Analysis.* Applications in Mechanics and Inverse Problems. 1996 ISBN 0-7923-3849-9
42. J. Menčik: *Mechanics of Components with Treated or Coated Surfaces.* 1996 ISBN 0-7923-3700-X
43. D. Bestle and W. Schiehlen (eds.): *IUTAM Symposium on Optimization of Mechanical Systems.* Proceedings of the IUTAM Symposium held in Stuttgart, Germany. 1996 ISBN 0-7923-3830-8

Kluwer Academic Publishers – Dordrecht / Boston / London

Mechanics

SOLID MECHANICS AND ITS APPLICATIONS

Series Editor: G.M.L. Gladwell

44. D.A. Hills, P.A. Kelly, D.N. Dai and A.M. Korsunsky: *Solution of Crack Problems.* The Distributed Dislocation Technique. 1996 ISBN 0-7923-3848-0
45. V.A. Squire, R.J. Hosking, A.D. Kerr and P.J. Langhorne: *Moving Loads on Ice Plates.* 1996 ISBN 0-7923-3953-3
46. A. Pineau and A. Zaoui (eds.): *IUTAM Symposium on Micromechanics of Plasticity and Damage of Multiphase Materials.* Proceedings of the IUTAM Symposium held in Sèvres, Paris, France. 1996 ISBN 0-7923-4188-0
47. A. Naess and S. Krenk (eds.): *IUTAM Symposium on Advances in Nonlinear Stochastic Mechanics.* Proceedings of the IUTAM Symposium held in Trondheim, Norway. 1996 ISBN 0-7923-4193-7
48. D. Ieşan and A. Scalia: *Thermoelastic Deformations.* 1996 ISBN 0-7923-4230-5
49. J. R. Willis (ed.): *IUTAM Symposium on Nonlinear Analysis of Fracture.* Proceedings of the IUTAM Symposium held in Cambridge, U.K. 1997 ISBN 0-7923-4378-6
50. A. Preumont: *Vibration Control of Active Structures.* An Introduction. 1997 ISBN 0-7923-4392-1
51. G.P. Cherepanov: *Methods of Fracture Mechanics: Solid Matter Physics.* 1997 ISBN 0-7923-4408-1
52. D.H. van Campen (ed.): *IUTAM Symposium on Interaction between Dynamics and Control in Advanced Mechanical Systems.* Proceedings of the IUTAM Symposium held in Eindhoven, The Netherlands. 1997 ISBN 0-7923-4429-4
53. N.A. Fleck and A.C.F. Cocks (eds.): *IUTAM Symposium on Mechanics of Granular and Porous Materials.* Proceedings of the IUTAM Symposium held in Cambridge, U.K. 1997 ISBN 0-7923-4553-3
54. J. Roorda and N.K. Srivastava (eds.): *Trends in Structural Mechanics.* Theory, Practice, Education. 1997 ISBN 0-7923-4603-3
55. Yu. A. Mitropolskii and N. Van Dao: *Applied Asymptotic Methods in Nonlinear Oscillations.* 1997 ISBN 0-7923-4605-X
56. C. Guedes Soares (ed.): *Probabilistic Methods for Structural Design.* 1997 ISBN 0-7923-4670-X
57. D. François, A. Pineau and A. Zaoui: *Mechanical Behaviour of Materials.* Volume I: Elasticity and Plasticity. 1998 ISBN 0-7923-4894-X
58. D. François, A. Pineau and A. Zaoui: *Mechanical Behaviour of Materials.* Volume II: Viscoplasticity, Damage, Fracture and Contact Mechanics. 1998 ISBN 0-7923-4895-8
59. L. T. Tenek and J. Argyris: *Finite Element Analysis for Composite Structures.* 1998 ISBN 0-7923-4899-0

Kluwer Academic Publishers – Dordrecht / Boston / London

Mechanics

FLUID MECHANICS AND ITS APPLICATIONS

Series Editor: R. Moreau

Aims and Scope of the Series

The purpose of this series is to focus on subjects in which fluid mechanics plays a fundamental role. As well as the more traditional applications of aeronautics, hydraulics, heat and mass transfer etc., books will be published dealing with topics which are currently in a state of rapid development, such as turbulence, suspensions and multiphase fluids, super and hypersonic flows and numerical modelling techniques. It is a widely held view that it is the interdisciplinary subjects that will receive intense scientific attention, bringing them to the forefront of technological advancement. Fluids have the ability to transport matter and its properties as well as transmit force, therefore fluid mechanics is a subject that is particularly open to cross fertilisation with other sciences and disciplines of engineering. The subject of fluid mechanics will be highly relevant in domains such as chemical, metallurgical, biological and ecological engineering. This series is particularly open to such new multidisciplinary domains.

1. M. Lesieur: *Turbulence in Fluids.* 2nd rev. ed., 1990 ISBN 0-7923-0645-7
2. O. Métais and M. Lesieur (eds.): *Turbulence and Coherent Structures.* 1991 ISBN 0-7923-0646-5
3. R. Moreau: *Magnetohydrodynamics.* 1990 ISBN 0-7923-0937-5
4. E. Coustols (ed.): *Turbulence Control by Passive Means.* 1990 ISBN 0-7923-1020-9
5. A.A. Borissov (ed.): *Dynamic Structure of Detonation in Gaseous and Dispersed Media.* 1991 ISBN 0-7923-1340-2
6. K.-S. Choi (ed.): *Recent Developments in Turbulence Management.* 1991 ISBN 0-7923-1477-8
7. E.P. Evans and B. Coulbeck (eds.): *Pipeline Systems.* 1992 ISBN 0-7923-1668-1
8. B. Nau (ed.): *Fluid Sealing.* 1992 ISBN 0-7923-1669-X
9. T.K.S. Murthy (ed.): *Computational Methods in Hypersonic Aerodynamics.* 1992 ISBN 0-7923-1673-8
10. R. King (ed.): *Fluid Mechanics of Mixing.* Modelling, Operations and Experimental Techniques. 1992 ISBN 0-7923-1720-3
11. Z. Han and X. Yin: *Shock Dynamics.* 1993 ISBN 0-7923-1746-7
12. L. Svarovsky and M.T. Thew (eds.): *Hydroclones.* Analysis and Applications. 1992 ISBN 0-7923-1876-5
13. A. Lichtarowicz (ed.): *Jet Cutting Technology.* 1992 ISBN 0-7923-1979-6
14. F.T.M. Nieuwstadt (ed.): *Flow Visualization and Image Analysis.* 1993 ISBN 0-7923-1994-X
15. A.J. Saul (ed.): *Floods and Flood Management.* 1992 ISBN 0-7923-2078-6
16. D.E. Ashpis, T.B. Gatski and R. Hirsh (eds.): *Instabilities and Turbulence in Engineering Flows.* 1993 ISBN 0-7923-2161-8
17. R.S. Azad: *The Atmospheric Boundary Layer for Engineers.* 1993 ISBN 0-7923-2187-1
18. F.T.M. Nieuwstadt (ed.): *Advances in Turbulence IV.* 1993 ISBN 0-7923-2282-7
19. K.K. Prasad (ed.): *Further Developments in Turbulence Management.* 1993 ISBN 0-7923-2291-6
20. Y.A. Tatarchenko: *Shaped Crystal Growth.* 1993 ISBN 0-7923-2419-6

Kluwer Academic Publishers – Dordrecht / Boston / London

Mechanics

FLUID MECHANICS AND ITS APPLICATIONS

Series Editor: R. Moreau

21. J.P. Bonnet and M.N. Glauser (eds.): *Eddy Structure Identification in Free Turbulent Shear Flows.* 1993 ISBN 0-7923-2449-8
22. R.S. Srivastava: *Interaction of Shock Waves.* 1994 ISBN 0-7923-2920-1
23. J.R. Blake, J.M. Boulton-Stone and N.H. Thomas (eds.): *Bubble Dynamics and Interface Phenomena.* 1994 ISBN 0-7923-3008-0
24. R. Benzi (ed.): *Advances in Turbulence V.* 1995 ISBN 0-7923-3032-3
25. B.I. Rabinovich, V.G. Lebedev and A.I. Mytarev: *Vortex Processes and Solid Body Dynamics.* The Dynamic Problems of Spacecrafts and Magnetic Levitation Systems. 1994 ISBN 0-7923-3092-7
26. P.R. Voke, L. Kleiser and J.-P. Chollet (eds.): *Direct and Large-Eddy Simulation I.* Selected papers from the First ERCOFTAC Workshop on Direct and Large-Eddy Simulation. 1994 ISBN 0-7923-3106-0
27. J.A. Sparenberg: *Hydrodynamic Propulsion and its Optimization.* Analytic Theory. 1995 ISBN 0-7923-3201-6
28. J.F. Dijksman and G.D.C. Kuiken (eds.): *IUTAM Symposium on Numerical Simulation of Non-Isothermal Flow of Viscoelastic Liquids.* Proceedings of an IUTAM Symposium held in Kerkrade, The Netherlands. 1995 ISBN 0-7923-3262-8
29. B.M. Boubnov and G.S. Golitsyn: *Convection in Rotating Fluids.* 1995 ISBN 0-7923-3371-3
30. S.I. Green (ed.): *Fluid Vortices.* 1995 ISBN 0-7923-3376-4
31. S. Morioka and L. van Wijngaarden (eds.): *IUTAM Symposium on Waves in Liquid/Gas and Liquid/Vapour Two-Phase Systems.* 1995 ISBN 0-7923-3424-8
32. A. Gyr and H.-W. Bewersdorff: *Drag Reduction of Turbulent Flows by Additives.* 1995 ISBN 0-7923-3485-X
33. Y.P. Golovachov: *Numerical Simulation of Viscous Shock Layer Flows.* 1995 ISBN 0-7923-3626-7
34. J. Grue, B. Gjevik and J.E. Weber (eds.): *Waves and Nonlinear Processes in Hydrodynamics.* 1996 ISBN 0-7923-4031-0
35. P.W. Duck and P. Hall (eds.): *IUTAM Symposium on Nonlinear Instability and Transition in Three-Dimensional Boundary Layers.* 1996 ISBN 0-7923-4079-5
36. S. Gavrilakis, L. Machiels and P.A. Monkewitz (eds.): *Advances in Turbulence VI.* Proceedings of the 6th European Turbulence Conference. 1996 ISBN 0-7923-4132-5
37. K. Gersten (ed.): *IUTAM Symposium on Asymptotic Methods for Turbulent Shear Flows at High Reynolds Numbers.* Proceedings of the IUTAM Symposium held in Bochum, Germany. 1996 ISBN 0-7923-4138-4
38. J. Verhás: *Thermodynamics and Rheology.* 1997 ISBN 0-7923-4251-8
39. M. Champion and B. Deshaies (eds.): *IUTAM Symposium on Combustion in Supersonic Flows.* Proceedings of the IUTAM Symposium held in Poitiers, France. 1997 ISBN 0-7923-4313-1
40. M. Lesieur: *Turbulence in Fluids.* Third Revised and Enlarged Edition. 1997 ISBN 0-7923-4415-4; Pb: 0-7923-4416-2

Kluwer Academic Publishers – Dordrecht / Boston / London

Mechanics

***FLUID* MECHANICS AND ITS APPLICATIONS**

Series Editor: R. Moreau

41. L. Fulachier, J.L. Lumley and F. Anselmet (eds.): *IUTAM Symposium on Variable Density Low-Speed Turbulent Flows*. Proceedings of the IUTAM Symposium held in Marseille, France. 1997 ISBN 0-7923-4602-5
42. B.K. Shivamoggi: *Nonlinear Dynamics and Chaotic Phenomena*. An Introduction. 1997 ISBN 0-7923-4772-2
43. H. Ramkissoon, *IUTAM Symposium on Lubricated Transport of Viscous Materials*. Proceedings of the IUTAM Symposium held in Tobago, West Indies. 1998 ISBN 0-7923-4897-4

Kluwer Academic Publishers – Dordrecht / Boston / London

ERCOFTAC SERIES

1. A. Gyr and F.-S. Rys (eds.): *Diffusion and Transport of Pollutants in Atmospheric Mesoscale Flow Fields*. 1995 ISBN 0-7923-3260-1
2. M. Hallbäck, D.S. Henningson, A.V. Johansson and P.H. Alfredsson (eds.): *Turbulence and Transition Modelling*. Lecture Notes from the ERCOFTAC/IUTAM Summerschool held in Stockholm. 1996 ISBN 0-7923-4060-4
3. P. Wesseling (ed.): *High Performance Computing in Fluid Dynamics*. Proceedings of the Summerschool held in Delft, The Netherlands. 1996 ISBN 0-7923-4063-9
4. Th. Dracos (ed.): *Three-Dimensional Velocity and Vorticity Measuring and Image Analysis Techniques*. Lecture Notes from the Short Course held in Zürich, Switzerland. 1996 ISBN 0-7923-4256-9
5. J.-P. Chollet, P.R. Voke and L. Kleiser (eds.): *Direct and Large-Eddy Simulation II*. Proceedings of the ERCOFTAC Workshop held in Grenoble, France. 1997 ISBN 0-7923-4687-4

KLUWER ACADEMIC PUBLISHERS – DORDRECHT / BOSTON / LONDON

ICASE/LaRC Interdisciplinary Series in Science and Engineering

1. J. Buckmaster, T.L. Jackson and A. Kumar (eds.): *Combustion in High-Speed Flows.* 1994 ISBN 0-7923-2086-X
2. M.Y. Hussaini, T.B. Gatski and T.L. Jackson (eds.): *Transition, Turbulence and Combustion.* Volume I: Transition. 1994 ISBN 0-7923-3084-6; set 0-7923-3086-2
3. M.Y. Hussaini, T.B. Gatski and T.L. Jackson (eds.): *Transition, Turbulence and Combustion.* Volume II: Turbulence and Combustion. 1994 ISBN 0-7923-3085-4; set 0-7923-3086-2
4. D.E. Keyes, A. Sameh and V. Venkatakrishnan (eds): *Parallel Numerical Algorithms.* 1997 ISBN 0-7923-4282-8
5. T.G. Campbell, R.A. Nicolaides and M.D. Salas (eds.): *Computational Electromagnetics and Its Applications.* 1997 ISBN 0-7923-4733-1
6. V. Venkatakrishnan, M.D. Salas and S.R. Chakravarthy (eds.): *Barriers and Challenges in Computational Fluid Dynamics.* 1998 ISBN 0-7923-4855-9

KLUWER ACADEMIC PUBLISHERS – DORDRECHT / BOSTON / LONDON

MIX
Papier aus verantwortungsvollen Quellen
Paper from responsible sources
FSC® C105338

If you have any concerns about our products,
you can contact us on
ProductSafety@springernature.com

In case Publisher is established outside the EU,
the EU authorized representative is:
Springer Nature Customer Service Center GmbH
Europaplatz 3, 69115 Heidelberg, Germany

Printed by Libri Plureos GmbH
in Hamburg, Germany